Optical Fiber Amplifiers: Design and System Applications

The Artech House Optoelectronics Library

Brian Culshaw, Alan Rogers, and Henry Taylor, *Series Editors*

Acousto-Optic Signal Processing: Fundamentals and Applications, Pankaj Das

Amorphous and Microcrystalline Semiconductor Devices, Optoelectronic Devices, Jerzy Kanicki, editor

Electro-Optical Systems Performance Modeling, Gary Waldman and John Wootton

The Fiber-Optic Gyroscope, Hervé Lefèvre

Field Theory of Acousto-Optic Signal Processing Devices, Craig Scott

Highly Coherent Semiconductor Lasers, Motoichi Ohtsu

Introduction to Electro-Optical Imaging and Tracking Systems, Khalil Seyrafi and S. A. Hovanessian

Introduction to Glass Integrated Optics, S. Iraj Najafi

Optical Control of Microwave Devices, Rainee N. Simons

Optical Fiber Sensors, Volume I: Principles and Components, John Dakin and Brian Culshaw, editors

Optical Fiber Sensors, Volume II: Systems and Applicatons, Brian Culshaw and John Dakin, editors

Optical Network Theory, Yitzhak Weissman

Principles of Modern Optical Systems, Volume I, I. Andonovic and D. Uttamchandani, editors

Principles of Modern Optical Systems, Volume II, I. Andonovic and D. Uttamchandani, editors

Reliability and Degradation of LEDs and Semiconductor Lasers, Mitsuo Fukuda

Single-Mode Optical Fiber Measurements: Characterization and Sensing, Giovanni Cancellieri

Optical Fiber Amplifiers: Design and System Applications

Anders Bjarklev

Artech House
Boston • London

Library of Congress Cataloging-in-Publication Data

Bjarklev, Anders
Optical Fiber Amplifiers: Design and System Applications/Anders Bjarklev
Includes bibliographical references and index.
ISBN 0-89006-659-0
1. Lasers 2. Optical Amplifiers 3. Fiber optics
4. Optical communications I. Title
TA1677.B54 1993 93-28368
621.382'75–dc20 CIP

685 Canton Street
Norwood, MA 02062

International Standard Book Number: 0-89006-659-0
Library of Congress Catalog Card Number: 93-28368

10 9 8 7 6 5 4 3 2 1

To my wife Araceli, my son Kristian, and our family

Contents

Preface

Few components have revolutionized the area of optical communications as the fiber-optic amplifier has. The simple technology of light-pumped erbium-doped silica fibers has today provided commercially available amplifiers with high gain and low-noise properties at 1550 nm. Furthermore, recent developments in neodymium- and praseodymium-doped fluoride glass amplifiers have shown promising results for the 1300-nm wavelength band.

This book presents a review of the properties of optical fiber amplifiers, with a specific focus on the telecommunication engineering aspects. Highly accurate models for erbium-doped fiber amplifiers are shown to be very useful tools for manufacturers developing optimized 1550-nm fiber amplifiers for various system applications. The merits of optimized erbium-doped fiber amplifiers are compared both experimentally and theoretically for the relevant 800-, 980- and 1480-nm pump bands, where semiconductor laser pump diodes are obtainable. Fiber designs are evaluated and practical designs are given with emphasis on applications in digital direct detection systems. Trunk systems with lumped and distributed gain are described, and there is a short review of amplifier applications in systems employing analog modulation. Amplifiers operating around 1300 nm are also described, and the attainment of significant improvements for neodymium- and praseodymium-doped fiber amplifiers is discussed.

Acknowledgments

I would like to express my thanks to my wife for her full support and understanding while I was writing this book, and for her invaluable help with the preparation of the illustrations. I would also like to thank Thomas Rasmussen for all of his efforts in reviewing and commenting on the drafts of the manuscript. Thanks also go to Ole Lumholt for his help during the revision of the book manuscript. These two colleagues, together with Karsten Rottwitt, have also been very helpful during many technical discussions, and I would like to thank them for their valuable suggestions and support.

In addition, I would like to thank Jørn Hedegaard Povlsen, Ejner Nicolaisen, and all my other colleagues at the Electromagnetics Institute, Technical University of Denmark, for their help and the fruitful discussions I had with them.

It is fair to say that many other colleagues have been very helpful during the time I spent on related work immediately prior to the writing of this book. I would especially like to mention here my colleagues at Jydsk Telefon, LYCOM A/S, NKT Research Center, NKT Electronics, and Telecom Denmark, Ltd. I would also like to thank the Danish Technical Research Council and the National Agency of Industry and Trade in Denmark for their financial support during that work.

Anders Bjarklev
Lyngby, Denmark
July 1993

Chapter 1
Introduction

Optical communications was introduced as a concept only 20 years ago, when it became possible to reduce the attenuation in silica glass to a few decibels per kilometer. Since then, optoelectronics and silica fibers have been the subject of largescale worldwide research and product development. As a result, optical communications is today established as one of the most promising technologies within the area of short- and long-distance data transmission.

A major thrust of research activity within the area of optical communications is to increase the capacity of optical systems. The ultimate capacity limits have until very recently been determined by the spectral bandwidth of the signal source and of the fundamental fiber parameters: attenuation and dispersion. In order to characterize a given transmission line, terms such as *loss-limited* or *dispersion-limited* systems have indicated the limiting property. In reviewing the development of optical systems, it can be seen that there has been an alternation between these two extremes.

Ten years ago, when most of the systems were made using multimode fiber, the difference of propagation constants between the fiber modes caused considerable pulse broadening, and the optical communication systems were clearly dispersion-limited. As reliable semiconductor lasers with high intensity and spectral purity became commercially available, an important turning point in the development of optical communication systems was reached. Namely, multimode fiber was superseded by single-mode fiber, in which only a much weaker chromatic dispersion can result in pulse broadening. At a specific signal wavelength (i.e., 1310 nm), standard silica fibers have no chromatic dispersion. This wavelength therefore became the most dominant, and the fibers were optimized for 1300 nm operation. Using these "standard fibers," the systems were, consequently, loss-limited. However, the minimal intrinsic loss of silica fibers is not at the zero dispersion wavelength but at 1550 nm, where the chromatic dispersion, on the other hand, is nonnegligible: 15 ps/(km·nm).

Many resources in the mid 1980s were guided towards the elimination of chromatic dispersion at 1550 nm. This was done by compensating for chromatic dispersion with an opposite and equally large waveguide dispersion through a carefully controlled design and manufacturing process. However, fibers of this type, normally called *dispersion-shifted* fibers, cannot be made without the introduction of additional losses in the fibers. A large-scale spreading of the dispersion-shifted fiber into commercial systems, however, did not immediately take place, partly because no urgent needs for zero-dispersion systems were present, since the ultralong transmission distances were still out of reach. This slightly restrained view of system improvement by dispersion-shifted fibers was supported by rapid component development, making dispersion less important as more spectrally pure signal sources were developed. Intensive international research on coherent communications, in addition, which is based specifically on signal sources of ultranarrow linewidth, indicated that installation of dispersion-shifted fibers with their somewhat poorer loss properties gave only little long-term guarantee that the best system was chosen.

There were, therefore, a number of practical problems that had to be overcome before a clear picture of future optical communication systems could be formed. Among the important subjects for clarifying this was the research on semiconductor amplifiers, which, however, suffered from problems such as cross-talk and polarization sensitivity in addition to difficult fiber-to-amplifier coupling.

At the same time, a parallel development took place, which in only three years would be shown to have a much larger impact on optical communication systems than that mentioned above. With a point of reference in work on rare-earth-doped glass lasers initiated as early as 1963 [1,2], the first fiber amplifiers (as possible useful devices for telecommunication applications [3]) were demonstrated in 1987. Progress since then has multiplied to the extent that amplifiers today offer far-reaching new opportunities in telecommunication networks [4–6]. The erbium-doped silica fiber amplifier for the 1550-nm telecommunication window has now become a well-established research laboratory tool. The fiber amplifier has been used both in system demonstrations on land and under the sea, and operational systems are commercially available from a number of manufacturers. In distribution experiments, splits of up to 39 million subpaths have been demonstrated for only two stages of amplification [7]. In short, this amplifier has revolutionized thinking on future optical fiber networks.

The main reason for the very strong impact that the optical amplifiers have and will have on data communication systems is to be found in the fact that they are unique in two respects: they are useful for amplifying input signals of different bit rates or formats, and they can be used to achieve simultaneous amplification of multiwavelength optical signals over a wide spectral region. These multiwavelength optical signals can carry different signal formats, including digital data, digital video, and analog video, allowing flexible upgrading in a broadband distribution network by adding extra *wavelength division multiplexed* (WDM) channels as needed. With optical amplifiers, therefore, the bottleneck of narrow and fixed bandwidths of electrical repeaters is avoided.

The outstanding properties offered by optical amplifiers are:

- High gain;
- High power conversion efficiency;
- Low noise;
- Low crosstalk;
- High saturation power;
- Polarization insensitivity;
- Broad spectral bandwidth;
- Very low coupling losses;
- Low cost.

This short list of fiber amplifier properties underlines the huge potential of these components, which most likely will extend to a large variety of applications in the years to come. The focus of this book will be, however, on the primary application, which is the optical amplifier for telecommunication systems, and the goal is to give a coherent description of the perspectives and limitations of fiber optical amplifiers. The organization of the book is aimed at providing the expert with a general reference source and the non-expert with a guide to the subject of rare-earth-doped fiber amplifiers.

First, it is necessary to understand in detail the physical properties of rare-earth-doped optical fiber. It is important to understand today's technological limitations, which are clarified through a description of the manufacturing process of rare-earth-doped optical fiber presented in Chapter 2. In the following chapters, the properties of rare-earth ions in glass host materials are discussed thoroughly. Chapter 3 presents the basic physical properties of rare-earth ions in silica and fluoride glass host materials, as well as the influence of different index raising co-dopants, and clarifies the most important mutual interaction between rare-earth ions. For all technical developments, it is essential to be able to characterize the components through measurements, and Chapter 4 is therefore focused on measurement techniques.

To date, the bulk of system work has centered on 1550-nm communication systems and has used erbium-doped silica-based fibers. In Chapters 5, 6, 7, and 8, the specific features and properties of *erbium-doped fiber amplifier* (EDFA) modules are discussed extensively. This naturally includes the use of other optical components, such as high-power semiconductor pump sources, fiber couplers, one or more isolators, and possibly optical filters. It is, however, not the purpose of this book to go into a thorough description of the physical properties of these additional components, since they are well characterized and described elsewhere. The focus will therefore be on erbium-doped fiber, which in Chapter 5 will be theoretically treated generally and in detail. Full-scale models of active erbium-doped fibers are presented, and related numerical problems in the solution of the governing differential equations are clarified and unraveled. A comprehensive analysis of the fiber design is performed in Chapter 6, and the results are used to optimize the performance of the EDFA used as in-line and preamplifiers, as well as power booster amplifiers. In Chapter 7, the different possibilities of pump wavelength choice are evaluated, and optimal properties are pointed out for various amplifier applications. Finally, the description of the EDFA component is extended in Chapter 8 by prescribing ways of achieving improved amplifier performance with respect to gain, noise, or gain efficiency. This is done by using more advanced amplifier configurations or by adding additional optics in the amplifier.

With the great potential of the EDFA, it is unfortunate that the vast majority of current operational systems are deployed in the 1300-nm band, and some fiber designs may prohibit upgrading to 1550 nm. An amplifier similar to the erbium-doped fiber, but for the 1300-nm band, is therefore very attractive, although it has been elusive until lately. However, very recently the 1300-nm amplifier has received renewed interest and stimulus [8–10] due to results on neodymium-doped and especially praseodymium-doped fluoride glass fibers. In Chapter 9, a presentation of these 1300 nm amplifiers is presented, showing theoretical limitations and basic experimental results. Furthermore, guidelines for the improvement and optimization of 1300-nm amplifiers will be given.

The increase in transmission distance or repeater spacing through the use of fiber amplifiers is not only the answer to many technological limitations. It will also reduce transmission equipment costs, improve line reliability, and simplify maintenance and operation functions of optical communication systems, as well as allow ultrahigh-speed transmission in direct detection systems. The advantages of coherent detection systems are no longer evident compared to direct detection schemes [11]. Chapter 10 is therefore devoted to a description of the potential of fiber amplifiers in direct detection systems, which, with the inclusion of fiber amplifiers, are expected to be the dominant technology for a number of years to come. The main interest is in erbium-doped fibers, and differences between booster, in-line, and preamplifier applications will be compared and evaluated, first by considering system upgrading possibilities by a single amplifier, and second by describing long transmission spans consisting of cascaded links of amplifier and passive fiber. Subjects such as gain optimization and *bit error rate* (BER) optimization of the active fiber length are described, as well as more specialized applications, such as remote pumped in-line amplifiers.

Direct detection will also be advantageous in soliton-based communication systems, where the existence of chromatic dispersion in combination with nonlinear effects can be used constructively for transmission of pulses, which will maintain their shape over extreme distances. For this type of high-speed intercontinental fiber connection without electrical signal regeneration, especially, the distributed fiber amplifier is a candidate [12]. The optimum use and design of such amplifiers are consequently the subject of Chapter 11, where the application in local area networks is also discussed. The distributed amplifier is here used as a power compensating backbone in a subscriber supply network with a large number of fiber taps along the distributed amplifier.

Analog *frequency division multiplexing* (FDM) optical transmission systems are attractive because the established analog FDM technology is combined with effective optical technology. For video transmission applications especially, fiber optical amplification of *amplitude modulated* (AM) signals is of increasing interest. Chapter 12 will therefore be focused on new possibilities that have become attainable with the use of fiber amplifiers.

REFERENCES

[1] Snitzer, E., "Neodymium Glass Laser," *Proceeding of 3rd International Conference on Quantum Electronics*, Paris, 1963, pp. 999–1019.

[2] Koester, C. J., and E. Snitzer, "Amplification in a Fiber Laser," *Applied Optics*, Vol. 3, 1964, pp. 1182–1186.

[3] Mears, R. J., L. Reekie, I. M. Jauncy, and D. N. Payne, "High-Gain Rare-Earth-Doped Fiber Amplifier Operating at 1.54 μm," *Optical Fiber Communications OFC '87*, Washington, D.C., 1987, Proc. Vol. 3, p.167.

[4] Simpson, J., "Fiber-Based Amplifiers: Progress in 1990," *Optical Fiber Communications OFC '91*, San Diego, 1991, Proc. Vol. 2, invited paper FA1, pp. 191–193.

[5] Millar, C. A., "Future Prospects for Active Fibre Devices," *European Conference on Optical Communication ECOC '90*, Amsterdam, 1990, Proc. Vol. II, invited paper, pp. 717–724.

[6] Ainslie, B. J., "Recent Advances in Discrete and Distributed Fibre Amplifiers," *European Conference on Optical Communication ECOC '91*, Paris, 1991, Proc. Vol. 2, invited paper, Vol. 2, pp. 47–50.

[7] Hill, A. M., R. Wyatt, J. F. Massicott, K. J. Blyth, D. S. Forrester, R. A. Lobbett, P. J. Smith, and D. B. Payne, "39.5 Million-Way WDM Broadcast Network Employing Two Stages of Erbium-Doped Fibre Amplifiers," *IEE Electronics Letters*, Vol. 26, No. 22, 1990, pp. 1882–1884.

[8] Miyajima, Y., "Progress in 1.3 μm Fluoride Fiber Amplifiers," *Techn. Dig. of Optical Amplifiers and Their Applications OAA '92*, Santa Fe, 24–26 June 1992, invited paper WB1, pp. 4–7.

[9] Durteste, Y., M. Monerie, J. Y. Allain, and H. Poignant, "Amplification and Lasing at 1.3 μm in Praseodymium-Doped Fluorozirconate Fibres," *IEE Electronics Letters*, Vol. 27, No.8, 1991, pp. 626–628.

[10] Carter, S. F., D. Szebesta, S. T. Davey, R. Wyatt, M. C. Brierley, and P. W. France, "Amplification at 1.3 μm in a Pr^{3+}Doped Single-Mode Fluorozirconate Fibre," *IEE Electronics Letters*, Vol. 27, No. 8, 1991, pp. 628–629.

[11] Tonguz, O. K., and R. E. Wagner, "Preamplified Direct Detection Versus Heterodyne Detection: Can They Be Equivalent?" *European Conference on Optical Communication ECOC '91*, Paris, 1991, Proc. Vol. II, paper WeC9-4, pp. 593–596.

[12] Simpson, J. R., H. T. Shang, L. F. Mollenauer, N. A. Olsson, P. C. Becker, K. S. Kranz, P. J. Lemaire, and M. J. Neubelt, "Performance of a Distributed Erbium-Doped Dispersion-Shifted Fiber Amplifier," *IEEE Journal of Lightwave Technology*, Vol. 9, No. 2, 1991, pp. 228–233.

Chapter 2
Manufacturing of Active Fibers

The first condition for the successful evolution of optical-fiber amplifiers was the development of controllable and reproducible fiber manufacturing processes. However, even though early reports on applications of rare-earth-doped fibers date back to the 1960s [1,2], the real breakthrough first happened almost two decades later, when single-mode fiber technology reached maturity. At this point, the theoretical limitations of passive optical direct detection system links were close to being met, making the need for an optical amplifier obvious. At the same time, it became possible to fulfill one of the basic requirements of fiber amplifiers: to achieve high pump light absorption at the same time as low loss at the signal wavelength. The key point of this new development was the demonstration and control [3] of low rare-earth dopant levels, whereby long interaction lengths and very low losses became inherent in the fibers.

In order to understand this, we must look at how rare-earth ions are included in glasses, as described in [4]. For simplicity, the explanation is at first illustrated by silica-based glasses, which are made up of network formers (e.g., SiO_2 and GeO_2). These form a covalently bonded structure, unlike conventional silicate glasses, which also contain network modifiers (e.g., Na^+ and Ca^{2+}), which allow modest processing temperatures to be realized by breaking up the covalent network. The rare-earth ions are usually trivalent, such as Er^{3+}, and they can replace a network modifier. In conventional laser glasses, a high concentration of nonbridging oxygen groups is important to allow for the incorporation of a few weight percents of rare-earth ions [5]. Since the pump light in this situation is absorbed over a relatively short path length, a high rare-earth ion concentration becomes necessary, and we find here one of the reasons for the development of multicomponent silicate and phosphate laser glasses. However, in glasses where network modifiers are absent, such as in optical fibers of silica-based glasses, a rigid structure exists and there is a lack of nonbridging $Si\text{-}O^-$ groups. This makes the coordination of Er^{3+} difficult and it contributes to a lack of solubility, which leads to clustering in the glass [6]. As we shall see in Chapter 3, this clustering is extremely harmful, since it leads to depopulation of the upper laser level of the erbium ions. The technical challenge was therefore to keep the rare-earth concentration low enough to avoid considerable clustering effects, or to modify the core glass to allow increased rare-earth ion solubility.

It was, however, not an easy task to overcome the problems of rare-earth ion clustering. This becomes clear when we look at one of the basic problems of including rare-earth ions in the optical fibers. The point of reference is taken from the well-developed methods for fabricating low-loss silica (SiO_2) fibers for communication purposes. These methods are all based on the oxidation or hydrolysis of silicon tetrachloride, and index-raising dopant ions (such as germanium, aluminum, or phosphorus) are introduced into a reaction stream as halide vapors [7]. Also, index-lowering dopants such as boron and fluorine are included in a similar way. The halide compounds of rare-earth ions have, however, generally lower vapor pressure (are not as volatile) than the commonly used chlorides and fluorides, thereby requiring vaporization and delivering temperatures of several hundred degrees (see Figure 2.1). The limitations imposed by high temperatures (500° to 1000°C) required for reasonable vapor pressure have caused a number of vapor and liquid phase methods to evolve.

The fundamental approaches of these vapor and liquid phase methods represent a central point in the understanding of the possibilities and restricting factors of fiber amplifiers. It is therefore the aim of this chapter to review the different methods for the manufacturing of rare-earth-doped fibers and to point out some of the important differences between these and standard fiber fabrication. It must also be noted that a complete guide to all the chemical and mechanical problems

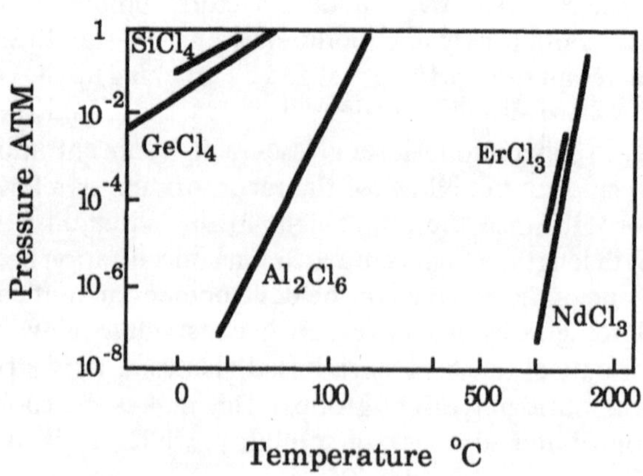

Figure 2.1 Vapor pressures of reactant halides. (Data from [3,7].)

of fiber manufacturing is not attempted, but merely to point out the major trends in the new development.

Appearing in the first part of this chapter are the fabrication methods for silica fibers, which represent adaptations of processes used to fabricate long-haul communication fiber with low attenuation. Special attention is also given to the fabrication of fluoride glass host fibers, which, as we shall see in Chapter 9, are of importance for the development of 1300-nm amplifiers. The requirements of small core diameters have also brought traditional rod-in-tube techniques [7] into focus, and their latest developments are described. Demands for low-noise amplifiers (as described in Chapter 8) added new requirements to the fiber drawing process, which will be discussed in the final part of this chapter.

2.1 RARE-EARTH IONS IN SiO_2 GLASS HOSTS

Due to the problems of rare-earth ion clustering, extensive studies of host materials have been performed [8,9] with the aim of finding a host-glass composition, which could alleviate the clustering problem. The majority of these studies have, however, still been within the confines of providing high-silica-based fibers, since these can be manufactured on the basis of the mature technology of fabricating low-loss silica fibers for communications. Through the use of well-established methods, controllability of the waveguide structures can be preserved.

The importance of maintaining the properties of passive optical fibers becomes clear because the optical waveguide enhances the performance of active devices primarily by increasing the pump energy flux in the vicinity of the rare-earth dopant. As we shall see in Chapter 6, the requirements of the waveguide structures are not weaker for active fibers than for passive, but they are different on certain points, at least as long as we do not consider distributed amplifiers. On one hand, the need for very low transmission losses at the signal wavelength is not in focus for active fibers, since only relatively short lengths are in use. On the other hand, do requirements for optimal performance direct the manufacturer towards fibers with a much higher *numerical aperture* (NA) than that needed for standard passive optical fibers? It is, however, still too early to prescribe standardization requirements for active optical fibers. Many system constructors do believe that the active fiber design should be nearly the same as that for telecommunication fibers to promote compatibility with the long-haul fiber systems. One way to find the best compromise between these points of view might be to fabricate fibers with the rare earth concentrated in the center of the core. This approach has already been demonstrated to lead to a more efficient use of pump power [10]. A more quantitative description of these possibilities is presented in Chapter 6.

Three different manufacturing processes are of interest for the design of active fibers: the *modified chemical vapor deposition* (MCVD) method [11], the *vapor axial deposition* (VAD) method [12], and the *outside vapor deposition* (OVD) method [13]. These mainstream fiber fabrication methods can be employed together with either *vapor-phase techniques* or *liquid-phase techniques*, as we shall discuss in the following sections.

2.1.1 Incorporation of Rare-Earth Ions in SiO_2 Fibers Using the MCVD Method

The vapor pressure of the halide starting materials used in telecommunication fibers is several tens of torrs at room temperature [4]. However, as discussed above, the much lower vapor pressure of the rare-earth halides require the heating of the reactants to several hundred degrees Celsius to approach a working vapor concentration.

Solutions to the low vapor pressure problem as applied to the MCVD method are shown for the configurations employed in Figure 2.2. Here, examples of heated sponge, heated source, and heated injector are shown to deliver rare-earth chloride vapor for reaction downstream by oxidation along with other standard reactants [14–16].

In all of these methods the low vapor pressure reactant is accommodated by bringing the vapor pressure source close to the reaction zone and immediately diluting the source by mixing it with other reactants.

In the case of the *dopant chamber* approach (as shown schematically in Figure 2.2(a)), a standard silica substrate tube has been modified just upstream from the reactant input end by adding a dopant chamber, inside which a rare-earth salt has been fused to the wall [3]. The required dopant, which, for example, could be $NdCl_36H_2O$ (the melting point of a 99.9% pure salt is 758°C), is during this process also dehydrated by heating it under a chlorine atmosphere. The next step is to clean the inside of the deposition tube by using SF_6 to remove any dopant deposited during the drying process. The cleaning is important, because anhydrous $NdCl_3$ crystals would form bubbles in the subsequently deposited cladding class. In the process of core deposition, during which $SiCl_4$, $GeCl_4$, and O_2 are fed down the tube, the dopant carrier chamber is heated to around 1000°C by a stationary second burner in order to produce small quantities of $NdCl_3$ vapor. The vapor is carried downstream by the reactant flow, where it is oxidized to Nd_2O_3 in the hot zone formed by the deposition burner and incorporated into the core. Note also that some control of the rare-earth concentration is possible by adjusting the temperature of the dopant chamber. As reported in [3], the first drying stage did not dehydrate the $NdCl_3$ sufficiently to produce low-loss fibers [17]. Consequently, a second drying process must be used, in which the core, consisting of SiO_2, GeO_2, and a small amount of Nd_2O_3, is deposited unfused at a low temperature. The porous core layer on the inside of the deposition tube is subsequently dried by

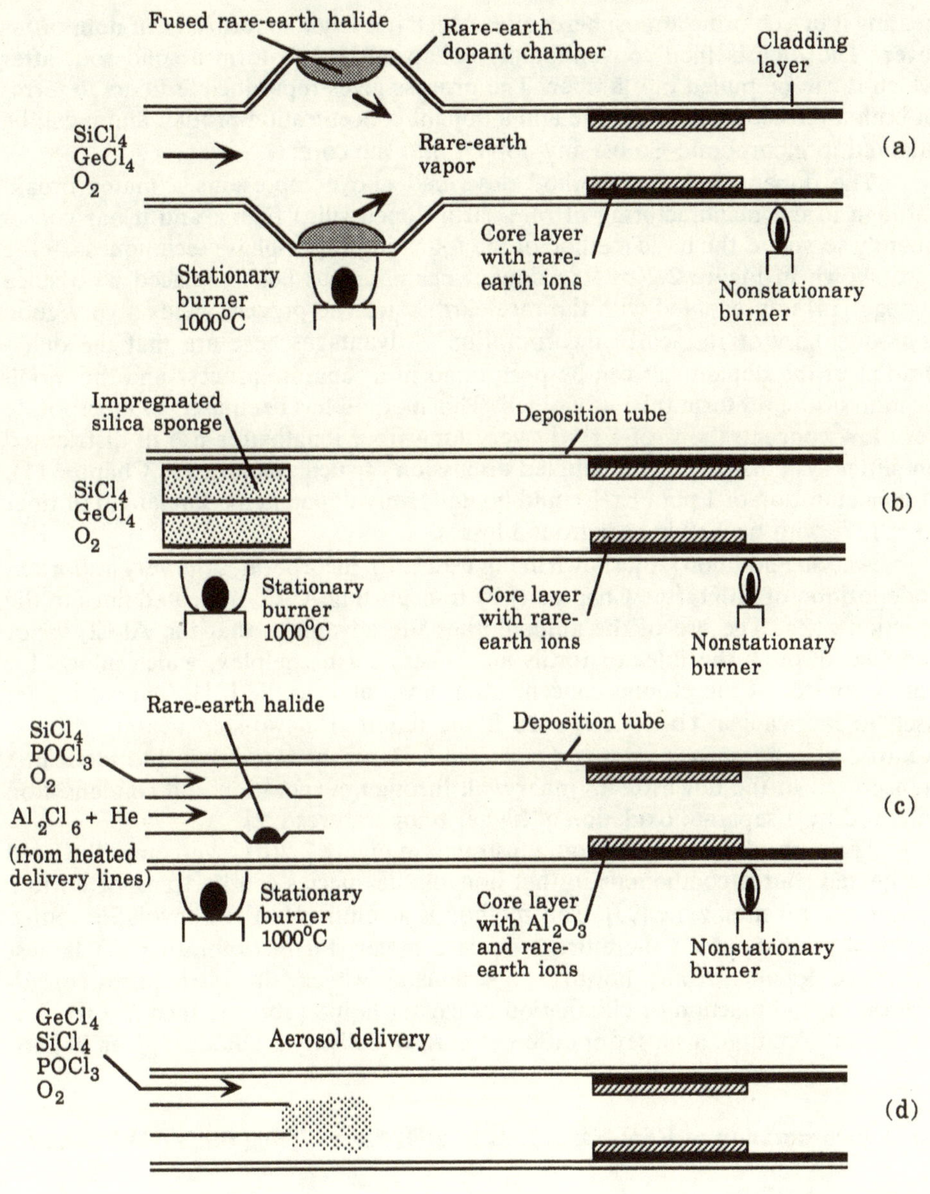

Figure 2.2 Low vapor pressure rare-earth dopant delivery approaches applied to the MCVD fabrication method (layout reproduced from [7]): (a) rare-earth halides placed in heated dopant chamber; (b) an impregnated silica sponge is heated; (c) incorporated Al_2O_3 deposition from Al_2Cl_6; (d) aerosol delivery.

heating it in a chlorine atmosphere, after which it is fused to form a clear nonporous layer. The tube is then conventionally collapsed [18] to form a solid rod, after which it can be pulled into a fiber. The process gives reproducible fibers in terms of both a refractive index profile and a dopant concentration profile, and it can be adapted to incorporate almost any dopant into the core.

The dopant chamber method described above represents a major breakthrough in the manufacturing of rare-earth-doped silica fibers, and it has consequently served as the basic method of the following vapor-phase techniques. In the case shown in Figure 2.2(b), the dopant chamber has been replaced by a silica sponge [14] impregnated with the rare-earth salt. The process yields a very good reproducibility of rare-earth incorporation. Advantages here are that the dehydration of the dopant salt can be performed in a separate process and the modification of the preform tube is avoided. This method has been used to incorporate very low concentrations of Er^{3+} in very long fiber lengths for use in distributed amplifier systems [19] (for a detailed discussion of such systems, see Chapter 11). A concentration of 1 ppm Er^{3+} could be uniformly doped in a 6-km length of fiber to $\pm 12\%$ with negligible background losses.

Another addition [20], shown in Figure 2.2(c), incorporates the very important codeposition of Al_2O_3 (see Chapter 4) by transporting Al_2Cl_6 in heated lines to the reaction zone. The use of the ampoule has the advantage that the Al_2Cl_6 vapor supplied through the injector forms an Al-rare-earth complex, which allows for better control of the erbium concentration level and profile [21]. One variation, used to fabricate a Yb_2O_3-SiO_2 core fiber, required a two-step heating process referred to as a *transport-and-oxidize method*. Here, the rare-earth chloride is first transported to the downstream inner wall through evaporation and condensation followed by a separate oxidation at higher temperatures [22].

The aerosol delivery method, illustrated in Figure 2.2(d), overcomes the need for heated source compounds in that nonvolatile species can be transported in a mist to the reaction zone [23]. This method is a solution to the nonvolatile source problem, and it should therefore provide a means for incorporating the largest variety of dopants. It has, however, a serious drawback, since it requires consideration for the reaction or elimination of carrier liquid products such as OH. For further information, a survey of different aerosol delivery methods is given in [24].

2.1.2 Incorporation of Rare-Earth Ions in SiO_2 Fibers Using the VAD Method

Rare-earth vapor reactions using the VAD hydrolysis process have also been demonstrated. Since the preforms manufactured by the VAD method are porous, the method is well suited for solution doping techniques, as described in Section 2.1.4.

However, to avoid the stage of impregnation of porous preforms, direct rare-earth incorporation methods have been developed.

In the VAD method, the doped glass is deposited using either a single flame burner or a more complex process, such as the double flame burner approach, which is illustrated in Figure 2.3. Here, the glass particles appear after chemical processes in the burners and they are then deposited axially by rotation of the soot preform. High rate deposition is accomplished by a double flame burner as described in [25]. The rare-earth doping of these soot preforms should in principle be possible by applying organometallic sources [4] to the center burner. As a result, β-diketonate complexes of rare-earth metals [26], which are solid substances with a melting temperature of approximately 100°C, are formed and deposited together with the other vapor source components.

The soot boules generated by the VAD method undergo a secondary drying and sintering process before being consolidated under very high temperatures (1400° to 1500°C) with fully sintered silica preforms. Experiments of this consolidation in an atmosphere of a rare earth have been attempted with some success with Nd^{3+} [28]. From these experiments, which were performed using the OVD technique described in the next section, it was found that the choice of consolidating atmosphere was important, because oxidizing conditions caused a buildup of Nd_2O_3 on the outer surface of the boule. A uniform distribution resulted when using helium as a sintering gas. This technique could also be extended to Er^{3+} incorporation, but in all cases a partly separate manufacturing of core and cladding must

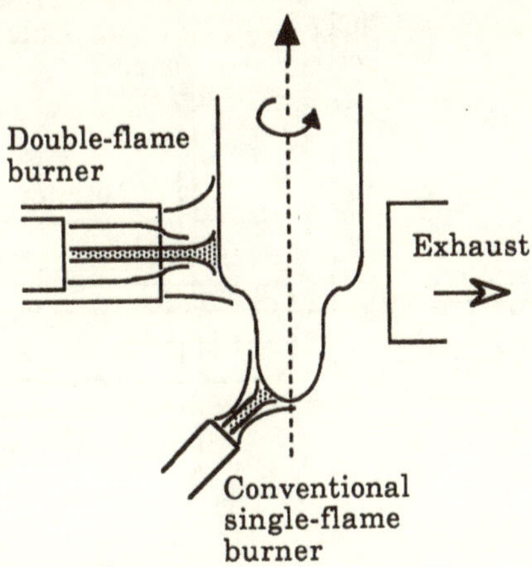

Figure 2.3 Fabrication of fiber preforms using the VAD method [25,27]. Glass particles are deposited axially on the rotating preform by using a double flame burner.

be considered. The importance of this will become clearer in Chapter 6, where the advantages of confining the erbium to the fiber core center are discussed.

Other approaches for combining conventional VAD techniques are likely, but they are yet to be seen. As an example, it can be mentioned that although there are no reported rare-earth solution aerosol flame methods, delivery of a nebulized aqueous solution of lead nitrate has been reported [29]. Likewise, according to [7], there are no reports of rare-earth chloride vapor flame reactions, even though this seems to be a likely method.

2.1.3 Incorporation of Rare-Earth Ions in SiO_2 Fibers Using the OVD Method

In the OVD method, a porous boule is formed on the surface of a rotating mandrel as shown in Figure 2.4. The deposition is done layer by layer as the single flame burner is moved back and forth in the longitudinal direction of the rotating soot preform. Refractive index differences between core and cladding compositions are produced by varying the reactants fed into the burner, as in the MCVD method. Finally, the central mandrel is removed and the porous preform is dried and fused in a He-Cl_2-O_2 gas mixture at temperatures between 1400° and 1600°C.

The incorporation of cerium, neodymium, and erbium has been accomplished in the OVD method by introducing organic vapors containing a rare earth into the reaction flame [26,28]. For the introduction of cerium, the organic source was

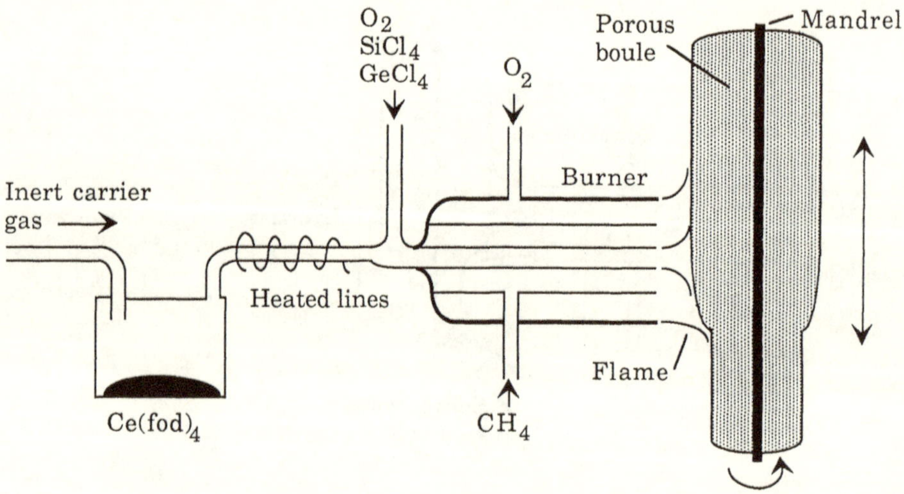

Figure 2.4 Rare-earth doping by an organo metallic source applied to the OVD process. (Figure layout from [4].)

cerium β-diketonate (Ce(fod)$_4$), which allowed delivery to the reaction flame by a more traditional bubbler carrier system with heated delivery lines [26]. This was possible because the (Ce(fod)$_4$) possessed a vapor pressure of about one torr for temperatures around 150° to 200°C.

As is the case for VAD preforms, a secondary drying and sintering process must be applied, which provides another opportunity for dopant incorporation; namely, rare-earth dopant vapors can be incorporated in the glass through a post-deposition diffusion process during sintering [30].

2.1.4 Solution Doping of Rare-Earth Ions

All the vapor phase methods for the manufacturing of rare-earth-doped fibers include some means of overcoming the low vapor pressure of the reactant rare-earth halides. As we have seen, this is normally accomplished by preheating the reactants under highly accurate temperature control, whereby a prescribed concentration of rare-earth ions is achieved. Because of the extra complexity this adds to the fabrication process, alternative methods become very interesting.

Therefore, we will discuss in this section the liquid-phase methods, which are used today to a large extent for commercial manufacturing of rare-earth-doped fibers. The first reports of the incorporation of low-volatility halide ions into high-purity fiber preforms were in the early 1970s and were called the *solution doping technique* [31] or *soot impregnation method* [32].

The principle of the solution doping technique is demonstrated in Figure 2.5 in the specific case in which it is combined with the MCVD method. As the figure shows, the first step is to deposit a cladding layer inside a substrate tube in the normal way for standard preform fabrication. After this, the core deposition is carried out at a lower temperature so that the layers are not fully sintered. The substrate tube appears at this stage with a white layer on the inside. This porous layer is then soaked in an aqueous or alcoholic solution of the rare-earth salt, whereby impregnation of the pores of the core layer will take place. Soaking time of an hour or more is typical [7]. The impregnated layer is then dried by a Cl_2-O_2 flow and fused into a clear glassy layer. Finally, the tube is collapsed into a preform in the usual manner. It is important to note that the Cl_2 drying process must be carefully performed, especially if the soaking took place in an aqueous solution. Incomplete drying will leave water in the fiber core, which is a well-known cause of highly increased fiber losses in the 1.55-μm wavelength band. Note also that the solution doping technique can be used to co-dope with Al_2O_3 [33] as well as with the rare earth. This will be outlined in the following chapter.

The principle of liquid-phase doping can be applied for any porous silica-based material, and, as mentioned earlier, adaptation to the VAD method is not complicated; only a two-step process becomes necessary. The procedure is illus-

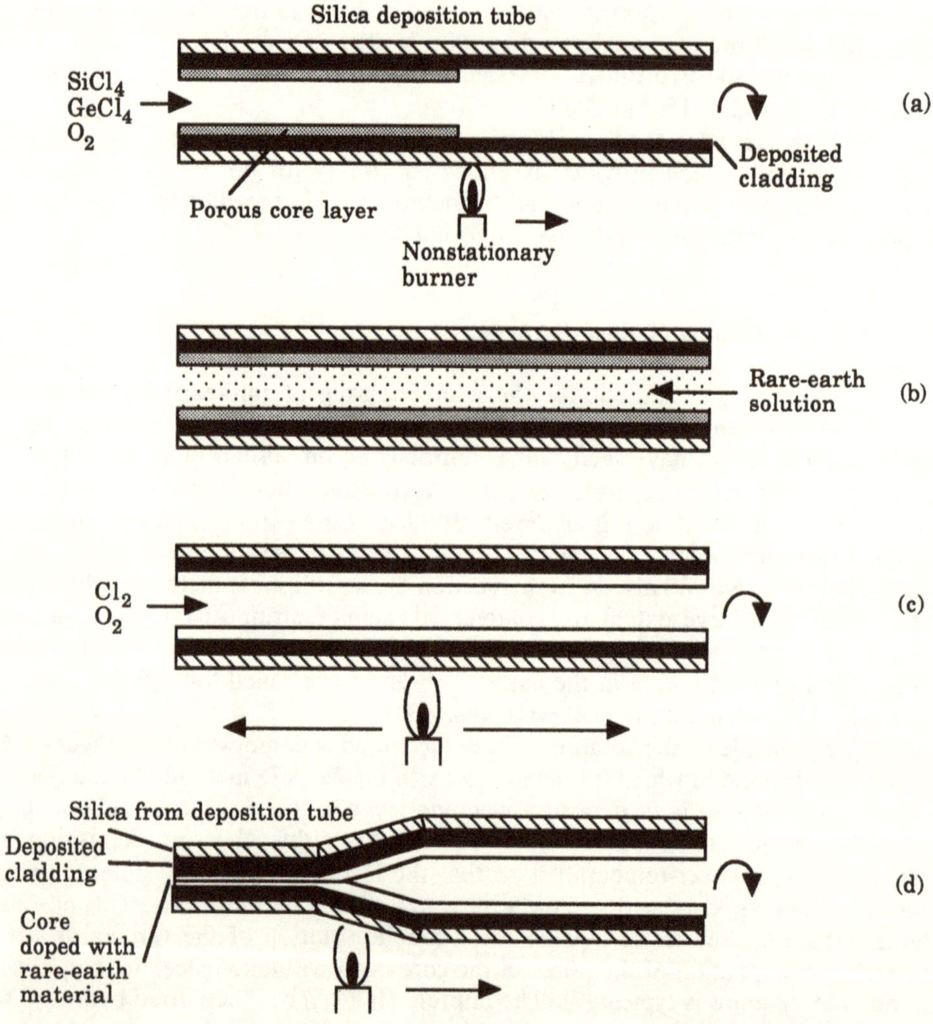

Figure 2.5 Solution doping applied to the MCVD method (illustration combined from descriptions in [4,7]): (a) deposition of a porous core layer; (b) soaking the porous core layer in a solution of rare-earth salt; (c) drying the impregnated porous layer; (d) fusing the porous layer and collapsing the tube into a solid rod preform.

trated in Figure 2.6. A pure silica boule is first fabricated by flame hydrolysis, and this boule is then immersed in a methanol solution of the dopant salt for typically one hour [32], after which it is dried for 24 hours. The dried boule is sintered to

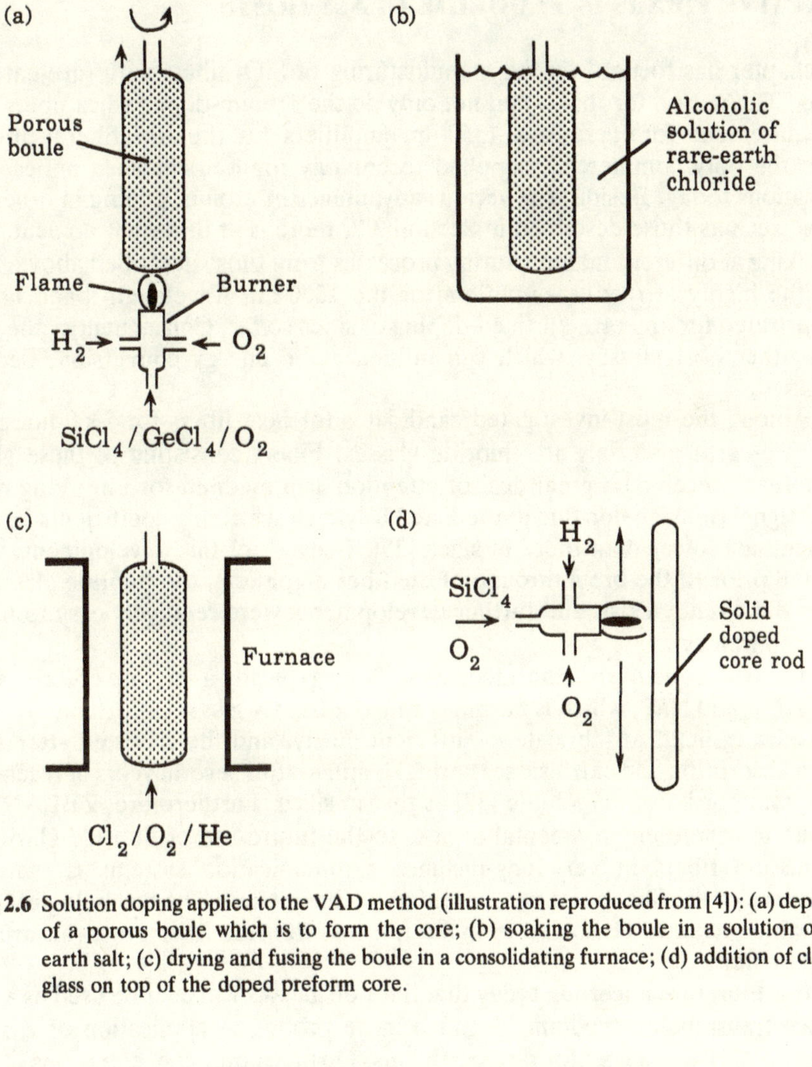

Figure 2.6 Solution doping applied to the VAD method (illustration reproduced from [4]): (a) deposition of a porous boule which is to form the core; (b) soaking the boule in a solution of rare-earth salt; (c) drying and fusing the boule in a consolidating furnace; (d) addition of cladding glass on top of the doped preform core.

a bubble-free glass rod, which has an approximately homogenous dopant concentration controlled by the ion concentration in the solution. However, the process still needs its second step to be finished by adding cladding glass on top of the rare-earth-doped core glass. The porous cladding is finally sintered before the preform can be drawn into a fiber. The method referred to as *molecular stuffing* [34] can generally be used, and it can also be applied to the OVD method.

It is also important to note here that the same strict requirements for drying are necessary, as in the case for solution doping of MCVD preform tubes.

2.2 ACTIVE FIBERS IN FLUORIDE GLASS HOSTS

This chapter has focused on the manufacturing of SiO_2 fibers with rare-earth co-doping. The reason for this is that not only do the erbium-doped silica fibers fulfill the requirements for operational 1550-nm amplifiers, but the silica fiber technology is also the only commercially applied technology for long-distance optical communications today. Besides the various advantages of erbium doping in other host glasses, such as those described in Section 3.2, there is at first sight no acute need for looking at different manufacturing processes from those described above. However, the highly attractive amplifiers for the 1300-nm wavelength band are not demonstrated for any rare-earth co-doping of silica so far. Consequently, the interest in other host glasses, which can influence the energy conversion, becomes intensified.

Among the most-investigated candidates for host fibers for 1300-nm amplifying rare-earth materials are fluoride glasses. Fibers consisting of these glasses have already received a great deal of attention as a medium for ultra long repeaterless signal transmission due to predicted Rayleigh scattering coefficients an order of magnitude lower than those in silica [35]. Because of this development, which occurred prior to the breakthrough of the fiber amplifiers, the fluoride glass technology did already exist, and further developments were relatively easy to handle in the laboratory.

The backbone of the fluoride glasses [36] is typically a mixture of ZrF_4, BaF_2, LaF_3, AlF_3, and NaF, which is normally called ZBLAN glass. Unfortunately, these glasses are difficult to fabricate in sufficient purity, and the reported attenuation (due to absorption and extrinsic scattering) is still—after several years of research—on the same order of magnitude [37] as that in silica. Furthermore, ZBLAN glass is prone to severe environmental attack, so the future use of fluoride glasses for transmission fibers in very-long-distance communication systems is, with the appearance of the fiber amplifier, considerably postponed. The technological problems, which need to be solved to make low-loss fluoride fiber practical, are very considerable.

It is therefore uncertain today that fluoride glasses will ever be used as a long-distance transmission medium. A much more promising application of ZBLAN glass is therefore as a host for rare-earth ions. Furthermore, concentrations of rare-earth ions up to several percent are possible by substituting LaF_3 for the desired rare-earth fluoride [21].

In particular, fluoride glasses can be more efficient hosts for rare-earth ions than silica, since the lifetime for nonradiative decay is longer. This is due to the weaker glass network and because excited state absorption is greatly diminished for some ions [38] (for further discussion of these issues, refer to Chapter 3).

2.2.1 The Built-in Casting Process

The standard method for fabricating ZBLAN fibers is called *rotational casting* and is illustrated in Figure 2.7. The first step is to melt the cladding material of the appropriate composition, and it is important to note that the melting point of 850°C is considerably lower than that for SiO_2 glass. The melt is then poured into a mold, which is spun around its axis to form a tube of the cladding material. The core composition is then poured into the resulting hole and the ensemble is drawn into a fiber.

The standard process described above is only suitable for making multimode fibers [39] because the very small core diameters required for single-mode fiber operation cannot be made in a controlled way. The multimode fibers have, however, a number of drawbacks, since the structures do not allow a sufficiently high pump

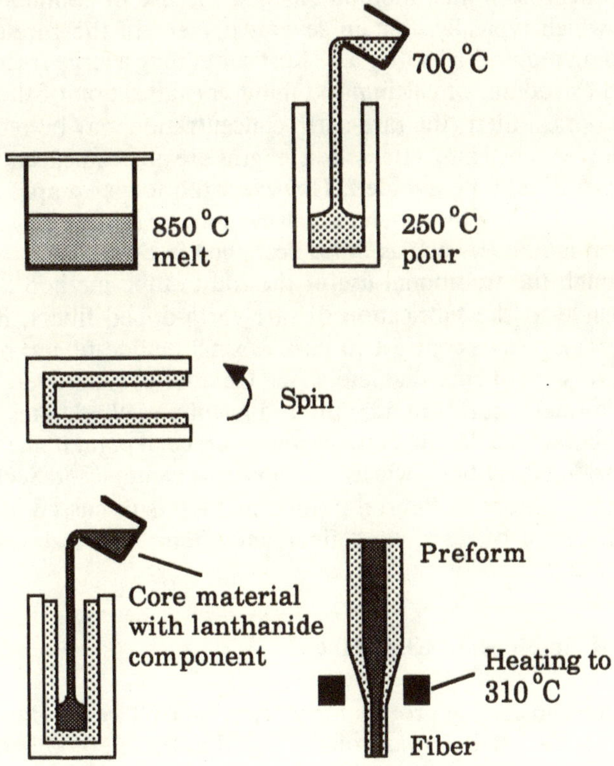

Figure 2.7 Standard process for the manufacturing of rare-earth-doped fluoride glass fibers. (Illustration reproduced from [38].)

power intensity, and there are severe problems with the coupling from the active fiber to a standard single-mode fiber. As we shall see in the following section, solutions to these problems are prescribed, but before the discussion of the rotating casting method is left, it is important to note that it allows construction of fibers with more than one core layer. Namely, circular rings can be made by adding more steps to the deposition process, and this becomes relevant for new and highly effective 1300 nm amplifiers, as described in Chapter 9.

2.3 PREFORM OVERCLADDING TECHNIQUES OR ROD-IN-TUBE METHODS

One of the simplest techniques for fabricating rare-earth-doped fibers consists of inserting a rod containing the rare-earth ions into a tube. The tube becomes the cladding once the ensemble is drawn into the fiber [40,41]. At an early stage of development, this *rod-in-tube* method enabled the use of commercially available laser glasses, which typically contain several percent of the rare-earth material. Such high loading requires a "soft-glass" host containing a large fraction of network modifiers such as sodium or calcium to inhibit crystallization of the rare earth. In a single-mode optical fiber, the rare-earth concentration may be reduced to below 1,000 ppm or lower, since long interaction lengths are no obstacle and crystallization or clustering can hereby be avoided. However, the longer amplifier lengths also set the requirement of low-loss fibers, which can only be achieved by current optical-fiber fabrication methods, such as those described in Section 2.1.

Even though the traditional use of the rod-in-tube method is therefore not directly applicable to the fabrication of rare-earth-doped fibers, it is still highly relevant. For example, in contrast to passive single-mode fibers, optimum active fibers require very small core diameters (see Chapter 6), and overcladdings of the preforms are normally performed by the rod-in-tube method. This is not only the case for silica fibers: overcladding also becomes a crucial point in the manufacturing process of fluoride glass fibers, which are chosen as examples in Section 2.3.1. The second new development of the rod-in-tube method is discussed in Section 2.3.2. Here, the fabrication of distributed fiber amplifiers by the so-called *seed fiber technique* is presented.

2.3.1. Single-Mode Fluoride Glass Fibers

The standard built-in casting process for fabrication of fluoride glass preforms was presented in Section 2.2.1. It was pointed out that the method is limited to multimode fiber fabrication because of the problem of controlling the small core radii. In the early stages of the development of rare-earth-doped fiber, this made fluoride glass an even weaker candidate as a host material. The reason was that the multimode nature was an extra difficulty on top of problems such as the fluoride fibers' vulnerability to environmental attacks.

Adaptations to the standard rotational casting method therefore had to be developed, and it was a remarkable step in the development of active fluoride glass fibers when the *stretch and overclad* process was demonstrated [39]. Here, normal multimode preforms, which are fabricated as illustrated in Figure 2.7, are "stretched" from about 10 mm to a final diameter of about 1 mm. The stretched preform is then overclad with a cladding tube in order to reduce the core-to-clad ratio to single-mode dimensions in the final fiber.

Alternatively, a double-crucible arrangement, as illustrated in Figure 2.8, allows codrawing of the core and cladding melts simultaneously [21]. This process, which also represents a rod-in-tube method, often needs, however, extra over-cladding by the stretch and overclad approach.

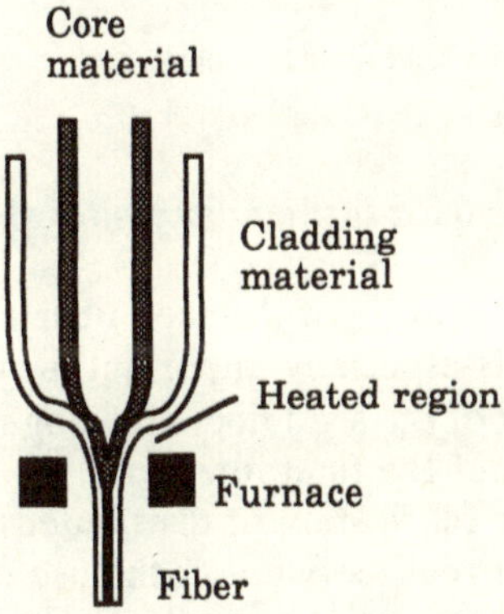

Figure 2.8 Double-crucible method for drawing soft-glass fibers. (Illustration based on [21].)

2.3.2 Manufacturing of Distributed Fiber Amplifiers

One of the first reports on active fibers with a very low concentration of rare-earth ions was made by Simpson et al. [42], who presented a method in which the dopant was introduced by a rod, in fiber form, analogous to a scaled down rod-in-tube method. A doped silica preform was first fabricated by the MCVD method, interrupting the collapse step to leave a small bore of 1 mm in diameter. As illustrated in Figure 2.9, a "seed" fiber was then introduced by inserting it into the small preform bore. The collapse process was thereafter resumed, bringing the preform to a solid rod as in a conventional MCVD technique. From accurate control of this process, a uniform dopant level was ensured with a higher concentration near

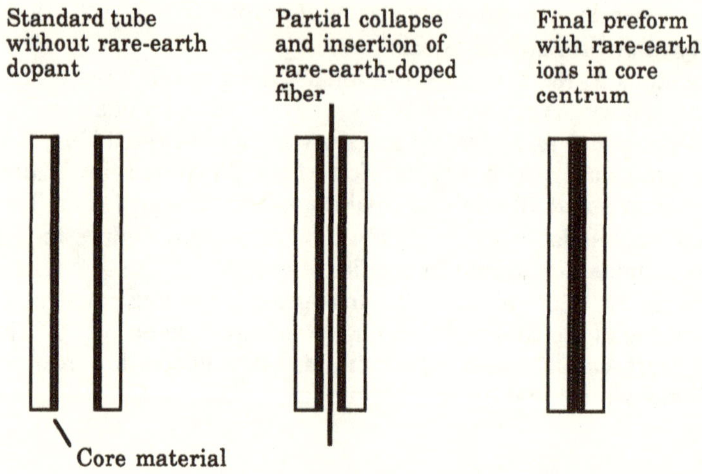

Figure 2.9 The principal steps in the seed fiber technique.

the core center. See Chapter 6 for further discussion of the advantages of this rare-earth ion placement.

As reported in [42], the purity of the seed fiber and the care with which it is placed in the support tube is especially important in realizing a low-extrinsic-loss distributed fiber. The purity of the seed fiber cladding is also important in that this material will become a part of the final fiber core.

In connection with the fabrication of distributed fiber amplifiers, it is worth noting that they can also be realized without the use of the seed fiber technique. This requires, however, concentrations of erbium oxide as low as 10 to 100 ppb [42], corresponding to a maximum fiber attenuation of around 0.2 to 0.3 dB/km. The low concentrations allow for just enough gain per unit length to cancel the fiber loss. To relate this concentration to the cases of other erbium-doped fibers, it can be mentioned that amplifiers with erbium oxide concentrations of up to 1400 ppm (corresponding to a maximum fiber attenuation in the relevant wavelength interval of 38 dB/m) have been reported [42].

2.4 NEW DEVELOPMENTS IN FIBER DRAWING TECHNIQUES

The subject of fiber drawing techniques is not normally described in much detail in literature concerning fabrication of rare-earth-doped fibers. The basic reason for this is that the drawing techniques represent mature processes, which over the years have been developed to obtain high reproducibility in the fabrication of optical fibers. Even more important is that the rare-earth-doped fibers did not at first require new development in drawing techniques, since standard equipment has been sufficient in this last step of the fiber manufacturing process.

A special aspect of the fiber drawing techniques is, however, included in this chapter because development of advanced and highly effective active fibers suggest a new challenge for the fiber manufacturer. This challenge is to fabricate long, tapered fibers, which have special advantages for low-noise applications of fiber amplifiers.

The traditional way of performing the fiber drawing process is to heat the preform with a circular graphite furnace placed just below the vertically fixed preform. The furnace has a temperature above 2000°C for SiO_2 fibers, which enables the glass to flow, and a fiber with a constant radius can then be drawn by controlling the force (or drawing speed) applied to the fiber. The diameter of the fiber is controlled by an optical diameter monitor placed below the furnace, and the output from this monitor is then servo-linked to the winding drum speed controller. This enables a diameter control of better than $\pm 2\ \mu m$. The fiber is finally coated using a cup applicator as shown in Figure 2.10. With this technique, one or more polymer coating layers, which are cured by ultraviolet light, can be added.

As illustrated in Figure 2.10, this method has been improved further, so very long tapered fibers can be manufactured. This is accomplished by a highly accurate

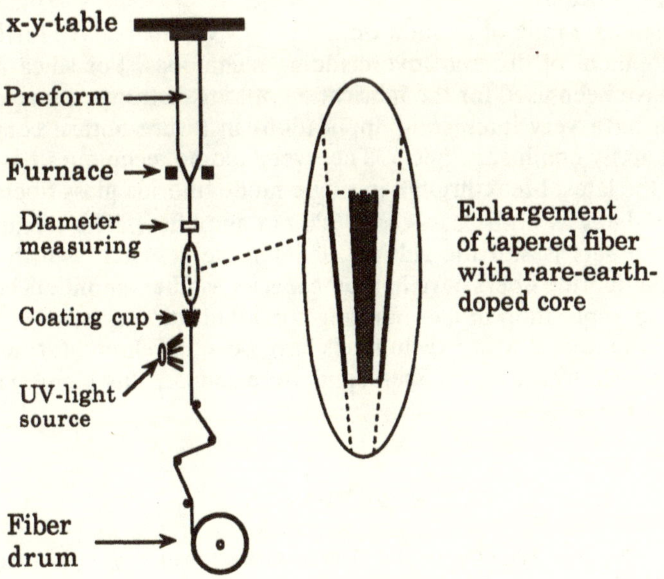

Figure 2.10 New advances in drawing process control made relevant by the development of fiber amplifiers.

control of the drawing speed and furnace temperature. It allows the manufacturing of fibers, whose outer diameter only varies a few tenths of a micron per meter of fiber.

2.5 SUMMARY

Although they were demonstrated more than two decades ago, rare-earth-doped silica fibers have only been fully recognized during the last few years. A main reason is that single-mode systems have only recently been installed in networks, so the requirements and applications of fiber based optical amplifiers were not present during the 1960s and 70s. Among the additional reasons for this "delay" has been that the fiber manufacturer had to face the problem of incorporating dopant halides with vapor pressures much lower than those for the standard reactants. Solutions were first to develop gas phase methods, which applied preheating of the rare-earth dopant reactants, and secondly to redevelop liquid-phase techniques, which for this reason have had a renaissance during the last few years.

Another challenge has been the development of fibers with NAs more than or around twice the values seen for standard passive optical fibers. However, there is the advantage that much larger intrinsic fiber loss is acceptable in active fibers, and index-raising co-dopants such as aluminum have found their way back to the manufacturing process.

The small core radii of the new optimal high-NA fibers have furthermore led to the development of different overcladding techniques. For silica fibers, these techniques have been used for the fabrication of long distributed rare-earth-doped fibers, which have very interesting applications in future optical communication systems that apply nonlinear effects. The overcladding techniques have also been essential for the latest breakthrough in single-mode fluoride glass fibers, which are the best candidates as host glasses for 1300-nm amplifiers. The manufacturing of fluoride glass fibers is still the subject of intensive research, which is far from ended, but the fluoride fibers have in their capacity as fiber amplifiers been brought much closer to application in an operating communication system.

Finally, the fiber drawing techniques have been developed further due to the requirements given by the active fibers, and, for example, long tapered active fibers are fabricated today.

REFERENCES

[1] Snitzer, E., "Proposed Fibre Cavities for Optical Masers," *Journal of Applied Physics*, Vol. 32, 1961, pp. 36–39.

[2] Koester, C. J., and E. Snitzer, "Amplification in a Fibre Laser," *Applied Optics*, Vol. 3, 1964, pp. 1182–1186.

[3] Poole, S. B., D. N. Payne, and M. E. Fermann, "Fabrication of Low-Loss Optical Fibres Containing Rare-Earth Ions," *IEE Electronics Letters*, Vol. 21, No. 17, 1985, pp. 737–738.

[4] Ainslie, B. J., "A Review of the Fabrication and Properties of Erbium-Doped Fibers for Optical Amplifiers," *IEEE Journal of Lightwave Technology*, Vol. 9, No. 2, 1991, pp. 220–227.

[5] Patek, K., *Glass Lasers*, London: Butterworth, 1970.

[6] Snitzer, E., "Glass Lasers," *Applied Optics*, Vol. 5, No. 10, 1966, pp. 1487–1499.

[7] Simpson, J., "Fabrication of Rare Earth Doped Glass Fibers," *Proceedings of Society of Photo-Optical Instrumentation Engineers Conference, OE/Fibers '89*, part: Fiber Laser Sources and Amplifiers, Boston, Sept. 1989, Vol. 1171, pp. 2–9.

[8] Miniscalco, W. J., "Erbium-Doped Glasses for Fiber Amplifiers at 1500 nm," *IEEE Journal of Lightwave Technology*, Vol. 9, No. 2, 1991, pp. 234–250.

[9] Monerie, M., D. Ronarch, and F. Auzel, "Rare-earth doped fluoride glass fibre amplifiers," *European Conference on Optical Communication ECOC '91*, Paris, 1991, Proc. Vol. II, invited paper, pp. 9–16.

[10] Ainslie, B. J., J. R. Armitage, S. P. Craig, and B. Wakefield, "Fabrication and Optimization of the Erbium Distribution in Silica Based Doped Fibres," *European Conference on Optical Communication ECOC '88*, Brighton, England, 1988, Proc. Vol. I, pp. 62–65.

[11] Smithgall, D. H., T. J. Miller, and R. E. Frazee, Jr., "A Novel MCVD Process Control Technique," *IEEE Journal of Lightwave Technology*, Vol. 4, No. 9, 1986, pp. 1360–1366.

[12] Ohmori, Y., F. Hanawa, and M. Nakahara, "Fabrication of Low-Loss Al_2O_3-Doped Silica Fibres," *IEE Electronics Letters*, Vol. 19, 1983, pp. 261–262.

[13] Keck, D. B., P. C. Schultz, and F. W. Zimar, U.S. Patent 3,737,393.

[14] Ainslie, B. J., S. P. Craig, and S. T. Davey, "The Fabrication and Optical Assessment of Nd^{3+} in Silica-Based Optical Fibers," *Material Letters*, Vol. 5, No. 4, 1987, pp. 143–146.

[15] Poole, S. B., D. N. Payne, R. J. Mears, M. E. Fermann, and R. I. Laming, "Fabrication and Characterization of Low-Loss Optical Fibers Containing Rare-Earth Ions," *IEEE Journal of Lightwave Technology*, Vol. 4, No. 7, 1986, pp. 870–876.

[16] Simpson, J. R., and J. B. MacChesney,"Alternate Dopants for Silica Waveguides," *Optical Fiber Communications '82*, 1982, paper TUCC5.

[17] Stone, J., and G. E. Walrafen, "Overtone Vibrations of OH Groups in Fused Silica Optical Fibers," *Journal of Chemical Physics*, Vol. 76, No. 4, 1982, pp. 1712–1722.

[18] Kao, C. K., *Optical Fibre*, London: Peter Peregrinus, 1988.

[19] Craig-Ryan, S. P., B. J. Ainslie, and C. A. Millar, "Fabrication of Long Lengths of Low Excess Loss Erbium Doped Optical Fibre," *IEE Electronics Letters*, Vol. 26, No. 3, 1990, pp. 185–186.

[20] MacChesney, J. B., and J. Simpson, "Optical Waveguides With Novel Compositions," *Optical Fiber Communications OFC '85*, paper WH5, 1982.

[21] DiGiovanni, D. J., "Fabrication of Rare Earth Doped Optical Fiber," *Proceedings of Society of Photo-Optical Instrumentation Engineers Conference, OE/Fibers '90*, part: Fiber Laser Sources and Amplifiers II, San Jose, Sept. 1990, Vol. 1373, pp. 2–8.

[22] Watanabe, M., H. Yokota, and M. Hoshikawa, "Fabrication of Yb_2O_3-SiO_2 Core Fiber by a New Process," *European Conference on Optical Communication ECOC '85*, Venice, 1985, Proc. Vol. I, pp. 15–18.

[23] Laoulacine, R., T. F. Morse, P. Charilaou, and J. W. Cipolla, "Aerosol Delivery of Non-Volatile Dopants in the MCVD System," *Extended Abstracts of the Amer. Inst. Chem. Eng. Annual Meeting*, Washington, D.C., paper 62B, 1988.

[24] Morse, T. F., L. Reinhart, A. Kilian, W. Risen, Jr., and J. W. Cipolla, Jr., "Aerosol Doping Technique for MCVD and OVD," *Proceedings of Society of Photo-Optical Instrumentation Engineers Conference, OE/Fibers '89*, part: Fiber Laser Sources and Amplifiers, Boston, Sept. 1989, pp. 72–79.

[25] Suda, H., S. Shibata, and M. Nakahara, "Double-Flame VAD Process for High-Rate Optical Preform Fabrication," *IEE Electronics Letters*, Vol. 21, No. 1, 1985, pp. 29–30.

[26] Thompson, D. A., P. L. Bocko, and J. R. Gannon, "New Source Compounds for Fabrication of Doped Optical Waveguide Fibres," *Proc. of Fibre Optics Adverse Environment II*, No. 506, SPIE, 1984, pp. 170–173.

[27] Suda, H., S. Shibata, and M. Nakahara, "High-Rate Fabrication of Wholly Synthesized Fiber Preforms by the Multi-Flame VAD Method With SiHCl3 Raw Materials," *European Conference on Optical Communication ECOC '85*, Venice, 1985, Proc. Vol. I, pp. 65–68.

[28] Bocko, P. L., "Rare-Earth Doped Optical Fibres by the Outside Vapour Deposition Process," *Optical Fiber Communications OFC '89*, Houston, 1989, paper TuG2.

[29] Sanada, K., T. Shioda, T. Moriyama, K. Inada, S. Takahashi, and M. Kawachi, "PbO Doped High Silica Fiber Fabricated by Modified VAD," *European Conference on Optical Communication ECOC '80*, 1980, Proc. Vol. I, pp. 14–17.

[30] Li, T., ed., *Optical Fiber Communications, Vol. 1, Fiber Fabrication*, Academic Press, 1985.

[31] Stone, J., and C. A. Burrus, "Nd^{3+} Doped SiO_2 Lasers in End Pumped Fibre Geometry," *Applied Physics Letters*, Vol. 23, 1973, pp. 388–389.

[32] Schultz, P. C., "Optical Absorption of the Transition Elements in Vitreous Silica," *Journal of American Chemical Society*, 1974, pp. 309–313.

[33] Poole, S. B., "Fabrication of Al_2O_3 co-Doped Optical Fibres by a Solution-Doping Technique," *European Conference on Optical Communication ECOC '88*, Brighton, England, 1988, Proc. Vol. I, pp. 433–436.

[34] Gozen, T., Y. Kikukawa, M. Yoshida, H. Tanaka, and T. Shintani, "Development of High Nd^{3+} Content VAD Single Mode Fibre by the Molecular Stuffing Technique," *Optical Fiber Communications OFC '88*, New Orleans, 1988, paper WQ1.

[35] Miyashita, T., and T. Manabe, "Infrared Optical Fibers," *IEEE Journal of Quantum Electronics*, Vol. 18, No. 10, 1982, pp. 1432–1450.

[36] Tran, D. C., G. H. Sigel, Jr., and B. Bendow, "Heavy Metal Fluoride Glasses and Fibers: A Review," *IEEE Journal of Lightwave Technology*, Vol. 2, No. 5, 1984, pp. 566–586.

[37] Carter, S. F., M. W. Moore, D. Szebesta, J. R. Williams, D. Ranson, and P. W. France, "Low Loss Fluoride Fibre by Reduced Pressure Casting," *IEE Electronics Letters*, Vol. 26, No. 25, 1990, pp. 2115–2117.

[38] Urquhart, P., "Review of Rare Earth Doped Fiber Lasers and Amplifiers," *IEE Proceedings J.*, Vol. 135, No. 6, 1988, pp. 385–407.

[39] France, P. W., and M. C. Brierley, "Progress in Fluoride Fibre Lasers and Amplifiers," *Proceedings of Society of Photo-Optical Instrumentation Engineers Conference, OE/Fibers '90*, part: Fiber Laser Sources and Amplifiers II, San Jose, Sept. 1990, Vol. 1373, pp. 33–39.

[40] Snitzer, E., and R. Tumminelli, "SiO_2-Clad Fibers With Selectively Volatilized Soft Glass Cores," *Optics Letters*, Vol. 14, No. 14, 1989, pp. 757–759.

[41] Yamashita, T., S. Amano, I. Masuda, T. Izumitani, and A. Ikushima, "Nd and Er Doped Phosphate Glass Fiber Lasers," *Proc. of Conference on Lasers and Electro-Optics CLEO '88*, 1988, paper ThH2.

[42] Simpson, J. R., L. F. Mollenauer, K. S. Kranz, P. J. Lemaire, N. A. Olsson, H. T. Shang, and P. C. Becker, "A Distributed Erbium Doped Fiber Amplifier," *Techn. Dig. of Optical Amplifiers and Their Applications OAA '90*, Monterey, 1990, paper PD19-1.

Chapter 3
Properties of Rare-Earth Ions in Glass Host Materials

In order to achieve a basic understanding of fiber amplifiers, it is necessary to study the properties of the active materials (i.e., the rare earths, or lanthanides) when optical fibers are doped with them. The most important host materials are therefore the SiO_2-based glasses, which are used for commercial fiber manufacturing. Second, another type of host material comes into question, fluoride glasses, which for more than a decade have been studied with the purpose of obtaining very low loss fibers (with theoretical losses $\approx$ 0.01 dB/km). As we will see later, it is, however, not the low loss that is the attractive property in this connection, but the fact that host glass composition significantly affects the optical properties of rare-earth-doped fibers. The fluoride glass is here an example of a host material, which can lead to superior performance in some applications of the active fibers.

However, before we discuss these properties, let us look at Figure 3.1(a), where the lanthanides are listed. This shows a list of 15 elements ranging from lanthanum (La), with an atomic number of 57, to lutetium (Lu), with an atomic number of 71. All the rare-earth atoms have the same outer electronic structure of $5s^25p^66s^2$, which are filled shells [1]. In Figure 3.1(a), the electronic structure is also listed, and here [Xe] designates the configuration of xenon, which includes the 5s and 5p electrons. The optical characteristics of the rare earths will, as described in [2], be dictated by the number of electrons occupying the inner 4f shell. Lanthanum has zero and Lutetium has fourteen 4f electrons, and since optical absorption and emission cause transitions within the 4f shell, these two materials will therefore have special properties compared with the rest.

Ionization of the rare earths usually takes place to form a trivalent state. Two of the 6s electrons and one of the 4f electrons are removed, but the outer 5s and 5p shells remain intact. The remaining 4f electrons are therefore partially shielded from perturbation by external fields. The fluorescence and absorption wavelengths

Lanthanum	57 La	[Xe] $(6s^2\,5d)$
Cerium	58 Ce	[Xe] $(6s^2\,4f\,5d)$
Praseodymium	59 Pr	[Xe] $(6s^2\,4f^3)$
Neodymium	60 Nd	[Xe] $(6s^2\,4f^4)$
Promethium	61 Pr	[Xe] $(6s^2\,4f^5)$
Samarium	62 Sm	[Xe] $(6s^2\,4f^6)$
Europium	63 Eu	[Xe] $(6s^2\,4f^7)$
Gadolinium	64 Gd	[Xe] $(6s^2\,4f^7 5d)$
Terbium	65 Tb	[Xe] $(6s^2\,4f^9)$
Dysprosium	66 Dy	[Xe] $(6s^2\,4f^{10})$
Holmium	67 Ho	[Xe] $(6s^2\,4f^{11})$
Erbium	68 Er	[Xe] $(6s^2\,4f^{12})$
Thulium	69 Tm	[Xe] $(6s^2\,4f^{13})$
Ytterbium	70 Yb	[Xe] $(6s^2\,4f^{14})$
Lutetium	71 Lu	[Xe] $(6s^2\,4f^{14}5d)$

Figure 3.1 (a) The rare-earth materials with indication of atom number and electron configuration. [Xe] stands for the electronic configuration of xenon. (Figure layout from [1].)

are consequently less dependent on external electric fields than other ions of the transition elements, which do not experience similar electronic shielding [1].

Important parameters in establishing the use of rare-earth-doped fibers in amplifier applications are the strength and energy of absorption and fluorescence. The reason is that the absorption spectrum holds accurate information about the location of possible pump wavelengths and the ability to excite the rare-earth ions to higher energy levels. From this higher energy level the electron can relax, transferring its packet of energy to the ground state either radiatively or nonradiatively. The nonradiative deexcitation involves the creation of phonons (i.e., a quantized vibration of the surrounding medium), whereas the radiative decay to

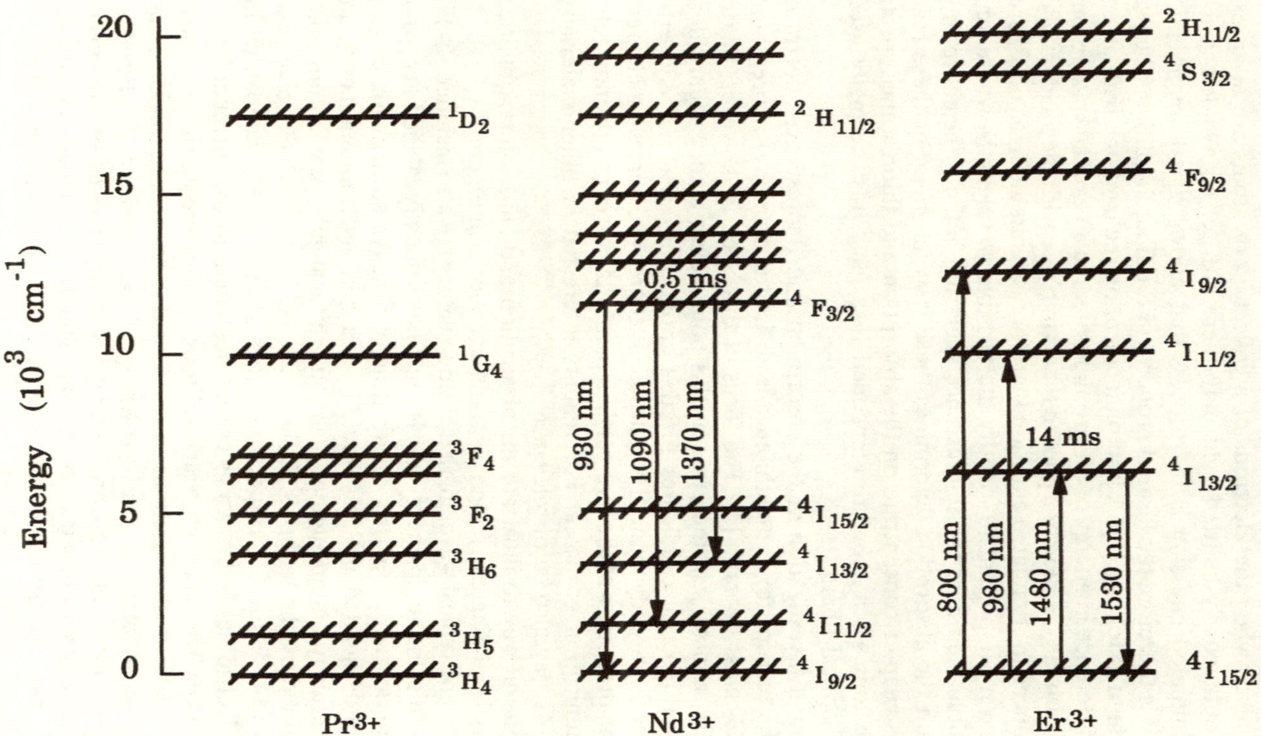

Figure 3.1 Continued. (b) Energy level diagrams for Er^{3+}, Nd^{3+}, and Pr^{3+}. The wavelengths of the optical transitions in SiO_2 glass are shown, together with the average lifetime in the upper laser levels.

lower energy levels takes one of two forms. These are known as spontaneous and stimulated emission, and in both cases photons are emitted. Spontaneous emission always takes place when the electrons of a collection of atoms are in an excited state, and spontaneously emitted light may therefore not be avoided in a fiber amplifier. Stimulated emission is the process that allows signal amplification to take place and is therefore the desired property of the fiber amplifier. The process may be explained as follows [1]: A photon incident on the medium, with an energy equal to the difference in energy of the ground state and an excited state, promotes deexcitation, with the creation of a photon that is in phase with the incident photon. The light produced in stimulated emission therefore exhibits a degree of coherence. It is naturally very important to identify the possible wavelengths where amplification is possible, and here the fluorescence spectrum is the principal tool, since it together with the absorption determines these wavelengths. For these reasons, we will in this chapter concentrate on the absorption and fluorescence spectra of the rare-earth ions when they are embedded in a glass host normally used for fabrication of optical fibers.

A thorough investigation of the absorption and fluorescence spectra of all the rare earths (except for promethium due to its radioactivity) is presented for SiO_2GeO_2-P_2O_5 glass fibers in [3]. The focus of this book is, however, limited to fiber amplifiers, and we will therefore at this point select among them the three most relevant rare-earth candidates for this application area, namely, erbium, neodymium, and praseodymium. The purpose is thereby to achieve a clearer presentation of the properties of the rare-earth dopants that are currently most relevant for telecommunication purposes. For readers interested in other materials and applications, [4] may also be of relevance.

A very useful way to illustrate the absorption and emission possibilities are through energy level diagrams. Such diagrams are shown in Figure 3.1(b) for Er^{3+}, Nd^{3+}, and Pr^{3+}, and typical wavelength transition values are also included. In Figure 3.1(b), the atomic and ionic energy levels are conventionally labeled according to the angular momentum properties of the atom or ion. An atom or ion existing in an energy level designated as $^{2S+1}L_J$ would have a spin quantum number S, a total angular momentum quantum number J, and an orbital angular momentum quantum number L defined by the letters S, P, D, F, G, H, I, . . . , which correspond to $L = 0, 1, 2, 3, 4, 5, 6, \ldots$, respectively. In a glass host, the lanthanide ion is subjected to electric fields, known as crystal fields due to the surrounding atoms in the host lattice. This causes a Stark splitting of the rare-earth ion orbitals (J manifolds), and site-to-site variations of the field due to the amorphous nature of glass results in an inhomogeneous broadening of the transition [5]. Besides this broadening mechanism, phonon broadening (or homogeneous broadening) also takes place. This happens because energy exchange in the connection with a transition between two energy levels may include the creation or annihilation of a phonon. This broadening, which is temperature-dependent [6], is illustrated in Figure 3.1(b) with hatched lines.

When the ions are pumped to manifolds above the upper laser level, the population cascades down through the intervening levels through nonradiative transitions and eventually leads to a population of the upper laser level. This level is normally described as metastable, since transitions from this to lower manifolds only appear because the crystal field has broken the inversion symmetry of the ion's environment, permitting the electric dipole transitions to occur between Stark levels in different *LSJ* multiplets [4]. It is noteworthy that the involved lifetimes in the upper laser level are for this reason many orders of magnitude larger than the nonradiative lifetimes of the higher energy levels.

Another important property may be seen in Figure 3.1(b) by comparing the energy level diagrams of Er^{3+} and Nd^{3+}. The important difference is that Er^{3+} (for the lower pump wavelengths) works as a three-level laser system, whereas Nd^{3+} may be described as a four-level system (when considering the 1370 nm band). In a three-level system, the lower lasing level is either the ground state or a level so close to the ground state that it has a significant thermal population. In a four-level laser, there is still a further transition from the lower laser level down to the ground state, which is usually nonradiative in nature. In a three-level laser system, signal absorption from the ground state directly to the upper laser level takes place, causing competition with the stimulated emission of photons. This is also the case for the 1480-nm pumping of Er^{3+} fibers, at which wavelength the Er^{3+} works as a two-level system, meaning that the upper pump level and the upper laser level are coinciding. In a four-level laser system, the lower lasing level is depopulated rapidly, and the necessary pump power to obtain population inversion therefore becomes significantly smaller. This will, as we will see, result in the principal differences between EDFAs and *neodymiumdoped fiber amplifiers* (NDFA).

In order to specify the properties of rare-earth ions in different environments, the remainder of this chapter will focus on the emission and absorption properties of two types of host glasses. The first is fused silica glasses, which are specifically relevant for EDFAs. The second, with special reference to 1300-nm amplifiers, is fluoride glasses, whose rare-earth ion properties become relevant. We will here, in connection with the description of erbium in silica host materials, present results on the influence of refractive index-raising co-dopants. Furthermore, Section 3.1 will describe important additional properties such as *excited-state absorption* (ESA) and up-conversion for erbium in silica. Finally, a short review of fluorescence and absorption properties of praseodymium and neodymium in silica is given. Section 3.2 describes the properties of the most important rare-earth materials in fluoride glasses. Section 3.3 provides two cases that exemplify the advantages of co-doping with more than one lanthanide. First, co-doping with ytterbium is discussed for the purpose of extending the possible choices of pump wavelengths. Second, lanthanum co-doping is considered with the aim of increasing the maximum erbium concentration in silica fibers. Section 3.4 presents results on the influence of temperature on fluorescence and absorption spectra for erbium in silica glass hosts.

3.1 SILICA GLASS HOST MATERIALS

One of the reasons for the successful and rapid development of EDFAs has been that manufacturing technology could refer to mature silica fiber production methods. Furthermore, it is important that the users of erbium-doped fibers always had confidence in the handling requirements. Most important, however, was that Er^{3+}-doped silica fibers permit amplification at wavelengths which are perfectly located within the 1550-nm transmission band of optical communication systems. For this reason, it is natural to initiate the description of fluorescence and attenuation properties by examining Er^{3+} embedded in silica host fibers.

3.1.1 Fluorescence and Attenuation Properties of Erbium in SiO_2

A principal reason for the interest in the EDFA is its luminescence spectrum, as shown in Figure 3.2. Gain has been demonstrated over practically the entire emission spectrum, and values above 20 dB have been readily obtained for wavelengths between 1530 and 1560 nm [7]. A critical factor in the success of Er^{3+}-doped fiber

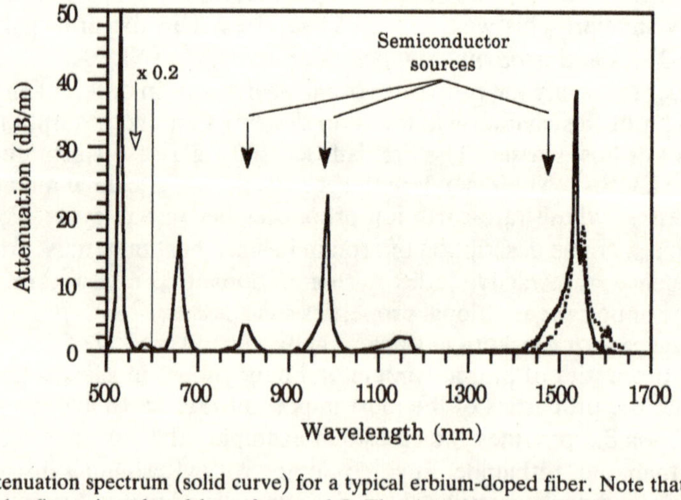

Figure 3.2 Attenuation spectrum (solid curve) for a typical erbium-doped fiber. Note that the left part of the figure is reduced by a factor of 5. The dotted curve is the fluorescence spectrum (measured using 650-nm pump light with a power of 100 mW). (Data from [8].)

amplifiers in optical communications is the long lifetime of the metastable state, which permits the required high-population inversions to be obtained under steady-state conditions using modest pump powers. Furthermore, an optically pumped device must have suitable pump bands to be practical, meaning that the EDFA absorption bands should preferably coincide with the emission wavelengths of compact and coupling effective pump sources such as semiconductor lasers or *light-emitting diodes* (LED). That this is the case can be seen from Figure 3.2, which also illustrates the absorption bands for an Er/Ge-doped silica fiber. Semiconductor pump sources have been successfully used to pump amplifiers in the 800-, 980-, and 1480-nm absorption bands (we shall discuss this in more detail in Chapter 7). Besides this, Figure 3.2 shows that pumping around 520, 590, and 660 nm is also applicable. However, semiconductor pump sources are not available at these wavelengths, and their importance as pump bands have been outlived within the area of optical communications. Note, however, that dye-laser pumping at wavelengths below 700 nm, for instance, played an important role in the early characterization of rare-earth-doped fibers.

3.1.2 Index-Raising Codopants

Silica-based single-mode optical fibers may be fabricated by adding germanium (GeO_2) and/or aluminum (Al_2O_3) in the fiber core. By doing this, the refractive index value is increased compared to the value in pure SiO_2. By exchanging GeO_2 with Al_2O_3 as the index-raising co-dopant, the performance of the EDFA is significantly influenced, and it is natural to ask what advantages are to be obtained by this. Part of the explanation is that the solubility of erbium in silica is affected positively by aluminum doping. The reason is that the four-coordinated germanium does not alter the tetrahedral silica network significantly. Germanium and Er^{3+} are therefore not tied to each other, and the clustering of Er^{3+} ions becomes likely to appear [7] (see also Section 6.2.3). The addition of aluminum has, on the other hand, been found to reduce the clustering problem, which is explained as follows: Al_2O_3 is dissolved in SiO_2, where it forms a solvation shell around the rare-earth ion, and the resulting complex is readily incorporated into the silica network [9]. This means that aluminum-doped fibers allow larger erbium concentrations, and so shorter fiber lengths are required.

The structural difference between GeO_2 and Al_2O_3 in the glass network also helps with the understanding of a related phenomenon: the diffusion properties of erbium ions during the fiber fabrication and drawing. Since the erbium ions are expected to form a complex with the Al-O groups, they are also more tightly bound at the location of the index-raising aluminum. A key point here is that, unlike GeO_2, the Al_2O_3 is not volatile during the preform collapse process. The additional benefit of controlling the radial doping profile of rare-earth ions is obtained because the Er^{3+} distribution follows the controllable distribution profile of aluminum. This is illustrated in Figure 3.3, where it is also found that diffusion may be a problem for germanium-doped EDFAs. In the present example, the MCVD preform fabrication process depletes the rare-earth dopants at the center of the core for a germanium-doped fiber [10]. The addition of aluminum, even when co-doped with germanium or phosphorus, prevents the depletion of the rare earth [10] and is believed to be a straightforward way to improve the efficiency of active fibers made from MCVD.

The effect of either germanium or aluminum co-doping on the fluorescence and absorption properties of erbium in silica-based glasses is shown in Figure 3.4. The spectra presented here have been determined experimentally in an Al/Er fiber (Figure 3.4(a)) [11] and in a Ge/Er fiber (Figure 3.4(b)) [12]. The emission peak wavelengths are 1530 nm for the Al/Er fibers and 1535 nm for the Ge/Er fibers. The respective fluorescence lifetimes have been measured to be 10.6 ms for the aluminum-co-doped and 12.5 ms for the germanium-co-doped fibers, and the absorption cross section at 980 nm was found from attenuation measurements to be 2.0×10^{-25} m^2 and 4.5×10^{-25} m^2, respectively. Today, many commercially available fibers are manufactured using a combination of aluminum and germanium doping. Depending on the mutual concentrations of the two index-raising co-dopants, the lifetime can normally be found in the interval between 10.6 ms and

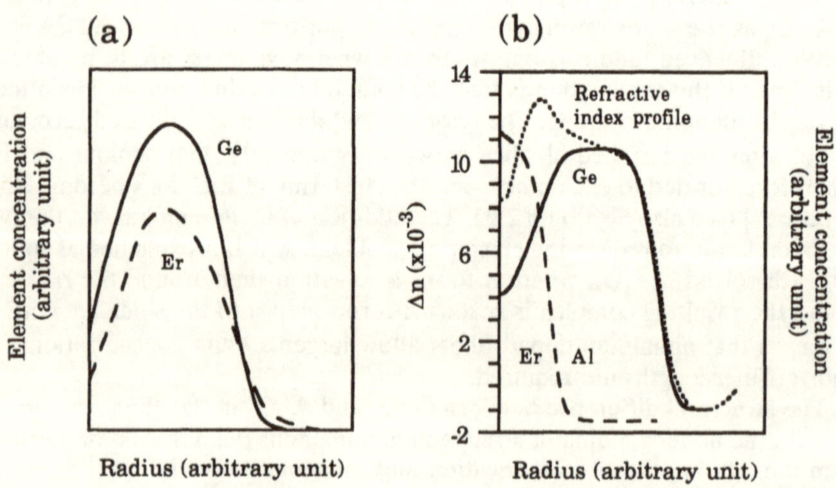

Figure 3.3 (a) The elemental distribution of Er^{3+} and Ge in a GeO_2-SiO_2 preform core. (b) Refractive index, Δn, and elemental distributions of Er^{3+}, Al, and Ge of a partially drawn Al_2O_3-SiO_2 preform core. The fiber is made with the erbium confined to the core center. (Data are averaged results from [10].)

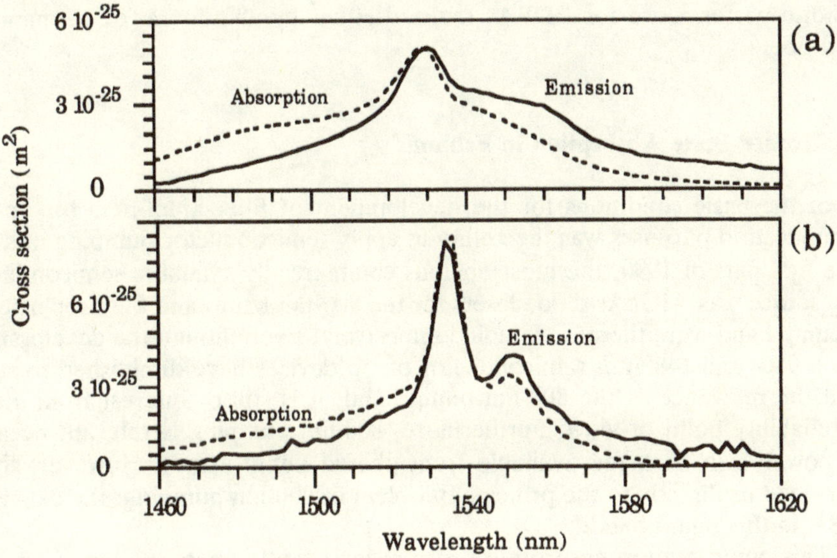

Figure 3.4 (a) Typical emission and absorption cross-section spectra in an erbium-doped fiber co-doped with aluminum. The erbium concentration is 2.0×10^{24} m^3. (b) Typical emission and absorption cross sections versus wavelength in an erbium-doped fiber co-doped with germanium. The erbium concentration is 4.6×10^{24} m^{-3}. (Data for all cross sections from [15].)

12.5 ms; as an example, a value of 11.4 ms has been published [13]. It should be noted here that all the specified lifetimes are valid only for erbium concentrations sufficiently low to avoid clustering effects (see Section 6.2.3). Examinations of the fluorescence linewidths have been made in [14] for core hosts spanning from GeO_2/SiO_2 to pure Al_2O_3/SiO_2. It was found here that the narrow 26.5-nm *full width half maximum* (FWHM) value of the GeO_2 was broadened right out to 43 nm for an Al:Ge ratio of 1:10, which although predominantly a GeO_2/SiO_2 host, gives a fluorescence spectrum very similar to an Al_2O_3/SiO_2 glass. It is only for an Al:Ge ratio as high as 1:33 that a glass with a FWHM fluorescence band midway between Al_2O_3/SiO_2 and GeO_2/SiO_2 is found. This also suggests that the Er^{3+} is preferentially forming an association with the aluminum in the glass matrix, even down to very low aluminum concentrations, rather than the germanium, which is present in much higher concentrations. It is reasonable that the coordination of rare-earth ions by Al-O^- is possible even at relatively low Al:Ge ratios, since the Er^{3+} concentration is always very low [14].

As indicated in Figure 3.4, the spectra in aluminum-doped fiber amplifiers are typically significantly broader than in germanium-doped fiber amplifiers. As we shall see in the following chapters, these properties will result in different gain and noise performance for EDFAs realized either by aluminum- or germanium-doped fibers.

3.1.3 Excited-State Absorption in Erbium

One of the basic conditions for the development of fiber amplifiers for optical communication purposes was the ability to apply semiconductor pumping. As late as the first part of 1990, the most obvious commercially available semiconductor pump source was AlGaAs diode lasers for the 800-nm band, and the properties of this pump band were therefore studied intensively. Even though the development of both 980- and 1480-nm semiconductor pump devices have diminished to some extent the relevance of the 800-nm pump band, it is still of interest from a cost and reliability point of view. Furthermore, 800-nm pumping is relevant because high powers can be made available from phased pump arrays. However, these factors lead us directly to the principal problem of 800-nm pumping: the existence of ESA in this pump band.

This pump photon absorption process significantly limits the amplifier performance for Er^{3+}-doped silica fibers pumped around 800 nm [16–19]. We will therefore define in this section its origin and relative strength compared to the *ground-state absorption* (GSA). It should be pointed out that the present description is limited to silica based host glass, even though other glasses have been predicted to be significantly better than silica in this respect [20,21]. The reason for this limitation is, however, that the principal idea may alone be understood from considering silica glass, and, more important, this host glass is the most likely material to be chosen by fiber manufacturers in the future.

In order to understand the importance of ESA, a simplified energy level diagram, as shown in Figure 3.5(a), may be advantageous. The 800-nm pump ESA signifies the process that an ion in the metastable $^4I_{13/2}$ state (the upper laser level) is excited to a still higher level through the absorption of a pump photon. Even though the ion almost momentarily returns to the $^4I_{13/2}$ level, a photon will be lost to unwanted phonon emission (heat in the fiber). Excited ions have an ESA spectrum which differs from their GSA spectrum due to the new energy gaps between the ground state and the higher lying states. This problem is important because the ESA in certain wavelength bands may be so strong that it dominates the GSA. This results in the situation in which the increase in absorption due to populating the metastable state more than offsets the decrease from depleting the ground state.

To quantify the particularly important implications of the $^4I_{13/2} \rightarrow {}^2H_{11/2}$ ESA transition on the $^4I_{15/2} \rightarrow {}^4I_{9/2}$ pump band at 800 nm, Figure 3.5(b) shows a comparison between ESA and GSA cross-section values for the wavelengths of interest. The present spectra are measured on an Al/P-doped silica fiber. The significance of the

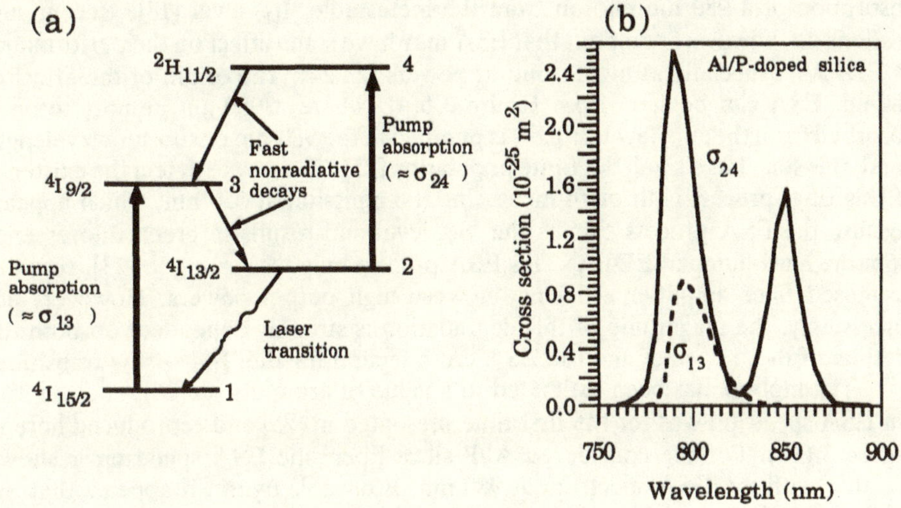

Figure 3.5 (a) Simplified energy level diagram illustrating pump ESA in a three-level laser medium. (b) Excited-state cross section (solid curve) and ground-state cross section (dashed curve) for Er^{3+} in an Al/P-doped SiO_2 fiber. (Data from [21].)

ESA problem can be characterized by the ratio of the ESA cross section to the GSA cross section [7]. This parameter, $R_p = \sigma_{ESA}/\sigma_{GSA}$, has been shown to be sensitive to the co-dopants [18,19]. As summarized in [7], at 810 nm, $R_p \approx 2$ for germanium-doped silica, while the addition of large amounts of phosphorus reduces it to ≤ 1. As shown in Figure 3.5(b), a value of ≈ 1 is typical for Al/P at this wavelength, and this has also been found realistic for Al-co-doped fibers [19]. Note that fluorozirconate fibers were reported to be significantly better than silica with $R_p < 1$ for all wavelengths larger than 800 nm [19].

For a typical aluminum-doped silica fiber, the spectra of Figure 3.5(b) are therefore representative, and it shows that the ESA cross section σ_{24} exceeds the GSA cross section σ_{13} over the complete pump band, except near 820 nm. This explains the high gain values realized for Al-silica amplifiers pumped at 820 nm, such as those reported in [22]. We will, however, return to the subject of pump wavelength choice in more detail in Chapter 7.

Though less significant, ESA may also be found at different wavelengths than around the 800-nm band, and not only pump but also signal ESA may appear. We will, however, wait a moment on the subject of signal ESA and first examine another possibility of pump light absorption, which is due to ESA in the 980-nm band. Studies of this effect were not considered relevant at an early stage of amplifier development because the positions of the Er^{3+} levels do not permit the absorption of a 980 nm photon from the metastable $^4I_{13/2}$ level [18]. Recent suggestions do, however, point out that ESA may have some effect on the performance of EDFAs, especially at higher output powers [23,24]. The origin of the effect of 980-nm ESA can be seen from Figure 3.6(a), where a 980-nm pump photon is absorbed from the $^4I_{11/2}$ level. If Er^{3+} is pumped in the 980-nm or shorter wavelength band, this state has a small but finite population [25]. The way to detect the existence of this ESA process is through measuring the emission at 550 nm, which appears because the ESA process excites the $^4S_{3/2}$ level and results in green fluorescence from the hard-pumped EDFA. This ESA process may, according to [23], result in decreased fiber amplifier efficiency at very high output powers. However, not surprisingly, the magnitude of this degradation is strongly dependent on both the lifetime of the $^4I_{11/2}$ state and the ESA cross section for the $^4I_{11/2} \rightarrow {}^4F_{7/2}$ transition.

The lifetime has been estimated to a value of around 1 μs [26], and the 980-nm ESA spectrum was for the first time presented in [25] and reproduced here in Figure 3.6(b). For the considered Al/P silica fiber, the ESA spectrum is shown together with the GSA spectrum at 980 nm. From this figure, it appears that the ESA peak is the same order of magnitude as the GSA peak, but is shifted to shorter wavelengths by about 10 nm. This spectral behavior becomes important as input

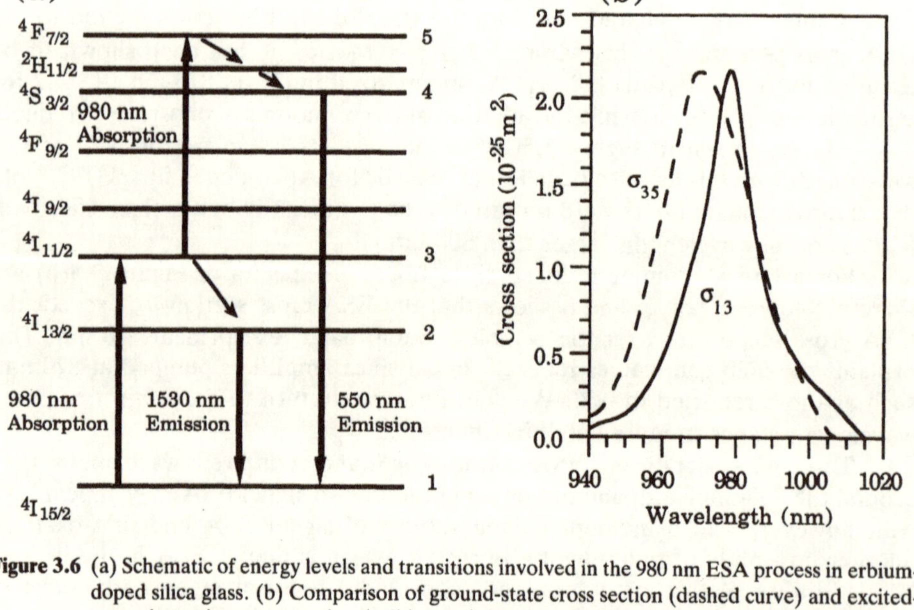

Figure 3.6 (a) Schematic of energy levels and transitions involved in the 980 nm ESA process in erbium-doped silica glass. (b) Comparison of ground-state cross section (dashed curve) and excited-state absorption cross section (solid curve) for Er/Al/P-doped silica. (Spectra from [25].)

to accurate descriptions of nonlinearities introduced into the EDFA by 980-nm pumping. It should be noted in this connection that other host materials such as fluorophosphate and fluoride glass show more narrow ESA spectra—around 980 nm [25].

As mentioned above, the signal photons may also be absorbed through an ESA process. This was considered in [27], where the cross section for absorption of signal photons through the transition $^4I_{13/2} \rightarrow {}^4I_{9/2}$ has been estimated. We will at this point, however, not go further into the discussion of this subject, but merely offer the conclusion that the cross section for signal ESA is insignificant in comparison with GSA and stimulated emission cross sections. ESA of signal photons will therefore in general not affect the amplifier performance significantly. It should, however, be noted that previous work has demonstrated ESA cross section values at 1480 nm, as well as at 1530 nm to be 5×10^{-26} m^2 [28]. This may be of relevance for future work in which a deeper understanding of the limiting factors of EDFAs is essential.

3.1.4 Up-Conversion in Erbium Fibers

As we shall see in Section 6.2.3, the lifetime in the metastable $^4I_{13/2}$ level will be reduced if the erbium concentration is made high enough. This is believed to be due to close proximity of the Er^{3+} ions, with mutual energy transfer and cross-relaxation as a consequence. This relaxation may be described by a two-particle upconversion process involving neighboring ions excited to the upper laser level $^4I_{13/2}$. As illustrated in Figure 3.7(a), energy is transferred nonradiatively from one atom (the donor) to the other atom (the acceptor). The energy transfer results in a de-excited donor atom and an acceptor atom being promoted to the $^4I_{9/2}$ state. The key point is that both ions in the process start in the metastable level, and the mechanism will therefore be dependent on pump power. This is the main mechanism restricting high erbium concentration in fibers, and early work within the area describes the up-conversion process [29,30]. According to [29,31], the acceptor atom will relax to the $^4I_{11/2}$ state within a few nanoseconds, and from here it is most likely to further decay to the upper laser level. However, roughly one out of 10,000 atoms relaxes down to the ground state under the emission of a photon with a wavelength of 980 nm. This 980-nm luminescence may not only be seen as evidence of the suggested process, but it is also quantitatively related to the gain degradation, which is the practical result of the up-conversion process. This is illustrated in Figure 3.7(b), which shows the maximum amplification of a 25 to 26 mW pump at 1480 nm as a function of the detected up-conversion luminescence at 980 nm. The curve is the averaged result of measurements presented in [31], where it was found by measuring the maximum gain for a number of fibers, each with a different erbium concentration below 1000 ppm by weight. Each fiber length was optimized

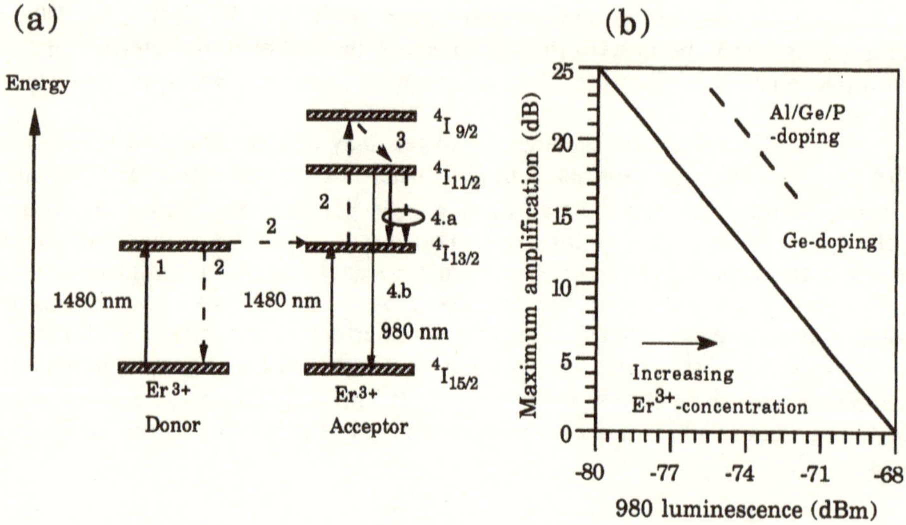

Figure 3.7 (a) Two-particle up-conversion in erbium-doped fibers. Nonradiative transitions are represented by dashed arrows, and the radiative transitions with solid arrows. The numbers indicate the succession of the processes. (b) Maximum amplification as a function of up-conversion at 980 nm with 25- to 26-mW pump power at 1480 nm. Each point on the curves represents a different fiber with a different erbium concentration and optimized fiber length for maximum amplification. (Data from [31].)

for maximum amplification at pump powers of 25 to 26 mW. Investigations were made both for erbium fibers solely co-doped with germanium and for fibers in which aluminum had also been used as an index-raising co-dopant. Based on the previous explanation that the Er^{3+} is most strongly bound to the aluminum atoms, we must expect the Er^{3+} to be more homogeneously distributed in Al/Er fibers. Therefore, higher concentrations of erbium may be accepted for such fiber, but at some point clustering may be seen anyway. This is also seen in Figure 3.7(b), where the slope of the maximum amplification versus the 980-nm luminescence is shown to be the same irrespective of the host material.

3.1.5 Neodymium and Praseodymium Doping in Silica Glasses

Among the first active optical fibers activities were investigations of the different lanthanides embedded in silica host glass. Here, especially, neodymium played a significant role in pioneering work [32], and based on these experiences the techniques were extended to incorporate all the rare earths, except for the radioactive promethium [3]. These historical reasons are, however, not the only ones justifying a description of lanthanides other than erbium. The additional motivation is partly found in the general wish to use silica glasses as host material for amplifiers. The important point here is to examine why neither neodymium nor praseodymium amplifiers can be made with silica hosts for 1300-nm operation.

To facilitate a better understanding of the properties of Nd^{3+} in a silica glass host, Figure 3.8(a) has been included. From this illustration, we find a number of possible pump absorption wavelengths between 500 and 1000 nm. Among the most significant are absorption peaks around the wavelengths 530, 590, 750, 810, and 900 nm. Note especially that the 810-nm peak is of interest for pumping with reliable, low-cost AlGaAs laser diodes. Also, the fluorescence spectrum of Nd^{3+} is shown in Figure 3.8(a), and we identify significant peaks at 900, 930, 1090, and 1370 nm. By comparison with the absorption spectrum, the clear difference between a three-level laser system and a four-level system appears. Namely, we find that no absorption is present around 1090 nm, where the Nd^{3+} ions work as a four-level system, whereas absorption is found at wavelengths around 930 nm, where three-level operation is the case. Finally, and probably most important, is the presence of a fluorescence peak located in the 1300-nm wavelength area and centered at 1370 nm. Even though this emission peak is located at wavelengths that are somewhat high for use in the 1300-nm telecommunications window, the width may at first cause optimism. However, what is not apparent from Figure 3.8(a) is

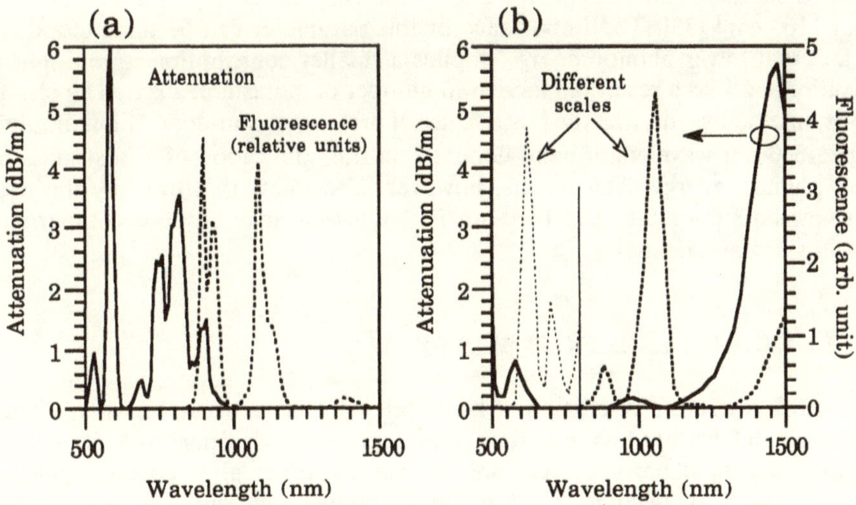

Figure 3.8 (a) Typical emission and absorption cross sections versus wavelength in a neodymium-doped silica fiber (data from [8]). (b) Emission and absorption cross-section spectra for a praseodymium-doped silica fiber. (Data from [3].)

the existence of a very strong ESA in this wavelength interval. This ESA will unfortunately completely rule out the application of Nd^{3+}doped silica fiber amplifiers for the 1300-nm wavelength interval.

After the clarification of this ESA problem in neodymium-doped silica, the only hope for a silica-based 1300-nm amplifier was by praseodymium doping. However, before we examine this possibility, it is worthwhile to note that neodymium-doped silica fibers have potential as a gain medium around 1090 nm, which, for instance, may be of relevance for laser manufacturers. Note also that the center wavelength of this emission peak may be changed by varying of the index-raising co-dopant.

The basic spectral properties of praseodymium-doped silica glasses are shown in Figure 3.8(b). From these data, which originate from [3], we find that the absorption bands in praseodymium are few and broad. It is very significant that no absorption is present between 670 and 900 nm, which rules out pumping with AlGaAs semiconductor sources [3]. However, a small ground-state absorption peak exists at 970 nm, but significant absorption is also present in the 1300-nm wavelength range. The fluorescence spectrum of Pr^{3+} shows peaks around 880 and 1050 nm, and, for the present interest in a broad wavelength interval, ranging from 1300 to 1700 nm. A major problem, however, exists with Pr^{3+}-doped silica fibers for 1300-nm amplification. The problem is the large nonradiative decay rate caused by the small energy level difference (about 2790 cm^{-1}) between the $^1G_4$ and the $^3F_4$ levels. This leads to complete quenching of the $^1G_4$ level in silica glasses, and no 1300-nm radiative emission can be expected because silica glass has high phonon energy of about 1100 cm^{-1} [33]. The importance of this parameter can be understood from the fact that lower phonon energy implies a smaller contribution to multiphonon relaxation, and, as a result, an increased number of metastable excited levels. This means, in other words, that the positive use of praseodymium-doped fiber amplifiers in the 1300-nm wavelength band depends on the application of a host glass with lower phonon energy. The results, however, also imply that even by the use of such new host materials, the 1300-nm Pr^{3+} emission may be poor compared the Er^{3+} in silica at 1530 nm.

3.2 FLUORIDE GLASS HOST MATERIALS

Section 3.1 contains a discussion of how major problems occurred when attempts were made to fabricate 1300-nm fiber amplifiers based on silica glass host materials. For this reason, it becomes relevant to examine other glass compositions, and fluoride glass fibers become an attractive alternative. One of the principal reasons is that they can fulfill the same functions as silica fibers, but due to their lower energy multiphonon edge, they yield more numerous metastable levels. Consequently, this opens up the possibility of using both neodymium- and praseodymium-doped fiber amplifiers for 1300-nm applications. The reason for the lower phonon energies involved in fluoride glasses is that they are composed of heavy metals, and the bonding between the atoms is relatively weak, which results in vibrational quantum energies that are lower than those of silica. Besides these obvious advantages, fibers fabricated in ZBLAN glass can be easily doped with rare-earth materials up to a few mol percent [34]. This is because LaF_3 is an integral component of a fluorozirconate glass matrix, and due to this property, short lengths of rare-earth-doped fiber may be allowed. It is, however, still an open question as to how short the fiber lengths may be, since the dependency on concentration effects (e.g., cooperative up-conversion) needs to be examined. Finally, in the following discussion of erbium-, neodymium-, and praseodymium-doped ZBLAN, we are going to see the results of the fact that the excited energy level depends not only on the electron orbit and spin, but also on the crystal field around the rare-earth ion [34]. Because of the different crystal fields around the rare-earth ions, we will therefore find spectroscopic characteristics such as absorption and fluorescence spectra in fluoride glass different from those of conventional silica glass.

3.2.1 Erbium Amplifiers in ZBLAN Glasses

On the subject of erbium-doped ZBLAN fiber properties, we will in this section focus on two different aspects. The first of these is the possibility of obtaining new emission wavelengths, which are nonaccessible in silica-based fibers. The second property is the spectral modification of emission and absorption cross sections of Er^{3+} ions that appear when the silica host material is interchanged with ZBLAN.

The aspect of obtaining new gain wavelengths is treated in Figure 3.9(a). Here, the energy level diagram is shown for an Er/ZBLANP fiber (P indicates that phosphorus is also included as a component of the host glass). In the specific situation of 476.5-nm pumping, it is indicated that light emission is possible not only at the wavelengths of 850, 988, and 1550 nm, but also at around 2700 nm. Note here that the erbium ions work as a four-level system for the 850- and 2700-nm emissions, whereas three-level operation is the case for the 988- and 1550-nm emissions. Furthermore, it is important to point out that the observed transition at 2700 nm has the potential that fiber laser sources may be manufactured for what has been described as the fourth-wavelength window for optical communications [33]. In this window, the fluoride fibers are used as a very-low-loss transmission medium for signal wavelengths around 2700 to 2800 nm. Based on the already documented capability of EDFA-enhanced 1550-nm optical communication systems, the application of this fourth window may, however, be postponed for an indefinite time. The long-wavelength laser sources are therefore more likely to find applications in other fields (e.g., in systems for medical treatment).

Another important aspect of fluoride glass fibers doped with erbium can be seen in Figure 3.9(b), where the fluorescence spectra for erbium ions in Ge/SiO_2 and in ZBLANP glass are shown. A clear difference can be seen in the broader

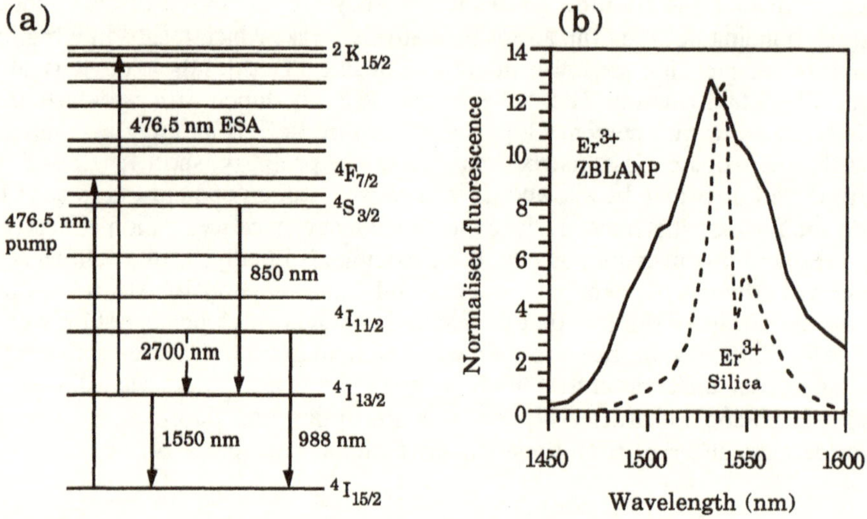

Figure 3.9 (a) Energy level diagram for erbium in a ZBLAN host material. The relevant pump wavelengths are shown together with the observed laser transitions. (b) Normalized fluorescence from Er^{3+}-doped ZBLANP and silica fibers. (Data from [35].)

emission spectrum of the fluorozirconate host. This observation has implications for the relative gain bandwidth and noise performance of amplifiers, as well as the continuous tuning range of fiber lasers in this material [35]. It is important to note that even though some of the later results of erbium-doped ZBLAN fiber amplifiers [33,36] only show bandwidths comparable with those obtained in silica-based EDFAs, recent comparisons [37] support the idea that increased bandwidths are obtainable in fluoride glass fibers. However, due to present problems in ZBLAN fiber manufacturing and handling, the bandwidth advantages should be more significant in order to make the erbium ZBLAN fiber a realistic candidate for practical applications.

3.2.2 Fluorescence and Attenuation in Nd- and Pr-Doped Fluoride Glass

In Section 3.1, we saw that neodymium cannot be used in 1300-nm silica fiber amplifiers. The reason is that the 1370-nm ($^4F_{3/2} \rightarrow {}^4I_{13/2}$) transition suffers from signal ESA from the $^4F_{3/2} \rightarrow {}^4G_{7/2}$ transition (see Figure 9.1), which limits the gain in silica fibers to wavelengths above 1360 nm [38]. However, in fluoride glass (ZBLAN) fibers, the ESA wavelength shifts to shorter values, allowing some gain to be obtained from 1310 to 1380 nm, with the peak occurring near 1345 nm [39,40].

Another factor that limits the 1300-nm performance is the existence of competitive transitions of the $^4F_{3/2} \rightarrow {}^4I_{11/2}$ transition corresponding to 1050-nm emission in ZBLAN glasses. The respective emission strengths in the different wavelength bands may be evaluated on the basis of measured spectral properties. However, for the sake of readability of the present presentation, these are presented in Chapter 9 (Figure 9.2), where a thorough discussion of the 1300-nm amplifier is given. The discussion in Chapter 9 will also concern other issues, such as fiber strength, durability, and the problems of coupling to standard telecommunication fibers.

The other rare-earth candidate for 1300-nm amplifiers is the praseodymium ion. As in the case for neodymium-doped silica fibers, praseodymium-doped silica also suffers from a major problem, though it is different in nature. The difficulty for silica-based praseodymium-doped glass is the large nonradiative decay rate, which eventually leads to complete quenching of the $^1G_4$ level in silica glasses. The lower phonon energy of ZBLAN glasses diminishes the phonon decay rate sufficiently to obtain gain in praseodymium-doped ZBLAN fibers. However, high pump powers are required because the quantum efficiency (i.e., the fraction of pump photons converted into signal photons) is estimated to be as low as 1.5% [41]. Methods of improving praseodymium-doped ZBLAN fiber amplifiers are discussed in detail in Chapter 9, where the spectral properties of the fibers are also shown (see Figure 9.12). We shall merely note here that praseodymium concentrations of less than 1000 ppm have been reported as being preferable in order to reduce the quenching due to cross-relaxation between Pr^{3+} ions [42].

3.3 FIBERS WITH MORE THAN ONE RARE-EARTH MATERIAL

We have already in the previous part of this chapter seen how the absorption and fluorescence properties of rare-earth ions are affected by the host material. It is therefore understandable that a large research effort has been focused on combining lanthanide and host glass composition optimally. Besides the already indicated possibilities of changing index-raising co-dopants or even host glass basis, also the inclusion of more than one rare-earth dopant may be of interest. To illustrate the potential use of such combined applications of more than one rare-earth material, we will in this section discuss two methods of modifying the properties of EDFAs.

3.3.1 Ytterbium Codoping of Erbium Amplifiers

Section 3.1 discusses how two-ion processes such as cooperative up-conversion and quenching appear in erbium- and praseodymium-doped silica glass. It is therefore not surprising that processes of energy transfer from neighboring but different rare-earth ions can take place. A constructive application of such an effect is made in

Er^{3+}-Yb^{3+}-doped silica fibers. The purpose was at first to utilize the broader absorption spectrum of ytterbium at wavelengths above 800 nm [43]. This is advantageous because the Yb^{3+} after absorption of the pump light undergoes a cross-relaxation process with adjacent ions of Er^{3+} [43,44], whereby the absorbed energy is transferred to the erbium system. This allows another route for pumping the erbium system, and the effective pump band of the fiber is thus extended, as can be seen in Figure 3.10(a). The relationship between the attenuation and the cross sections is discussed in Chapter 5.

The effect of ytterbium co-doping is very significant, but it should be understood that it also implies some drawbacks. Namely, the successful energy transfer requires that ytterbium and erbium ions be located close to each other, and consequently a high ytterbium concentration is necessary (a Yb^{3+}:Er^{3+} ratio of 30:1 is realistic). Besides the impact of this on fabrication techniques, it is therefore clear that a large number of ytterbium ions will be excited by the pump without being able to transfer this energy on to the erbium ions. Consequently, the pump efficiency must be expected to be very low for Yb-co-doped EDFAs, but this is the price that has to be paid to extend the pump wavelength band as illustrated. It should, however, also be pointed out that even though high concentrations (several weight percentages) of ytterbium can be incorporated into a silica host, it can still give an efficient amplifier.

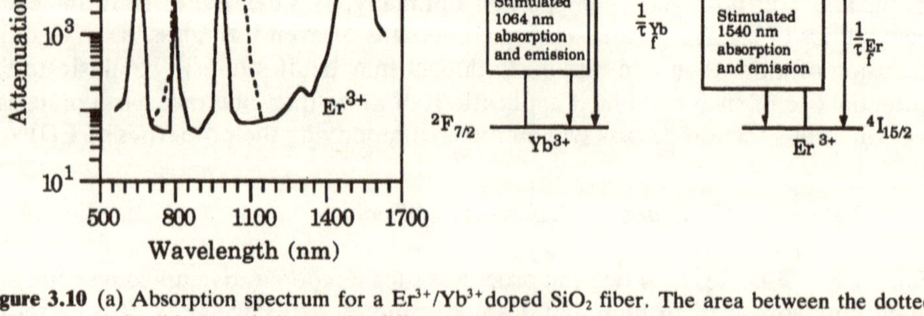

Figure 3.10 (a) Absorption spectrum for a Er^{3+}/Yb^{3+} doped SiO_2 fiber. The area between the dotted and solid curves indicates the difference between the actual fiber and a fiber doped solely with erbium (curves from [48]). (b) Erbium-ytterbium energy level diagram illustrating the 1064-nm pumping possibility.

Lately, the Er/Yb-doped fiber amplifiers have received renewed interest, since they have been demonstrated to provide high gain and efficient power amplification around 1540 nm using diode-pumped miniature Nd:YLF_4 (YLF) or Nd:$Y_3Al_5O_{12}$ (YAG) laser pump sources operating at 1064 nm [45,46]. According to [47], such demonstrations are significant in that high-power TEM_{00} mini Nd:YAG sources are highly developed, readily available, and, crucially, they use high reliability, low-brightness AlGaAs pump sources. It is especially this pump reliability question that is likely to become one of the most severe limitations of fiber amplifier systems.

Figure 3.10(b) illustrates the involved cross-relaxation process, which leads to the excitation of the erbium ions through pumping the ytterbium. Even though we must direct the reader to references such as [45–47] for further details, a few results shall, however, be mentioned here. It is noteworthy that small-signal gains of up to 51 dB and output saturation powers of +24.6 dBm have been demonstrated [46]. This has been possible because the diode-pumped Nd:YLF laser pump source has allowed the use of high-power, reliable AlGaAs diode laser arrays. From these, coupled pump powers of as much as 1.5W are obtainable.

3.3.2 Lanthanum-Doped Erbium Fibers

The erbium-doped aluminum, germanium, and Al/Ge silica fiber amplifiers have been the subjects of the most extensive studies with respect to host glass optimization of EDFAs. For further optimization, new co-doping materials are of interest. Important goals are to increase the NA (see Chapter 6), or to increase the erbium concentration, while still having effective amplifiers. Furthermore, it may for some applications be attractive to achieve further broadening or flattening of the gain spectrum. Among some of the interesting possibilities is the Er/Ca-doped fiber, which was fabricated with the purpose of shifting the ESA at 820 nm [49].

However, optical passive rare-earth materials may also be included in the EDFA, and in [50], Er/La-doped Al_2O_3-SiO_2 fibers were presented. Also, Er/La-doped GeO_2-SiO_2 fibers have been described in [51]. The selection of lanthanum as a co-dopant was based on several factors. First, the refractive index of rare-earth-doped aluminum-silicate glass decreases through the series of rare earths [52]. Since lanthanum has the lowest atom number, a fiber having a high NA should therefore be achieved when lanthanum is used as a co-doping material. Second, lanthanum is an optical passive rare-earth because it has no electrons in the 4f shell. This means that no significant changes are expected to occur to the absorption and emission cross sections of erbium when lanthanum is doped into the $ErAl_2O_3$-SiO_2 fiber. Still, chemically, lanthanum acts like any other rare-earth dopant and will therefore occupy an interstitial site in the silica network in the same way that erbium does. If the ratio between lanthanum and erbium ions is much larger than one, clusters—if formed—will contain many more lanthanum than erbium ions.

This separation of the erbium ions is expected to prevent the cooperative upconversion [14], which is a common problem, when clusters are formed in highly doped erbium fibers.

Among the principal results of the first fabricated Er/Ladoped Al_2O_3-SiO_2 fiber are the NA of 0.31 and the cutoff wavelength of 980 nm [50]. The erbium concentration was made as high as $23 \times 10^{18}\,cm^{-3}$, and the ratio between the number of lanthanum and erbium ions was 14:1. The cross sections shown in Figure 3.11 have been determined from fluorescence and absorption measurements on both an Er/La- and an erbium-doped Al_2O_3-SiO_2 fiber. The spectra show a large similarity, both in width, intensity, and in the position of the emission and absorption peaks. However, a small difference is observed in the peak intensity, showing a 14% larger value in the absorption cross section and an increase of 6% in the emission cross section for the lanthanum-doped fiber.

As expected, the measured lifetime of the Er/La-doped fiber is comparable with that of the erbium-doped fiber, and 10.4 ms and 10.6 ms were measured, respectively. In addition, a component with a short lifetime was observed in the fluorescence decay when the fiber was highly inverted. A similar fast component has been observed previously in highly erbium-doped germanium and aluminum fibers [12], and it was explained as cooperative up-conversion when clusters were

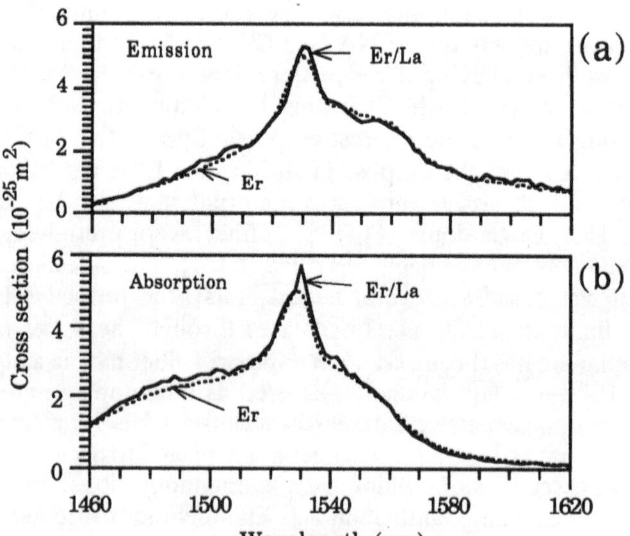

Figure 3.11 Emission (a) and absorption (b) cross sections as a function of the wavelength for an Er/Ladoped fiber and a fiber doped solely with erbium. (Measurements from [50].)

formed. As mentioned above, it is most unlikely, however, that this phenomenon will take place in the lanthanum fiber.

Lanthanum has therefore been shown to be an excellent index-raising codopant for the erbium-doped fiber. Gain higher than 30 dB and output power larger than 15 dBm was measured at very short lengths of the fiber (2m to 4m) when pumped at 1470 nm [50]. It is noteworthy that the short fiber lengths of a few meters (compared to maybe 20 m in fibers without lanthanum co-doping) is considered a considerable advantage by manufacturers of fiber amplifiers. Furthermore, the technique is interesting, since it is aimed at fabrication methods of reducing problems such as cooperative up-conversion in fiber amplifiers.

3.4 EXTERNAL FACTORS THAT AFFECT RARE-EARTH ION PROPERTIES

Lanthanum-doped fibers will, like any optical fiber, show transmission properties that depend to some degree on the environment in which it is placed. Among the important external factors that affect rare-earth-doped fiber are radiation, chemicals, mechanical stress, and temperature. However, many of these factors are either only faintly influential or the fiber may in the design of the amplifier module be protected against factors such as chemicals and mechanical stress. The means of providing such protection is therefore highly relevant in the product development phase, but is not the subject of this book. Therefore, the focus of this section will be limited to a discussion of temperature dependence of the spectral properties of the rare-earth dopant. Furthermore, the discussion will only deal with erbium-doped silica fibers, since they are by far the most relevant today.

3.4.1 Temperature Dependence of Spectral Properties

The temperature properties of EDFAs are interesting because the absorption and fluorescence spectra are affected. This is primarily due to the multilevel (Stark split) nature of the $^4I_{13/2}$ - $^4I_{15/2}$ Er/SiO_2 laser system, which, with respect to Stark level energies and related transitions, may be affected by temperature [6].

The manifolds at the ground state ($^4I_{15/2}$) and the first excited energy band ($^4I_{13/2}$) are found experimentally with a resolution of six and four Stark levels [6], respectively. This means that the ground-state $^4I_{15/2}$ covers an energy range of 270 cm^{-1} and the $^4I_{13/2}$ covers 230 cm^{-1}. The electrons in the ground-state manifold will populate the Stark levels according to a Boltzmann distribution:

$$p_i = Z \cdot \exp(-\Delta E_i / k \cdot T) \tag{3.1}$$

where p_i is the probability of finding an electron at the sublevel number i, Z is a normalization factor, k is the Boltzmann constant, T is the temperature, and ΔE_i is the energy difference between the sublevel i and the lowest lying sublevel in the manifold. A similar distribution will be observed among electrons in the Stark levels within the upper manifold when pumped to this band. As can be seen from (3.1), the population is strongly dependent on the temperature. In addition to the population distribution, the homogeneous linewidth for each transition is also temperature-dependent [6]. The fluorescence lifetime may also be temperature-dependent, though only small changes have been reported [53,54]. All together, these parameters determine the temperature variation of the spectrum, since the fluorescence spectrum, for a given fiber, is the total of the overlapping fluorescence line shapes from each of the transitions between the manifolds.

At small wavelengths (<1520 nm), the temperature-dependent changes of the spectrum can be predicted, since these transitions are strongly affected by the population decrease, and this will dominate changes due to variations in the other parameters. At cryogenic temperatures, the upper Stark levels of $^4I_{13/2}$ are empty and no fluorescence occurs in the region around 1480 nm. It is also important to note that the erbium laser system in silica-based fibers mainly acts as a homogeneously broadened two-level system [55,56]. Thus, the temperature variation of homogeneous linewidths, Boltzmann's distributions, and lifetime are directly expressed in the change of emission and absorption cross sections.

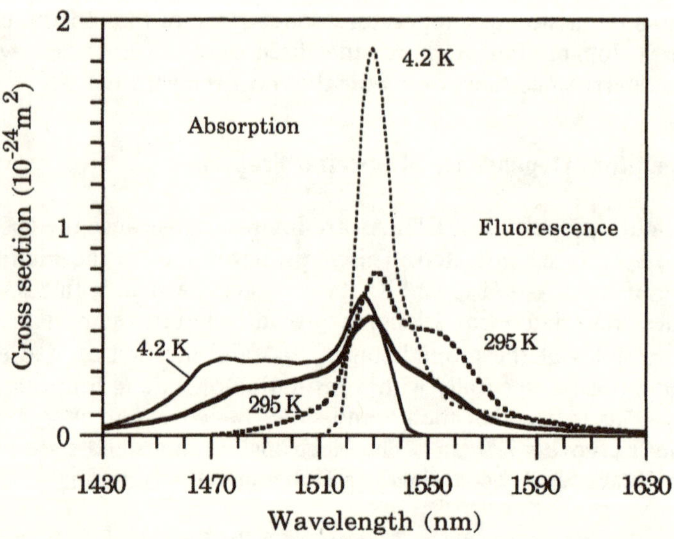

Figure 3.12 Absorption and fluorescence cross sections of an Er/Al-doped fiber sample calculated from spectral data recorded at temperatures T = 4.2K and T = 295K. (Data from [6].)

This is also apparent in Figure 3.12, which shows the absorption and emission cross sections at a temperature as low as 4.2 K and at normal room temperature. As expected, we find no emission at 1480 nm for the EDFA cooled down to 4.2 K, and the spectrum has a much more significant emission peak at 1530 nm compared to the spectrum measured at room temperature.

It is therefore evident that the spectral properties are temperature-dependent, but it is equally important to note that the significant spectral changes first appear at very low temperatures. As we shall see in Chapter 7, only small gain variations (−0.019 dB/K) will be observed within the temperature range for normal operation (down to −50°C) where stability is required [57]. The properties of erbium-doped fibers at cryogenic temperatures are, however, still relevant as an important tool for amplifier characterization [58].

3.5 SUMMARY

The basic properties of lanthanide-co-doped fibers have been presented in this chapter through the discussion of the most relevant dopant and host materials for fiber amplifier fabrication. The description has therefore primarily been devoted to praseodymium, neodymium, and erbium, and the two most significant types of host glass: silica and fluoride glass. The description includes discussions of relevant energy level diagrams, as well as absorption and fluorescence spectra of the involved rareearth materials.

The properties of erbium-doped silica were first discussed, and pump bands were identified around 520, 590, 660, 800, 980, and 1480 nm. The three highest wavelengths among these are of greatest interest because of the possibility of semiconductor pumping. The fluorescence data reveal that gain is obtainable for wavelengths from 1530 nm and up to above 1600 nm. The specific spectral properties were, however, shown to be closely related to the choice of refractive index-raising co-dopant. An important difference is that the spectra in aluminum-doped fibers are typically significantly broader than in germanium-doped fibers, and fluorescence lifetimes are changed from 10.6 ms in aluminum-doped fibers to 12.5 ms in germanium-doped fibers. Today, most fibers are manufactured using both aluminum and germanium as index raising co-dopants, whereby the broadband gain properties of aluminum co-doping is combined with the lower background loss of germanium-doped silica fibers.

Also, efficiency-degrading factors such as ESA have been described, and the pump absorption in the 800-nm band was shown to be significant. Among the primary conclusions are that Al-co-doped fibers are more efficient than Ge-co-doped EDFAs with respect to 800-nm pump ESA. Additionally, the preferred pump wavelength in the 800-nm band should be close to 820 nm, since the best compromise between GSA and ESA appears to be at this wavelength. Also, 980-

nm pump absorption due to ESA was discussed, even though the influence of this was found to be much smaller than ESA in the 800-nm band. The practical importance of 980-nm pump ESA is therefore not large, and this is also the case for pump ESA at 1480 nm and signal ESA at 1530 nm. Finally, another EDFA degrading factor was presented: a two particle up-conversion process in which energy is transferred from one ion to a neighbor ion. However, as long as the Er^{3+} can be homogeneously distributed throughout the glass, it is not considered that this process will limit the amplifier performance significantly.

The discussion of lanthanum-doped silica fibers was finally rounded out by a short description of neodymium and praseodymium properties in this host. It was found that Nd^{3+} has absorption peaks around the 530-, 590-, 750, 810-, and 900-nm wavelengths. Emission may be seen at 1060 to 1090 nm and around 1370 nm, but due to a very strong ESA, the possibility of 1300-nm amplification was ruled out. As expected, 1300-nm amplifiers based on Pr^{3+} in silica are not practical due to the large phonon energy of oxide glasses.

Applicable host materials for 1300-nm amplifiers are, however, found in the fluoride glass or ZBLAN composition, and here both the ESA problem of neodymium and the phonon energy problem of praseodymium are reduced. Also, erbium doping of ZBLAN glass has a number of advantages. Among the most significant advantages are that new laser transitions (such as 2700 nm) will be possible with Er/ZBLANP fibers. In addition to this, we also find indications that the cross-section spectra of Er^{3+} may be broadened by changing erbium from silica to a ZBLAN host.

The focus of the chapter was then returned to erbium-doped silica, but this time fibers with more than one rare-earth dopant were investigated. One of the relevant combinations is erbium fibers with a high additional co-dopant percentage of ytterbium. Such fibers are made with the purpose of extending the possible pump bands of erbium. This is possible because the Yb^{3+} ion after excitation may transfer its energy to a neighboring Er^{3+} ion. As a result, the 800-nm pump band may be extended, and efficient power amplification around 1540 nm is demonstrated using diode-pumped miniature Nd:YLF or Nd:YAG laser pump sources operating at 1064 nm. Note that Er^{3+}/Yb^{3+}-doped SiO_2 fibers are expected to show low power conversion efficiency, since the excitation of ytterbium ions, which have no neighboring erbium ions, adds an extra (but useless) pump absorption.

Another rare-earth co-dopant in EDFAs is lanthanum, which is optically passive, but may be used as refractive index raising material. More important, lanthanum may be used to replace erbium ions in formed rare-earth ion clusters, whereby the efficiency degrading effect of cross-relaxation is reduced. Lanthanum is therefore believed to be an attractive index-raising co-dopant for erbium-doped fiber.

Finally, the chapter briefly described the temperature dependence of the spectral properties of EDFAs. From this description it was evident that both emission and absorption spectra are changed as the temperature is lowered, but it is equally important to note that the significant spectral changes first appear at very low temperatures. For operation of fiber amplifiers in a normal temperature range, these effects are therefore not of serious concern.

REFERENCES

[1] Urquhart, P., "Review of Rare Earth Doped Fibre Lasers and Amplifiers," *IEE Proceedings, Pt. J*, Vol. 135, No. 6, 1988, pp. 385–407.

[2] France, P. W., "Optical Loss Mechanisms in Fluorozirconate Glasses and Infrared Fibres," Ph.D. thesis, University of Sheffield, U.K., 1987.

[3] Ainslie, B. J., S. P. Graig, and S. T. Davey, "The Absorption and Fluorescence Spectra of Rare Earth Ions in Silica-Based Monomode Fiber," *IEEE Journal of Lightwave Technology*, Vol. 6, No. 2, 1988, pp. 287–293.

[4] France, P. W., M. G. Drexhage, J. M. Parker, M. W. Moore, S. F. Carter, and J. V. Wright, *Fluoride Glass Optical Fibres*, Blackie and Son, Ltd. 1990.

[5] Desurvire, E., and J. R. Simpson, "Amplification of Spontaneous Emission in Erbium-Doped Single-Mode Fibers," *IEEE Journal of Lightwave Technology*, Vol. 7, No. 5, 1989, pp. 835–845.

[6] Desurvire, E., and J. R. Simpson, "Evaluation of $^4I_{15/2}$ and $^4I_{13/2}$ Stark-Levels Energies in Erbium-Doped Aluminosilicate Glass Fibers," *Optics Letters*, Vol. 15, No. 10, 1990, pp. 547–549.

[7] Miniscalco, W. J., "Erbium-Doped Glasses for Fiber Amplifiers at 1500 nm," *IEEE Journal of Lightwave Technology*, Vol. 9, No. 2, 1991, pp. 234–250.

[8] Larsen, C., and K. Dybdal, "Fibre Lasers and Amplifiers," *DOPS-NYT* (in Danish), Vol. 1, 1989, pp. 9–14.

[9] Arai, K., et al., "Aluminum or Phosphorus Co-doping Effects on the Fluorescence and Structural Properties of Neodymium-Doped Silica Glass," *Journal of Applied Physics*, Vol. 59, 1986, pp. 3430–3436.

[10] Ainslie, B. J., J. R. Armitage, S. P. Craig, and B. Wakefield, "Fabrication and Optimisation of the Erbium Distribution in Silica Based Doped Fibres," *Proc. of 14th European Conference on Optical Communication ECOC '88*, Brighton, 11–15 Sept. 1988, pp. 62–65.

[11] Pedersen, B., K. Dybdal, C. D. Hansen, A. Bjarklev, J. H. Povlsen, H. Vendeltorp-Pommer, and C. C. Larsen, "Detailed Theoretical and Experimental Investigation of High-Gain Erbium Doped Fiber Amplifier," *IEEE Photonics Technology Letters*, Vol. 2, No. 12, 1990, pp. 863–865.

[12] Dybdal, K., N. Bjerre, J. E. Petersen, and C. C. Larsen, "Spectroscopic Properties of Er-Doped Silica Fibers and Preforms," *Proceedings of SPIE OE/Fibers '89,* part: Fiber Laser Sources and Amplifiers I, Boston, Vol. 1171, 1989, pp. 209–218.

[13] Pedersen, B., A. Bjarklev, O. Lumholt, and J. H. Povlsen, "Detailed Design Analysis of Erbium-Doped Fibre Amplifiers," *IEEE Photonics Technology Letters*, Vol. 3, No. 6, 1991, pp. 548–550.

[14] Craig-Ryan, S. P., J. F. Massicott, M. Wilson, B. J. Ainslie, and R. Wyatt, "Optical Study of Low Concentration Er^{3+}-Fibres for Efficient Power Amplifiers," *Proc. of 16th European Conference on Optical Communication ECOC '90*, Amsterdam, 16–20 Sept. 1990, pp. 571–574.

[15] Pedersen, B., A. Bjarklev, J. H. Povlsen, K. Dybdal, and C. C. Larsen, "The Design of Erbium-Doped Fiber Amplifiers," *IEEE Journal of Lightwave Technology*, Vol. 9, No. 9, 1991, pp. 1105–1112.

[16] Nakazawa, M., Y. Kimura, and K. Suzuki, "High Gain Erbium Fibre Amplifier Pumped by 800 nm Band," *Electronics Letters*, Vol. 26, No. 8, 1990, pp. 548–550.

[17] Suzuki, K., Y. Kimura, and M. Nakazawa, "High Gain Er^{3+}-Doped Fibre Amplifier Pumped by 820 nm GaAlAs Laser Diodes," *Electronics Letters*, Vol. 26, No. 13, 1990, pp. 948–949.

[18] Laming, R. I., S. B. Poole, and E. J. Tarbox, "Pump Excited-State Absorption in Erbium-Doped Fibers," *Optics Letters*, Vol. 13, No. 12, 1988, pp. 1084–1086.

[19] Atkins, C. G., J. R. Armitage, R. Wyatt, B. J. Ainslie, and S. P. Craig-Ryan, "Pump Excited State Absorption in Er^{3+} Doped Optical Fibers," *Optics Communications*, Vol. 73, 1989, pp. 217–222.

[20] Andrews, L. J., W. J. Miniscalco, and T. Wei, "Excited State Absorption of Rare Earths in Fiber Amplifiers," *Proceedings of the First International School on Excited States of Transition Elements*, B. JezowskaTrzebiatowska, ed., World Scientific, Singapore, 1989, pp. 9–30.

[21] Zemon, S., G. Lambert, W. J. Miniscalco, R. W. Davies, B. T. Hall, R. C. Folweiler, T. Wei, L. J. Andrews, and M. P. Singh, "Excited State Cross Sections for Er-Doped Glasses," *Proceedings of SPIE OE/Fibers '90,* part: Fiber Laser Sources and Amplifiers II, San Jose, Vol. 1373, paper 3, 1990, pp. 21–32.

[22] Kimura, Y., K. Suzuki, and M. Nakazawa, "High Gain Erbium-Doped Fiber Amplifier Pumped in the 0.8 μm Pump Band," *Proc. of 16th European Conference on Optical Communication ECOC '90*, Amsterdam, 16–20 Sept. 1990, pp. 103–106.

[23] Quimby, R. S., "Output Saturation in a 980-nm Pumped Erbium-Doped Fiber Amplifier," *Applied Optics*, Vol. 30, No. 18, 1991, pp. 2546–2552.

[24] Krug, P. A., M. G. Sceats, G. R. Atkins, S. C. Guy, and S. B. Poole, "Intermediate Excited-State Absorption in Erbium-Doped Fiber Strongly Pumped at 980 nm," *Optics Letters*, Vol. 16, No. 24, 1991, pp. 1976–1978,

[25] Quimby, R. S., W. J. Miniscalco, and B. Thompson, "Excited State Absorption at 980 nm in Erbium-Doped Silica Glass," *Techn. Dig. of Optical Amplifiers and Their Applications OAA '92*, Santa Fe, 24–26 June 1992, paper WE3, pp. 67–70.

[26] Zyskind, J. L., "Erbium-Doped Fiber Amplifiers," *Proceedings of SPIE OE/Fibers '91,* part: Fiber Laser Sources and Amplifiers III, Boston, Vol. 1581, paper 2, 1991, pp. 14–23.

[27] Armitage, J. R., C. G. Atkins, R. Wyatt, B. J. Ainslie, and S. P. Graig, "Studies of Excited-State Absorption at 1.5 μm in Er^{3+} Doped Silica Fibres," *Conference on Lasers and Electro-Optics CLEO '89*, Washington, D.C., 1989, paper WF10, pp. 180–181.

[28] Blixt, P., J. Nilsson, J. Babonas, and B. Jaskorzynska, "Excited-State Absorption at 1.5 μm in Er^{3+}Doped Fiber Amplifiers," *Techn. Dig. of Optical Amplifiers and Their Applications OAA '92*, Santa Fe, 24–26 June 1992, paper WE2, pp. 63–66.

[29] Ainslie, B. J., S. P. Craig-Ryan, J. R. Davey, C. G. Atkins, and R. Wyatt, "Optical Analysis of Erbium Doped Fibres for Efficient Lasers and Amplifiers," *Proceeding of IOOC '89*, paper 20A3-2, 1989.

[30] Desurvire, E., J. L. Zyskind, and C. R. Giles, "Design Optimization for Efficient Erbium-Doped Fibre Amplifiers," *IEEE Journal of Lightwave Technology*, Vol. 8, No. 11, 1990, pp. 1730–1741.

[31] Blixt, P., J. Nilsson, T. Carlnäs, and B. Jaskorzynska, "Concentration-Dependent Upconversion in Er^{3+}-Doped Fiber Amplifiers: Experiments and Modeling," *IEEE Photonics Technology Letters*, Vol. 3, No. 11, 1991, pp. 996–998.

[32] Ainslie, B. J., S. P. Craig, and S. T. Davey, "The Fabrication and Optical Properties of Nd^{3+} in Silica-Based Optical Fibers," *Material Letters*, Vol. 5, 1987, p. 143.

[33] Monerie, M., D. Ronarc'h, and F. Auzel, "Rare-Earth Doped Fluoride Glass Fibre Amplifiers," *Proc. of 17th European Conference on Optical Communication ECOC '91*, Paris, 9–12 Sept. 1991, invited paper, pp. 9–16.

[34] Miyajima, Y., "Progress Towards a Practical 1.3μm Optical Fiber Amplifier," *Proc. of 18th European Conference on Optical Communication ECOC '92*, Berlin, 27 Sept.–1 Oct. 1992, invited paper, pp. 687–694.

[35] Millar, C. A., M. C. Brierley, and P. W. France, "Optical Amplification in an Erbium-Doped Fluorozirconate Fibre Between 1480 nm and 1600 nm," *Proc. of 14th European Conference on Optical Communication ECOC '88*, Brighton, 11–15 Sept. 1988, pp. 66–69.

[36] Sorel, Y., D. Ronarc'h, H. Ibrahim, J. F. Kerdiles, M. Guibert, and H. Poignant, "Performance of an Er3+ Doped ZBLAN Optical Fibre Amplifier Working at 1.54 μm in a 5 Gbit/s Transmission System," *Proc. of 17th European Conference on Optical Communication ECOC '91*, Paris, 9–12 Sept. 1991, paper WeC9-6, pp. 601–604.

[37] Miyajima, Y., "Erbium- and Neodymium-Doped Fluoride Glass Fiber Lasers and Amplifiers," *Proceedings of Society of Photo-Optical Instrumentation Engineers Conference, OE/Fibers '91*, part: Fiber Laser Sources and Amplifiers III, Boston, 5–6 September 1991, SPIE Vol. 1581, invited paper, pp. 304–315.

[38] Miniscalco, W. J., L. J. Andrews, B. A. Thompson, R. S. Quimby, L. J. B. Vacha, and M. G. Drexhage, "1.3 μm Fluoride Fibre Laser," *Electronics Letters*, Vol. 24, No. 1, 1988, pp. 28–29.

[39] Miyajima, Y., T. Komukai, T. Sugawa, and Y. Katsuyama, "Nd^{3+}-Doped Fluoro-Zirconate Fiber Amplifier Operated Around 1.3 μm," *OFC '90*, San Francisco, postdeadline paper PD-16.

[40] Brierley, M., S. Carter, P. France, and J. E. Pedersen, "Amplification in the 1300 nm Telecommunications Window in a Nd-Doped Fluoride Fibre," *Electronics Letters*, Vol. 26, No. 5, 1990, pp. 329–330.

[41] Carter, S. F., R. Wyatt, D. Szebesta, and S. T. Davey, "Quantum Efficiency and Amplification at 1.3 μm in a Pr^{3+} Doped Fluorozirconate Single Mode Fibre," *Proc. of 17th European Conference on Optical Communication ECOC '91*, Paris, 9–12 Sept. 1991, paper MoA2-3, pp. 21–24.

[42] Ohishi, Y., T. Kanamori, and S. Takahashi, "Concentration Quenching in Pr^{3+}-Doped Fluoride Fiber Amplifier Operating at 1.3 μm," *Proc. of 17th European Conference on Optical Communication ECOC '91*, Paris, Sept. 9–12, 1991, paper MoA2-2, pp. 17–20.

[43] Snitzer, E., and R. Woodcock, "Yb^{3+}-Er^{3+} Glass Laser," *Applied Physics Letters*, Vol. 6, 1965, p. 45.

[44] Hanna, D. C., R. M. Percival, J. R. Perry, R. G. Smart, and A. G. Trapper, "Efficient Operation of an Yb Sensitized Er Fiber Laser Pumped in the 0.8 μm region," *Electronics Letters*, Vol. 24, No. 17, 1988, pp. 1068–1069.

[45] Grubb, S. G., et al., "High-Power Sensitized Erbium Optical Fibre Amplifier," *Optical Fiber Communications OFC '91*, San Diego, 1991, pp. 31–34.

[46] Grubb, S. G., P. A. Leilabady, K. L. Sweeney, W. H. Humer, R. S. Cannon, and S. W. Vendetta, "High Output Power Er^{3+}/Yb^{3+} Co-doped Optical Amplifiers Pumped by Diode-Pumped Nd^{3+} Lasers," *Techn. Dig. of Optical Amplifiers and Their Applications OAA '92*, Santa Fe, 24–26 June 1992, paper FD1, pp. 202–205.

[47] Morkel, P. R., "Modelling Erbium/Ytterbium-Doped Fibre Amplifiers," *Techn. Dig. of Optical Amplifiers and Their Applications OAA '92*, Santa Fe, 24–26 June 1992, paper FD2, pp. 206–209.

[48] Barnes, W. L., S. B. Poole, J. E. Townsend, L. Reekie, D. J. Taylor, and D. N. Payne, "Er^{3+}-Yb^{3+} and Er^{3+} Doped Fiber Lasers," *IEEE Journal of Lightwave Technology*, Vol. 7, No. 10, 1989, pp. 1461–1465.

[49] Saifi, M. A., M. J. Andrejco, W. I. Way, A. V. Lehman, A. Y. Yan, C. Lin, F. Bilodeau, and K. O. Hill, "Er^{3+}-Doped GeO_2-CaO-Al_2O_3 Silica Core Fiber Amplifier Pumped at 813 nm," *Optical Fiber Communications OFC '91*, San Diego, 1991, paper FA6, p. 198.

[50] Lumholt, O., K. Dybdal, C. C. Larsen, S. Dahl-Petersen, K. Schüsler, A. Bjarklev, J. H. Povlsen, T. Rasmussen, and K. Rottwitt, "Er-La Doped Fibre Amplifier, Pumped at 1.47 μm," *Proc. of 17th European Conference on Optical Communication ECOC '91*, Paris, 9–12 Sept. 1991, paper TuPS-1, pp. 285–288.

[51] Nakazawa, M., and Y. Kimura, "Lanthanum Codoped Erbium Fibre Amplifier," *Electronics Letters*, Vol. 27, No. 12, 1991, pp. 1065–1067.

[52] Shelby, J. E., and J. T. Kohli, *Journal of the American Ceramic Society*, Vol. 73, No. 1, 1990, pp. 39–42.

[53] Kagi, N., A. Oyobe, and K. Nakamura, "Temperature Dependence of the Gain in Erbium-Doped Fibers," *IEEE Journal of Lightwave Technology*, Vol. 9, No. 2, 1991, pp. 261–265.

[54] Millar, C. A., T. J. Whitley, and S. C. Fleming, "Thermal Properties of an Erbium-Doped Fiber Amplifier," *IEE Proceedings J.*, Vol. 137, No. 3, 1990, pp. 155–162.

[55] Zyskind, J. L., E. Desurvire, J. W. Sulhoff, and D. J. Digiovanni, "Spectral Gain Hole-Burning in an Erbium-Doped Fiber Amplifier With GeO_2:SiO_2 Core," *Techn. Dig. of Optical Amplifiers and Their Applications OAA '90*, 6–8 August 1990, paper MD4, pp. 56–59.

[56] Giles, C. R., and E. Desurvire, "Modeling Erbium-Doped Fiber Amplifiers," *IEEE Journal of Lightwave Technology*, Vol. 9, No. 2, 1991, pp. 271–283.

[57] Lumholt, O., S. Dahl-Petersen, A. Bjarklev, C. C. Larsen, B. Pedersen, J. H. Povlsen, T. Rasmussen, and K. Rottwitt, "Gain Variations for an Erbium Doped Fiber Amplifier in a Temperature-Range From 45 K to 320 K," *Journal of Optical Communications*, Vol. 13, No. 3, 1992, pp. 114–116.

[58] Zemon, S., G. Lambert, L. J. Andrews, W. J. Miniscalco, B. T. Hall, T. Wei, and R. C. Folweiler, "Characterization of Er^{3+}-Doped Glasses by Fluorescence Line Narrowing," *Journal of Applied Physics*, May 1991.

Chapter 4
Characterization Measurements for Rare-Earth-Doped Fibers

The progress of any technological development depends on the possibilities of performing accurate measurements on the involved components, and the rare-earth-doped optical fiber is not different in this respect. We will therefore in this chapter focus on some of the central measurement techniques necessary for the accurate characterization of rare-earth-doped fibers. A description of the experimental characterization of pump sources, wavelength division multiplexers, filters, and isolators is given elsewhere [1,2]. Therefore, we will concentrate on the experimental description of the rare-earth-doped fibers, and use the erbium-doped silica fiber as a representative example.

The chapter is divided into three sections. The first describes the properties of the rare-earth-doped fiber that may be measured on the preform. The focus is on the special characteristics of measuring the rare-earth dopant distribution. The second section deals with spectroscopic properties measured on the active fiber, and the third describes some central experimental characterization methods for determining amplifier properties.

4.1 BASIC PROPERTIES MEASURED ON THE PREFORM

Before the rare-earth-doped fiber is drawn from the preform, a number of properties can be measured. These include measurement of the refractive index profile and, equally important, the erbium concentration profile. The significance of the latter becomes clear from the fact that an accurate knowledge of the ion concentration may not be determined on the optical fiber, since the small dimensions make any mapping procedure highly complicated and inaccurate. The major effort in this section is therefore aimed at a procedure for determining the rare-earth dopant, but a short discussion of the index profile measurement will also be given.

The reason is that even though the refractive index profile may be measured directly on the fiber, a comparison with preform measurements may still be of value. Such measurements may help us understand and control eventual diffusion processes occurring during the fiber drawing process.

4.1.1 Index Profile Measurements

An attractive and nondestructive way to measure the refractive index profile of optical-fiber preforms is by means of transverse illumination techniques [3]. The idea behind these methods is to determine the angle of refraction of a ray of light passing transversely through the preform. These measurements are performed as the ray is swept transversely across the preform, and a deflection function is obtained. The refractive index profile may be found by numerical transformation of this function. As described in [4], the implementations of the transverse illumination approach can differ crucially in their optical configurations, depending on whether collimated incoherent light or focused laser beams are used. We will not, however, go into a further discussion of these methods at this point, but just direct the reader to [5, 6] for additional information. The reason for this is that the index profile measurement is the same as that described in textbooks concerning passive optical fibers [7].

4.1.2 Radial Distribution of Rare-Earth Ions in the Preform

In this section, we will discuss methods applied in the determination of the radial distribution of rare-earth ions in the preform; specifically, the case of erbium-doped silica fibers will be considered. We first consider a method for measuring the absolute mean concentration of erbium, and then a method for determining the relative radial erbium distribution will be described.

The mean concentration of erbium ions in the core of the rare-earth-doped fibers may be determined by *Rutherford backscattering spectrometry* (RBS) on slices of the doped preforms [8,9]. We will briefly outline the method using the description in [9].

In RBS low mass, charged particles (in this case α-particles) are made to collide with a target (the preform core area) at a relatively high energy. The actual magnitude of this energy, which in the specific case is about 2 MeV, is chosen high enough for the atomic processes to be neglected. In other words, the energy is high enough for the electron cloud of the target to be completely penetrated. On the other hand, the energy must be low enough to avoid exciting nuclear reactions. With an appropriate choice of energy, the scattering from target nuclei becomes elastic and the differential scattering cross section, $d\sigma/d\Omega$, is given by Rutherford's formula [10]:

$$\frac{d\sigma}{d\Omega} = \left(\frac{q_1 q_2}{4E_0}\right)^2 \cdot \frac{1}{\sin^4(\theta/2)} \tag{4.1}$$

where q_1, q_2 are the electrostatic charges of the α-particle and erbium nuclei, respectively. E_0 is the impact energy of the α-particle (2 MeV) and θ is the scattering angle in the laboratory system. The differential cross section is the cross section per steradian (Ω) for scattering into the particular angle given by θ.

The energy E of the scattered particles from the surface of the target is determined by the kinematic factor K_{M2} by the relation $E = K_{M2} \cdot E_0$. The kinematic factor may be expressed as

$$K_{M2} = \left(\frac{\mathrm{M}_1 \cos(\theta) + \sqrt{M_2^2 - M_1^2 \sin^2(\theta)}}{M_1 + M_2}\right)^2 \tag{4.2}$$

where M_1 and M_2 are the masses of the α-particle and the target nuclei, respectively.

If all scattering processes took place at the surface only, the resulting energy spectrum would consist of narrow peaks. These would be positioned at energies corresponding to the kinematic factor for the masses involved, and their relative integral values would depend on the scattering cross section and the concentration of the elements in the target. However, this is not the case, since some of the α-particles will penetrate into the target, where they interact nonelastically with the surroundings and finally scatter with a lower impact energy. On their way out of the target material, they interact again, and in the end they will be detected with a lower energy than that given by (4.2). The measurement result will therefore not consist of narrow peaks, but of energy intervals, which may be identified as representing the different broadened contributions from the components of the target material. In a Ge/Er-doped SiO_2 fiber, we will be able to identify Rutherford backscattering from oxygen, silicon, germanium, and erbium. As described in [11], the relative measurements of the erbium and germanium to silicon concentration may be found. Since the atomic density of silicon in fused silica glass is known (2.2×10^{28} m^{-3}), the method is estimated to give 5% accuracy for the erbium mean concentration [8].

Now what is left is to specify the absolute erbium dopant profile by measuring the relative radial distribution of the Er^{3+} ions. The concentration profile can be measured along a diameter of a slice of preform by energy-dispersive x-ray analysis in a *scanning electron microscope* (SEM) [8, 12].

The function of the SEM is as follows: electrons (from a filament) are accelerated through an electric field (the voltage is typically up to 40 kV), and they are focused on the sample surface with a system of electromagnetic lenses. The focused electron beam can then be scanned across the sample surface. When the accelerated electrons interact with the sample material, a number of different signals are emitted and they may subsequently be detected. In the specific case, where the interesting property is the chemical composition of the fiber preform, it is relevant to measure the characteristic x-ray spectrum. Here, each element of the fiber may be identified by a specific distribution of energy levels. It is noteworthy that the spatial solution of this SEM measurement will typically be 1 to 2 μm, limited by the involved scattering processes. This is why the measurement is best performed on preforms rather than fiber samples. Figure 4.1 shows typical results of Erbium and Germanium, measured on a preform.

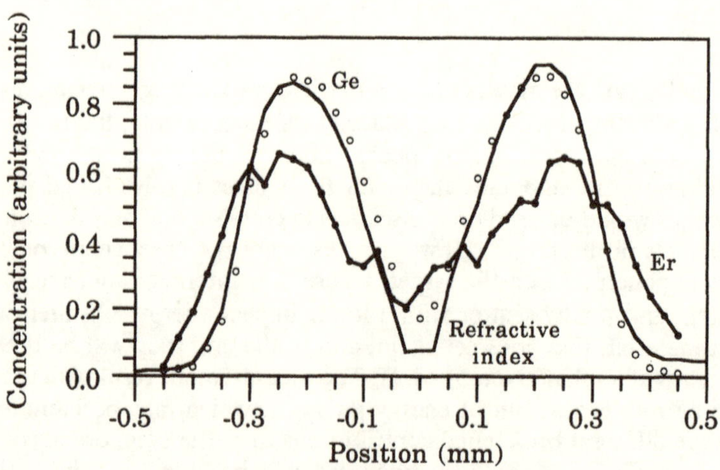

Figure 4.1 Concentration profiles of erbium and germanium measured on a preform. The measured refractive index profile is shown as a solid curve without dots. (Data from [8].)

4.1.3 Fluorescence Measurements

The fiber preform may also be used for measurements of the fluorescence spectrum, and, as described in [8], this may be measured after exciting the Er^{3+} ions in a preform slice with light from a 655-nm dye laser beam. The spectrum of the emitted light is then measured through a monochromator, which, in the example of [8], has a resolution of 1 nm. The spectrum is finally corrected for the spectral response of the measurement system.

The attractive property of this measurement is that the fluorescence is not spectrally deformed from additional processes of self-absorption and self-amplification, as well as stimulated de-excitation. The reason is that the emitted light only propagates through a very short length of pumped amplifier media, and line narrowing effects are therefore avoided. However, a serious drawback is the need for cutting a slice of the preform, which, from a practical application point of view, is very inconvenient. The process will therefore not be discussed in more detail in this section.

4.2 FUNDAMENTAL CHARACTERISTICS MEASURED ON THE FIBER

After drawing the rare-earth-doped preform into a fiber, a large number of fundamental properties may be measured. These naturally include both refractive index profile and cutoff wavelength, as in the case for passive standard optical fibers. However, as we shall see in the following section, additional aspects have to be considered when rare-earth-doped optical fibers are to be characterized from the well-known methods. In addition to this, Sections 4.2.2 to 4.2.4 will describe characterization measurements specific to the active fibers (i.e., absorption and fluorescence measurements, as well as radiative lifetime estimation).

4.2.1 Refractive Index Profile and Cutoff Wavelength

The experimental determination of the refractive index profile of lanthanide-doped optical fibers does not differ from established methods developed for passive optical fibers. It is therefore not the purpose of this book to go into a full description of these methods, but rather to briefly review one of the most well-known approaches.

A popular refractive index measurement method is the *refracted near field* (RNF) technique. In an instrument based on this principle, light is focused at the plane fiber end and refracted through the sides of the fiber [13]. The exit angles of the refracted rays depend on the refractive index of the fiber at the point at which the light is focused onto its end face, and also on the refractive index of the matching oil used to improve the coupling efficiency. Unlike preform profiling, where there is essentially a parallel light beam, the deflection of which is measured, the RNF method produces a cone of light. Instead of measuring angular changes, part of the light cone is obscured by an opaque disk, and changes in the power of the light passing the disk are measured. For small changes in refractive index and light distribution, which is approximately Lambertian, changes in power and refractive index are linearly related to the required accuracy [14]. One way to calibrate is by moving the light blocking disk along its axis by a known amount. Since this corresponds to an angular movement of the refracted rays, a change in light power can be related to a change in refractive index. According to [14], a typical resolution of the refractive index difference is ± 0.0001. A comparison with other refractive index measurement methods is also presented in [14], and we will refer to this description for further information.

The propagation properties of the fiber modes may be calculated based on the measured refractive index profile, and among the central parameters is the cutoff wavelength, which indicates where the fiber can be used in single-mode applications. In practice, this theoretical cutoff wavelength will differ slightly from the effective cutoff of the fiber, which is defined as the wavelength above which the LP_{11} (linearly polarized) mode is no longer transmitted. This difference is due to additional effects such as microbends, core diameter fluctuations, and bends, which together strongly increase the LP_{11} loss for wavelengths slightly below the theoretical cutoff of the fiber. Because of this minor difference and the complexity of calculations, it may alternatively be wished that the cutoff wavelength of a given fiber be measured. We will therefore briefly sketch the measurement method at this point.

The simplest measurement technique is, according to [15], combined with the spectral loss measurement (see Section 4.2.2). It consists of simply observing the somewhat abrupt transmitted power decrease when the wavelength increases and reaches the effective cutoff wavelength. More specifically, the attenuation is measured spectrally and examined for abrupt power decreases associated with the LP_{11} mode leakage. The nature of this becomes clearer when curves are measured in (at least) two cases: (1) when the fiber is only slightly bent, and (2) when the fiber experiences a significant bend (depending on the NA). By comparing the curves, a difference may be found in the loss gradient due to a more significant LP_{11} mode attenuation when the fiber sample was strongly bent.

However, a few final comments are relevant, and the first is related to the standard situation of passive fibers. It can generally be seen that the effective cutoff wavelength obtained from the described method depends on the curvature of the fiber sample, and a certain uncertainty is therefore unavoidable [15]. In addition, the rare-earth-doped fiber may be designed so that the cutoff wavelength is located within one of the absorption bands of the lanthanide ions, and the described method is in such cases highly inaccurate or maybe even inapplicable. If knowledge of the cutoff wavelength is required in these cases, the theoretical cutoff wavelength could be calculated.

4.2.2 Spectral Attenuation

As already has become clear from the description in Chapter 3, the attenuation spectrum of the rare-earth-doped fiber is an essential property in the characterization, since not only the location but also the strength of the pump bands may be found in this way. We will, therefore, at this point describe the standard procedure of attenuation measurement. The measurement setup is illustrated in Figure 4.2, and it is based on a standard cutback method [7]. The principle is to measure the transmitted optical power $P_B(\lambda)$ as a function of the optical wavelength for a given fiber length L_B. Hereafter, the fiber is cut back to a reference length L_A, and the wavelength-dependent transmitted power is measured again. As shown in Figure 4.2, the light source emits a white light spectrum (in the wavelength range of 300 to 2000 nm) from which the different spectral lines are chosen by a monochromator. Furthermore, it is noteworthy that the fundamental setup is similar to attenuation measurement equipment used for the spectral characterization of passive optical fibers. The attenuation, $A(\lambda)$ (in decibels per meter), may be calculated from the simple relation

$$A(\lambda) = \frac{10}{L_B - L_A} \cdot \log\left(\frac{P_A(\lambda)}{P_B(\lambda)}\right) \tag{4.3}$$

However, a number of special problems are related to the measurement on active fibers. These problems are due to the large attenuation difference between absorption peaks and wavelength intervals, where the rare-earth ions do not affect the fiber attenuation significantly. In order to secure an acceptable dynamic range for the measurement, the fiber length ($L_B - L_A$) should realistically be chosen so that the attenuation is around 10 dB at the absorption peaks. Such a requirement

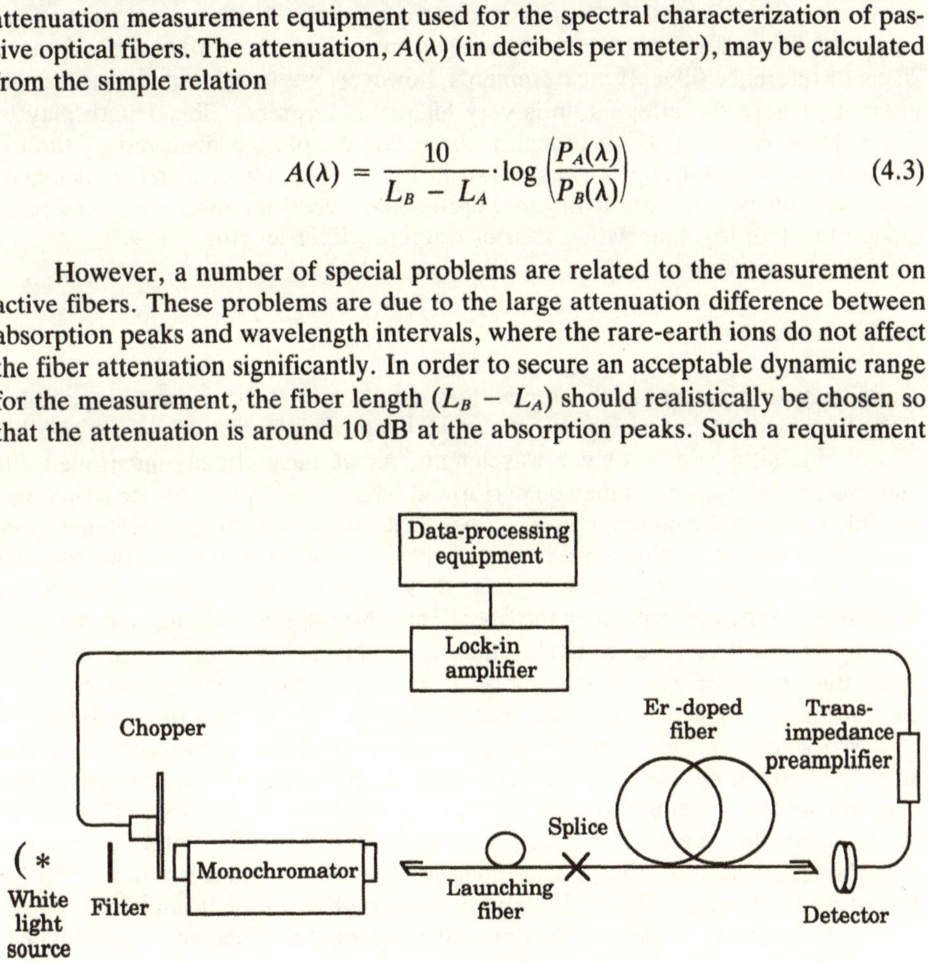

Figure 4.2 Setup for measurement of the spectral attenuation.

is, on the other hand, not generally obtainable, since the attenuation peak values may vary over several orders of magnitude (see Chapter 3). Consequently, measurements have to be made for different fiber lengths, depending on the wavelength interval of interest. This means that for fibers with high erbium concentrations, such as those described in Section 3.3.2, typical fiber lengths are on the order of 1 m, whereas maybe 100 m is needed for fibers with a low erbium concentration. Between the absorption peaks, where the attenuation is due to the intrinsic attenuation of SiO_2 fibers, it is often found that $P_A(\lambda) \approx P_B(\lambda)$, and the attenuation $A(\lambda)$ is therefore measured with low accuracy in these wavelength intervals.

Another problem is connected with the length of the reference fiber, which is necessary for eliminating the influence of higher order modes excited at the fiber input end. These higher order modes have normally leaked out after approximately 20 cm of reference fiber. If measurements, however, are to be made in wavelength intervals, where the attenuation is very high, this reference fiber length may be shown to be too long. Even though a controlled coupling is achieved by the use of the dispersion-shifted fiber, as shown in Figure 4.2, higher order modes may, however, still be present. It might therefore be necessary to choose a cladding mode filter [16] together with a shorter reference fiber length.

4.2.3. Fluorescence Measurements

A property as important as the attenuation spectrum of the rare-earth-doped fiber is the fluorescence or emission spectrum, which gives information on the possibilities of obtaining gain at a given wavelength. As we have already mentioned, the fluorescence measurement may be performed on the preform, but since this is only possible for the fiber manufacturer, a fiber measurement technique becomes more relevant. It should furthermore be noted that, according to [8], the fluorescence properties may differ significantly depending on whether the measurements are performed on the preform or on the fiber. This observation indicates that the Stark splitting is sensitive to even small changes in the Er^{3+} surroundings. From an application point of view, furthermore, it may lead to the recommendation that the fluorescence spectrum be measured on the fiber and not on the preform.

We will, therefore, describe an accurate method for fluorescence measurements performed on the fiber. The essential task here is to avoid the spontaneous emission spectrum being distorted by propagation through the fiber. This can be obtained by choosing a very short fiber length, and a practical rule for this is to keep the length so short that the attenuation at the gain peak is less than 1% [17]. The setup is shown in Figure 4.3, and it appears that the light emitted from the fiber is filtered by a monochromator and is thereafter detected by an InGaAs detector (for wavelengths relevant in the case of erbium-doped fibers). The signal is then amplified by a trans-impedance amplifier and fed to a lock-in amplifier,

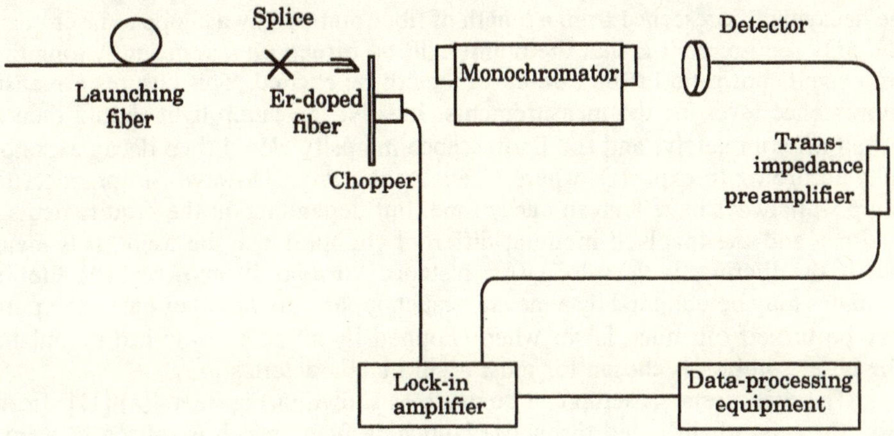

Figure 4.3 Setup for measurement of the fluorescence spectrum from rare-earth-doped fibers pumped through the dispersion-shifted fiber.

which is necessary because the fluorescence signal is very weak. The reference signal is obtained from the chopper, which is placed as close as possible to the fiber in order to minimize the effect of the divergence of the beam.

However, if the rare-earth dopant concentration, and thereby the attenuation, becomes large, the lengths of the rare-earth-doped fibers become impractically short (≤1 cm) with respect to handling in the laboratory. In this case, the fluorescence is instead viewed from the side of a short length of fiber (≈2 cm), which is pumped to high inversion by an appropriate pump mode. To avoid attenuation that is unaccounted for from the fiber coating, the coating is removed before the measurement.

The measured spectrum is normally distorted by the wavelength dependence of the detector and the efficiency of the monochromator, and a subsequent correction is therefore necessary before the final spectrum is obtained.

4.2.4 Lifetime in Upper Laser Level

Another very important parameter is the fluorescence lifetime, which is a key parameter not only for the accurate modeling of the fiber amplifier, but also for the qualitative understanding of its properties. Besides this, detailed measurements of the fluorescence from the upper laser level can provide information on the relative importance of ion-ion interactions such as up-conversion and quenching processes [18] (see Chapter 3).

The lifetime τ in the upper laser level may be measured simply by monitoring the decaying fluorescence from a length of fiber pumped by a chopped laser beam. The only requirement is that the pump light be turned on a sufficiently long time for a significant ratio of the rare-earth ions to be excited. This ensures a realistic fluorescence level for the measurements. Ideally, the pump light should then be turned off completely, and the fluorescence intensity would then decay exponentially according to $\exp(-t/\tau)$, where t denotes the time. However, in practice, the pump will always have a given decay time, but depending on the requirements of accuracy and the involved lifetime, different choppers may be used. This means that if the lifetime is very long (for instance, around 10 ms), realistic lifetime estimates may be obtained by a mechanical chopper. On the other hand, the pump may be turned off much faster when chopped by an acousto-optical modulator. The latter solution is chosen for most accurate characterizations.

The measurement setup can be made as shown in Figure 4.4(a) [17]. In this case, the pump light is fed through a launching fiber, which is spliced to a small piece of erbium-doped fiber. A dispersion-shifted launching fiber may often be the preferred choice, since better mode matching and, as a result, lower coupling losses can be obtained. This measurement fiber is then placed in an ellipsoidal mirror in such a way that a part of it coincides with one of the focal points. The fiber end is placed in a hollow needle to block the pump light and prevent it from reaching the detector, which is placed in the other focal point of the mirror. This setup ensures that a sufficient amount of fluorescence signal is collected and detected.

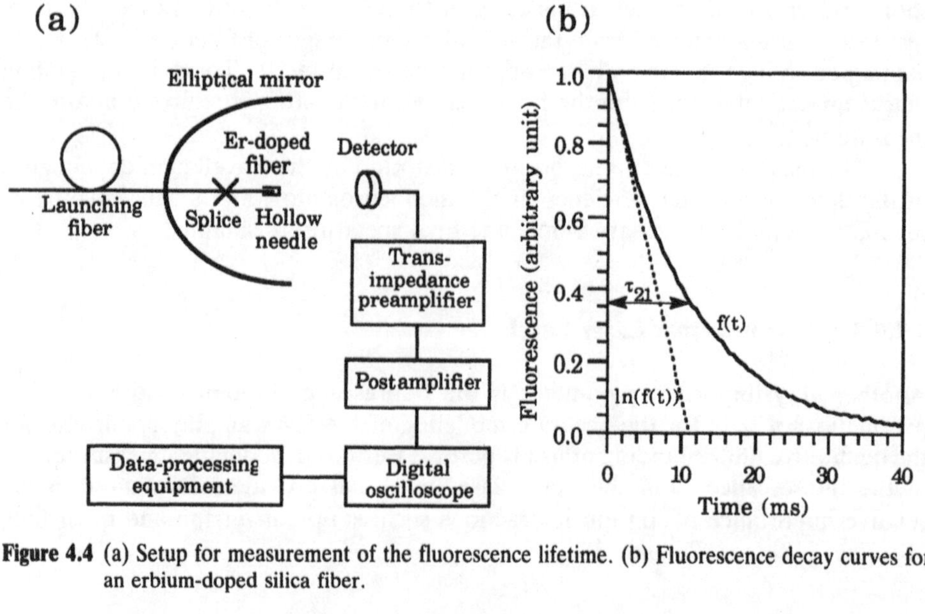

Figure 4.4 (a) Setup for measurement of the fluorescence lifetime. (b) Fluorescence decay curves for an erbium-doped silica fiber.

If the pump light is now rapidly turned off for a time interval, which is much longer than the lifetime in the upper laser level, results such as those illustrated in Figure 4.4(b) may be found.

4.2.5 Determination of Exited-State Absorption

As described in Section 3.1.3, several possibilities for ESA exist for the erbium-doped fiber amplifier at either the pump or the signal wavelength. To determine the strength of ESA experimentally, it is necessary to employ different methods, which may be found described in the references related to Section 3.1.3. However, a number of common properties may be found, and therefore at this point only one of these methods is referred to, which is developed in order to measure ESA in the 800-nm pump band (this method has been presented in [19, 20]). The relevance of measuring the exact shape and magnitude of the ESA spectrum is found in the fact that low-cost AlGaAs diodes may be used as pump sources in this wavelength band for erbium-doped fibers. We will bring forward here the central points of this measurement.

A basic problem with measuring ESA is obtaining a controlled population of an excited energy level of the involved rare-earth ion. The measurement described in [19] dealt with this problem in determining the difference in transmission of a probe beam when a pump beam was turned on and off. The transmission difference is denoted by $\Delta T = (T_{off} - T_{on})$, where T_{off} and T_{on} refer to the respective transmission factors in the two states of the pump power. To optimize the signal-to-noise ratio, the double modulation scheme was employed as illustrated in Figure 4.5. The pump beam (with wavelength $\lambda_p = 647.1$ nm) was chopped at 10 Hz, which is a very slow frequency with respect to the ≈ 10-ms lifetime of the metastable $^4I_{13/2}$ state. The pump (with power up to 0.5 W) is then reflected by a dichroic mirror and focused onto one end of the erbium-doped fiber. The probe beam is supplied by an incoherent 100 W lamp and chopped at 400 Hz. It was thereafter focused onto the other end of the fiber. The specific light source is chosen because it contains all the interesting spectral components in the 800-nm band. Note that excited cladding modes were stripped with index-matching fluid. The transmitted probe beam was collimated and focused into a spectrometer, and the selected wavelength component was then detected with a cooled germanium pin photodiode and the signal was filtered with a 400-Hz lock-in. The 10-Hz envelope was amplified by a second lock-in and the resulting data were recorded with a computer. To obtain ΔT from the raw data, a scaling factor (derived from accurate dc measurements) was used to correct for the averaging effect of the lock-in system. Finally, a scan was taken with the pump off to obtain T_{off}.

Saturation has in [20] been reported with excitation power as low as 0.2W. The procedure for deriving the cross sections for the erbium-doped fibers from the saturated value $(\Delta T/T_{off})_{sat}$ is as follows. First, the attenuation α was evaluated using

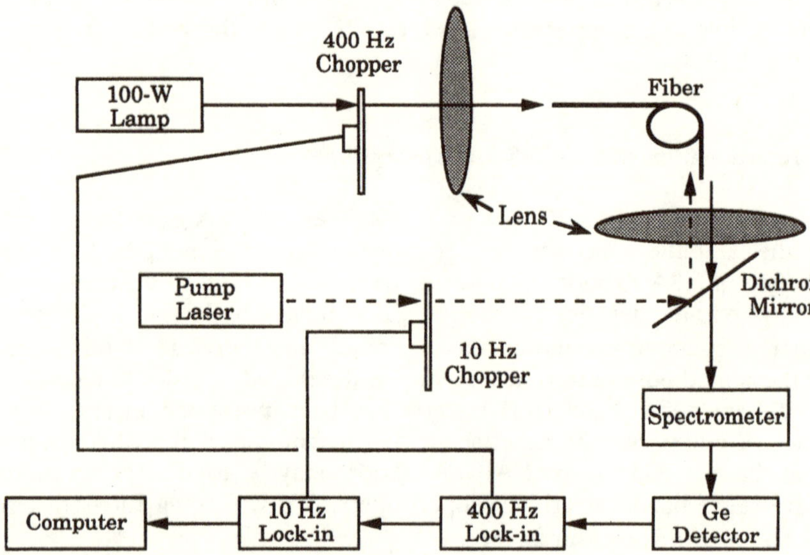

Figure 4.5 Sketch of the experimental system, which in [19] was used for measurement of the ESA spectrum in the 800 nm band.

$\alpha = \{\ln(1 - (\Delta T/T_{off})_{sat})\}/L_0$, where L_0 is the fiber length. To relate the measured attenuation to an actual cross section σ, the α data for the 980-nm band were compared to the GSA cross-section data, and a scaling constant C_0 was determined (i.e., $\sigma = C_0\alpha$. After this, the constant C_0 was used to translate the $(\alpha_{24} - \alpha_{13})$ data for the 800-nm band into the cross-section data $(\sigma_{24} - \sigma_{13})$ so that the ESA cross-section spectrum might be determined as $\sigma_{24} = C_0(\alpha_{24} - \alpha_{13}) + \sigma_{13}$. The last step requires that the GSA cross sections are determined separately, which may be done from cutback measurements (as described in Section 4.2.2) and knowledge of the Er^{3+} doping profile [21]. We will return to this analysis procedure in Chapter 5.

Note finally that in [18] it has been found that increasing phosphorous and the addition of aluminum are both beneficial for the reduction of ESA at 800 nm (and at 650 nm).

4.3 MEASUREMENTS OF AMPLIFIER CHARACTERISTICS

The final step in the experimental characterization of the rare-earth-doped fiber is to measure its properties as an amplifying medium. This means that the lanthanide ions (and we will again use Er^{3+} as an example) have to be pumped, and a probe signal must be included. The chief property in this respect is, not surprisingly, the amplifier gain, but determining the *amplified spontaneous emission* (ASE) intensity is also essential for the determination of the noise figure. It is the measurement of these properties that will be the focus of this chapter.

4.3.1 Measurement of Amplifier Gain and ASE

A basic setup used for gain measurements is shown in Figure 4.6 [22]. We have chosen to illustrate the technique at the pump wavelength λ_p = 654 nm. The coupled pump and signal (λ_s = 1530 nm) powers in the undoped launching fiber were accurately calibrated to the monitor detectors. The launching setup was included because the back-reflection into the erbium-doped fiber could thereby be reduced to −57 dB. The erbium-doped fiber was fusion spliced to the launching fiber and terminated with a cut at an oblique angle, whereby reflections of signal, pump, or amplified spontaneous emission were reduced. Splice losses at signal and pump

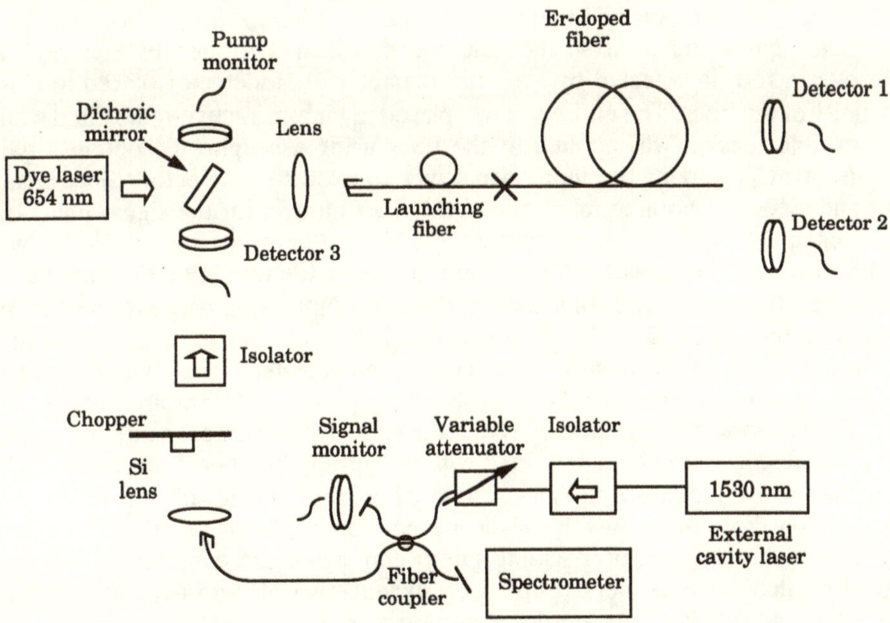

Figure 4.6 Er^{3+} fiber amplifier: experimental setup for measurement of gain-attenuated pump power and forward and backward ASE power. (From [22].)

wavelengths were measured after each series of measurements. In the setup described in [22], losses were low, typically 0.3 to 0.7 dB.

The residual pump power and forward- and backward-traveling ASE powers could now be measured in calibrated detectors, detector 1, detector 2, and detector 3, respectively. Removing detector 3 from the signal beam path, the output signal could be measured with detector 2 connected to a lock-in amplifier (input signal < -46 dBm). The results were estimated to be accurate to within ± 0.3 dB in [22] within the wavelength interval from 1530 to 1600 nm. This uncertainty estimate was based on the results of repeated measurements.

This method for experimental mapping of the amplifier gain is very useful and accurate for laboratory investigations, and it is commonly used for the characterization of fiber amplifiers. A typical application example is the case in which an optimal fiber length is sought. In such cases, gain properties at various lengths are determined by cutting back the erbium fiber (for a specified pump power) until the maximum gain is identified. However, this cutback method is time-consuming and it destroys the fiber amplifier being studied. For this reason, it is interesting to be able to perform gain measurements in an alternative way, which does not have the same drawbacks, and here the nondestructive method proposed in [23] becomes relevant.

The fundamental idea of this method appears in Figure 4.7, where it is illustrated how a short-radius bend on the fiber results in light escaping from the fiber core. The light scattered from the macrobend, which is formed by bending the fiber over a rod, is detected by a 10-mm diameter photodetector placed in close proximity of the fiber. The entire rod was placed in high-refractive-index fluid while reproducible tension was applied to the fiber using a weight. To obtain a gain measurement, a part of the input signal was coupled to a detector. Since pump light and ASE are both coupled out of the fiber along with the signal light, the input signal light was in [23] chopped at 50 kHz and phasesensitive detection was used to discriminate between them. Due to the long lifetime (10 ms) in the upper laser level, the modulation induced on the pump and ASE was expected to be attenuated by at least 20 dB at 50 kHz [24]. It should also be mentioned that a polarization controller is included to maximize the detected signal, whereby polarization dependence effects of the macrobend coupling coefficient are reduced.

The advantage of this method is that the gain, output power, pump efficiency, and optimal length may be determined without cutting the fiber. Furthermore, the setup may be used for absorption measurement by transmitting only one modulated signal. A few drawbacks must, however, be mentioned. One is that the fiber length has to be cut if counter-propagating pump and signal are considered. Another difficulty is that the macrobend coupling coefficient is wavelength-dependent, which may affect the relative error on the measured gain values.

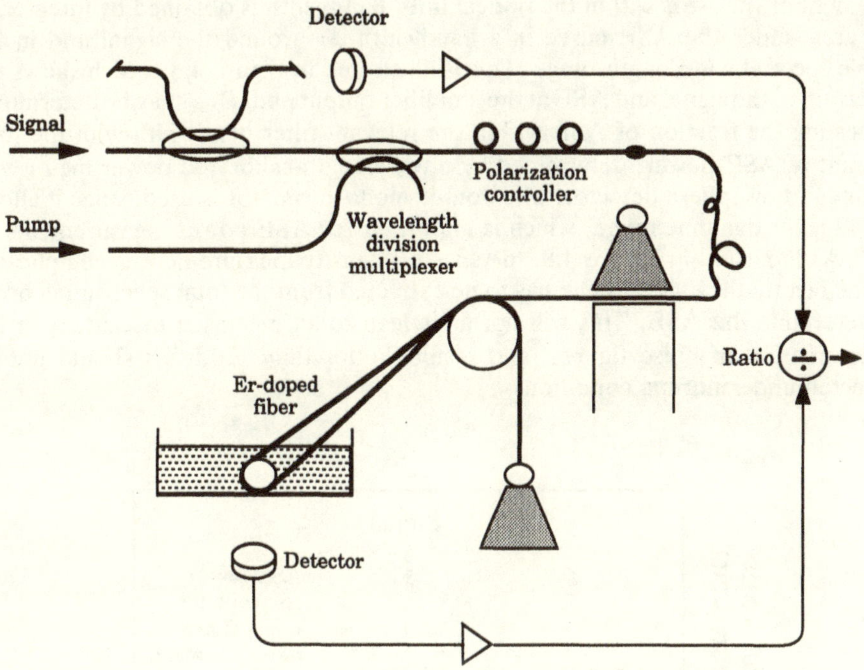

Figure 4.7 Apparatus used to nondestructively measure gain in fiber amplifiers. The erbium-doped fiber is fusion-spliced to standard single-mode fiber after the polarization controller. (From [23].)

4.3.2 Experimental Determination of Amplifier Noise Figure

The setup for evaluating the noise figure is somewhat similar to the one described in the previous section. The reason is that the noise figure is determined from corresponding measurements of gain G and ASE power P_{ASE} in a bandwidth $\Delta\nu$ around the signal frequency ν. The ASE power in question is the one co-propagating with the signal. As we shall see in Chapter 5, the noise figure F may be calculated from the following equation [25]:

$$F = \frac{1}{G} \cdot \left(\frac{P_{ASE}}{\Delta\nu \cdot \mathrm{h} \cdot \nu} + 1 \right)$$

where h is Planck's constant. The shape of the ASE spectrum may, for example, be recorded in 0.1 nm steps on a double monochromator [26]. A relative measurement of the ASE within the optical filter bandwidth is obtained by integrating the area under the ASE curve in a bandwidth $\Delta\nu$ around the signal and in the whole spectral wavelength range. This is illustrated in Figure 4.8, which shows the spectrum of the signal and ASE at the amplifier output end. P_{ASE} may be determined by scaling the fraction of ASE within the relevant filter bandwidth with the total amount of ASE power, determined by a wavelength-calibrated power meter with an integrating sphere detector. The double monochromator is used, since it allows for a higher dynamic range, which is important for ASE power measurements.

As indicated in Figure 4.8, the spectral density measurement is complicated by the fact that the signal peak has to be extracted from the total spectrum in order to determine the ASE. This will normally lead to an increased inaccuracy in the determination of noise figures, and results better than ±0.5 dB should not be expected under normal conditions.

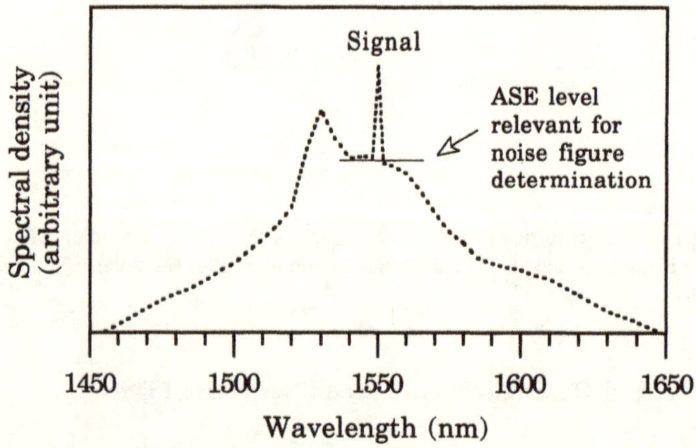

Figure 4.8 Spectrum of light intensity at the output end of an erbium-doped fiber amplifier. The peak at 1550 nm is due to the signal surrounded by the ASE spectrum.

4.4 SUMMARY

In order to obtain a complete understanding of the properties of a given fiber amplifier, it is necessary to characterize the amplifier components carefully. In this respect the rare-earth-doped fiber is a key component, and measurements may be performed already, since it is at the preform stage.

Besides the refractive index profiling, which is standard in many production facilities for passive optical fibers, the distribution of rare-earth ions is also of interest. A method for determining the rare-earth dopant profile has been reviewed in this chapter, and it was found that the mean lanthanide concentration may be estimated with an accuracy of 5% using Rutherford backscattering. Hereafter, the relative radial erbium distribution may be determined from x-ray analysis of a preform slice placed in a SEM. This measurement requires that the preform be sliced, and it may therefore be undesirable for standard characterization procedures. However, since the rare-earth distribution cannot be measured within the small dimensions of a drawn fiber, the preform method remains the only applicable way to achieve detailed information on the rare-earth distribution profile. The final preform measurement discussed in this context is the determination of the fluorescence spectrum. Here, line narrowing effects due to self-absorption and selfamplification may be avoided because the light only travels over a short distance within a preform slice. However, the attraction of this method is also questionable, since a part of the preform must be destroyed in order to prepare the measurement sample. In addition to this, fluorescence differences between preform and fiber measurements have been observed.

The primary characterization of the basic properties of the rare-earth-doped fiber is performed on the fiber. Among the fundamental parameters to be determined are the refractive index profile and the cutoff wavelength. The first of these may normally be found by using standard fiber characterization equipment, applying, for instance, the refracted near-field technique. However, it should be ensured that the dynamic range of the instrument is large enough to measure the very high NAs, which are often used in rare-earth-doped fibers. The increased attenuation associated with the leakage of the LP_{11} mode is normally used to determine the cutoff wavelength, and the method is also useful for EDFAs under the condition that the cutoff wavelength does not coincide with the attenuation peaks of the lanthanide ions. In this case, the increased attenuation due to the cutoff may be bleached by the high fiber attenuation. We may therefore conclude that the waveguide parameters for rare-earth-doped fibers may generally be measured using standard techniques, being aware of situations where these fail to be accurate.

More specifically for rare-earth-doped fibers are the experimental determination of spectral attenuation and fluorescence. The attenuation measurement is basically a normal cutback measurement, but it is complicated by the very large dynamic range of the attenuation. The solution is to operate with different and sometimes very short fiber lengths. In the latter cases, the higher order modes launched into the fiber may be removed by cladding mode filters. The fluorescence measurement is performed on a very short fiber length in order to minimize spectral distortion induced by self-absorption and amplification. For this reason, the fluorescence may be viewed from the side of a highly doped fiber.

Another key parameter is the fluorescence lifetime, which is determined from the decay of the fluorescence intensity after turning off the pump light. Also important is the ESA, which may be derived from comparative attenuation measurements on pumped and unpumped rare-earth-doped fibers. The pump power here must be large enough to significantly excite the considered energy level. Furthermore, the GSA cross-section values must be accurately determined by separate measurements, since they have to be extracted from the experimental data to achieve absolute values of the ESA cross sections.

The final step in the characterization of the rare-earth-doped fiber is to include it in an amplifier setup. A procedure for doing so has been described in this chapter, and it was pointed out that gain and ASE provide important information about the fiber. The ASE spectral density around the signal wavelength is especially important because it forms the basis for the calculation of the amplifier noise figure. However, though the gain may be practically determined with an accuracy of ± 0.3 dB, the noise figure accuracy is realistically limited to ± 0.5 dB.

REFERENCES

[1] Agrawal, G. P., and N. K. Dutta, *Long-Wavelength Semiconductor Lasers*, Van Nostrand Reinhold, 1986.

[2] Elion, G. R., and H. A. Elion, *Fiber Optics in Communication Systems*, New York and Basel: Marcel Dekker, 1978.

[3] Stewart, W. J., "Optical Fiber and Preform Profiling Technology," *IEEE Journal of Quantum Electronics*, Vol. 18, No. 10, 1982, pp. 1451–1466.

[4] Svendsen, D. A., "Resolution Model for Measurement of Optical Fibre Preforms by Focused Laser Transverse Illumination Technique," *IEE Proceedings J.*, Vol. 135, No. 3, 1988, pp. 196–201.

[5] Sasaki, I., D. N. Payne, and M. J. Adams, "Measurement of Refractive-Index Profiles in an Optical Fibre Preform," *Electronics Letters*, Vol. 16, No. 6, 1980, pp. 219–221.

[6] Watkins, L. S., "Laser Beam Refraction Traversely Through a Graded-Index Preform to Determine Refractive Index Ratio and Gradient Profile," *Applied Optics*, Vol. 18, No. 13, 1979, pp. 2214–2222.

[7] Marcuse, D., *Principles of Optical Fiber Measurements*, New York, San Francisco, London: Academic Press, 1981.

[8] Dybdal, K., N. Bjerre, J. E. Pedersen, and C. C. Larsen, "Spectroscopic Properties of Er-Doped Silica Fibers and Preforms," *Proceedings of Society of Photo-Optical Instrumentation Engineers Conference, OE/Fibers '89*, part: Fiber Laser Sources and Amplifiers, Boston, 6–8 Sept. 1989, SPIE Vol. 1171, pp. 209–217.

[9] Pedersen, J. E. "Rare Earth Doped Fibre Optical Amplifiers," Ph.D. Thesis EF249, Institute of Physics, University of Aarhus, March 1991.

[10] Wang, S., *Solid-State Electronics*, New York, San Francisco, St. Louis: McGraw Hill, 1966.

[11] Mayer, J. W., and E. Rimini, *Ion Beam Handbook for Material Analysis*, Academic Press, 1977.

[12] Ainslie, B. J., S. P. Craig, S. T. Davey, and B. Wakefield, "The Fabrication, Assessment and Optical Properties of High-Concentration Nd^{3+}- and Er^{3+}-Doped Silica-Based Fibres," *Materials Letters*, Vol. 6, Nos. 5 and 6, 1988, pp. 139–144.

[13] White, K. I., "Practical Application of the Refracted Near Field Technique for the Measurement of Optical Fibre Refractive Index Profiles," *Optical and Quantum Electronics*, Vol. 2, 1979, pp. 185–196.

[14] Raine, K. W., J. G. Baines, and R. J. King, "Comparison of Refractive Index Measurements of Optical Fibres by Three Methods," *IEE Proceedings J.*, Vol. 135, No. 3, 1988, pp. 190–195.

[15] Jeunhomme, L. B., *Single-Mode Fiber Optics*, New York and Basel: Marcel Dekker, 1983.

[16] Miller, S. E., and A. G. Chynoweth, *Optical Fiber Telecommunications*, New York, San Francisco, London: Academic Press, 1979.

[17] Dam-Hansen, C., "Measurements on Active Fibres," Master's Thesis E412 (in Danish), Electromagnetics Institute, Technical University of Denmark, 1990.

[18] Wyatt, R., "Spectroscopy of Rare Earth Doped Fibres," *Proceedings of Society of Photo-Optical Instrumentation Engineers Conference, OE/Fibers '89*, part: Fiber Laser Sources and Amplifiers, Boston, 6–8 Sept. 1989, SPIE Vol. 1171, invited paper, pp. 54–64.

[19] Zemon, S., G. Lambert, W. J. Miniscalco, R. W. Davies, B. T. Hall, R. C. Folweiler, T. Wei, L. J. Andrews, and M. P. Singh, "Excited State Cross Sections for Er-Doped Glasses," *Proceedings of Society of Photo-Optical Instrumentation Engineers Conference, OE/Fibers '90*, part: Fiber Laser Sources and Amplifiers II, San Jose, 18–19 Sept. 1990, SPIE Vol. 1373, pp. 21–32.

[20] Zemon, S., B. Pedersen, G. Lambert, W. J. Miniscalco, L. J. Andrews, R. W. Davies, and T. Wei, "Excited-State Absorption Cross Sections in the 800-nm Band for Er-Doped, Al/P-Silica Fibers: Measurements and Amplifier Modeling," *IEEE Photonics Technology Letters*, Vol. 3, No. 7, 1991, pp. 621–624.

[21] Singh, M. P., D. W. Oblas, J. O. Reese, W. J. Miniscalco, and T. Wei, "Measurement of the Spectral Dependence of Absorption Cross Section for Erbium-Doped Single-Mode Optical Fiber," *Symposium on Optical Fiber Measurements*, NIST Special Publication 792, 1990, pp. 93–96.

[22] Pedersen, B., K. Dybdal, C. Dam-Hansen, A. Bjarklev, J. H. Povlsen, H. Vendeltorp-Pommer, and C. C. Larsen, "Detailed Theoretical and Experimental Investigation of High-Gain Erbium-Doped Fiber Amplifier," *IEEE Photonics Technology Letters*, Vol. 2, No. 12, 1990, pp. 863–865.

[23] Evankow, J. D., and R. M. Jopson, "Nondestructive Measurement of Length Dependence of Gain Characteristics in Fiber Amplifiers," *IEEE Photonics Technology Letters*, Vol. 3, No. 11, 1991, pp. 993–995.

[24] Giles, C. R., and E. Desurvire, "Modeling Erbium-Doped Fiber Amplifiers," *IEEE Journal of Lightwave Technology*, Vol. 9, 1991, pp. 271–283.

[25] Pedersen, B., A. Bjarklev, J. H. Povlsen, K. Dybdal, and C. C. Larsen, "The Design of Erbium-Doped Fiber Amplifiers," *IEEE Journal of Lightwave Technology*, Vol. 9, No. 9, 1991, pp. 1105–1112.

[26] Lumholt, O., K. Schüsler, A. Grunnet-Jepsen, A. Bjarklev, S. Dahl-Petersen, J. H. Povlsen, T. Rasmussen, K. Rottwitt, and C. C. Larsen, "Low Noise High Gain Preamplifier Using an Isolator Within the Erbium Doped Fibre," *European Conference on Optical Communication ECOC '92*, Berlin, 1992, Proc. Vol. I, paper MoA3.5, pp. 93–96.

Chapter 5
Basics of 1550-nm Fiber Amplifiers

With the demonstration of high-gain erbium-doped amplifiers [1,2] and their applications in 1550-nm lightwave transmission systems [3,4], it became obvious that the potentials of optical communication took a large step forward. This initiated intense worldwide research and development effort, bringing the fiber amplifier from the laboratories and into operating communication links. In this process, a key point has been not only to develop technological and experimental facilities, but also to establish sufficiently accurate models for the EDFA. The models are important for predicting amplifier behavior and for yielding improved fiber designs. Furthermore, accurate models may serve as a useful tool in the process of understanding the properties of fiber amplifiers, since they may serve as test benches for new ideas on amplifier configuration. The scope of this chapter is therefore to describe the theory of the EDFA and to present some of the basic and central results of amplifier modeling.

It is at this point important to note that a number of models have been presented in the literature over the last few years. All these models originate from the same basic first-order rate and propagation equations, but differ with respect to complexity. These differences not only express model development over time, but they also depend on the specific problem considered in the context in which they appeared. This is motivated by the desire to limit model complexity.

The first theoretical papers presented on the EDFA had a point of reference in a plane wave description of the pump and signal light [2,5]. Furthermore, these early models did not include the frequency dependence of the ASE. These models were important because they provided simple expressions and formed the basis for more accurate descriptions of the EDFA. However, it was also obvious that more accurate results could only be obtained if it was taken into account that light propagates as modes in the EDFA. Furthermore, it was important to include that the ASE has a spectral dependence strongly dependent on amplifier configuration and power levels.

Among the first steps towards a complete description of the erbium-doped fiber amplifier was the work presented in [6]. For a part of this reference, a Gaussian dopant profile was assumed. This allowed simple analysis of the influence of signal and pump mode overlap with dopant ion distribution in the fiber core. A model that included accurate mode descriptions of not only the fundamental mode but also the higher order pump modes was presented in [7]. This model will be used to examine the influence of higher order pump modes on fiber amplifier performance later in this chapter. It must, however, be noted that this approach, in agreement with other contemporary [8] or successive models [9], replaces rather than resolves the ASE spectrum with an equivalent noise bandwidth (see the detailed description in Section 5.2.1). The models, which include accurate transverse mode representation, have their significance in the fact that the spatial representation is necessary for optimization of waveguide parameters and dopant distribution.

The second important step towards an accurate EDFA model was the inclusion of the spectral resolution of ASE as first described in [10]. In this model, the mode fields were represented by a Gaussian approximation of the fundamental LP_{01} mode. An important assumption was that the homogeneous broadening of the laser transitions between the two *J* manifolds (see Chapter 3 and Figure 5.2) far exceeds the inhomogeneous broadening due to the random glass structure. This assumption, later adopted in most amplifier models, is valid at room temperature for most rare-earth ions in glass because of the broadening effect of phonon coupling [11].

To relate the different models to each other, Figure 5.1 may be useful. From this diagram, it is obvious that most of the models have a point of reference in the homogeneous broadening assumption mentioned above. For these models, the first branching is between steady-state and time-varying descriptions. Again, most of the effort has been put in steady-state solutions, since these are fully sufficient for most amplifier applications. As it appears in Figure 5.1, the complexity of the spatial models varies significantly, and in each case further branching occurs between models that include ASE and those that ignore it. ASE terms are required for calculating amplifier noise and gain saturation caused by ASE. A partly sufficient description may therefore be given by modeling the ASE as optical power in an effective noise bandwidth.

The work on models, which includes time variation effects, also requires some comments. This effort was originally initiated because the researchers were concerned with potential problems of systems in which simultaneous amplification of more than one communication channel took place. The considered effects, which have been extensively studied in semiconductor optical amplifiers, are (1) transient gain fluctuations resulting from amplitude-modulated signals (i.e., the *saturation-induced crosstalk* (SIC)) [20,21], and (2) the gain modulation at the beat frequency of closely spaced signals (i.e., intermodulation distortion (IMD)) [22–24]. Two-

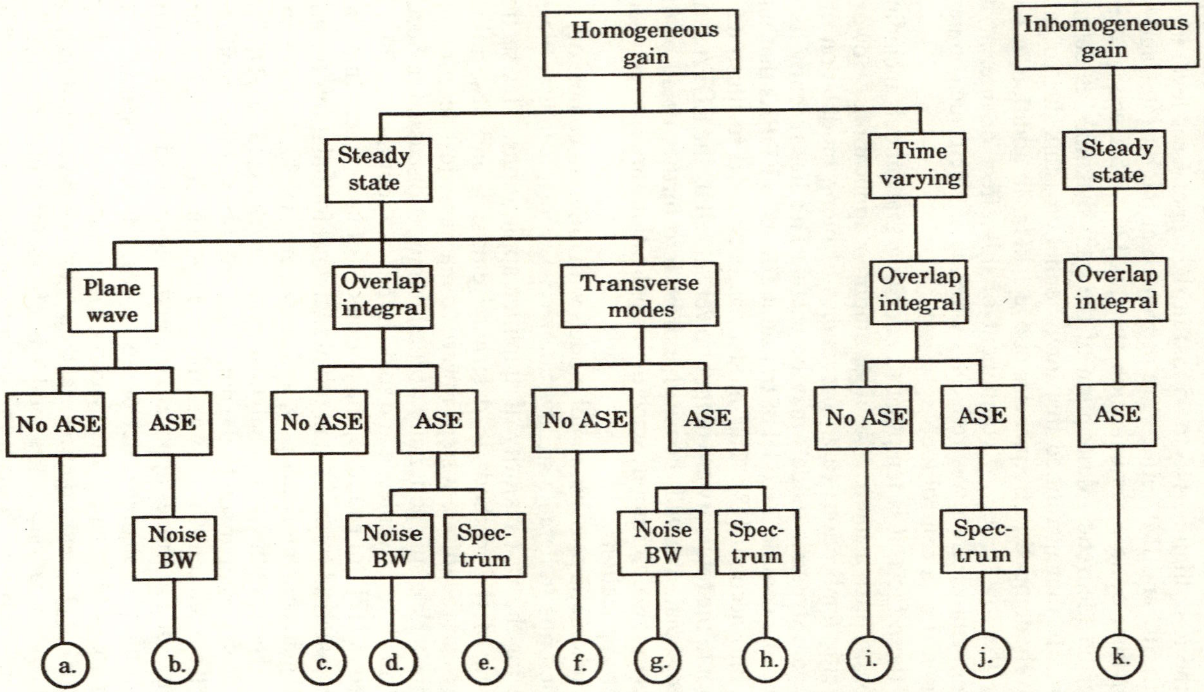

Figure 5.1 Tree diagram showing relationship among existing amplifier models. The letters in the different branches indicate references in which the models are described: a. [2], b. [5], c. [12–14], d. [9], e. [12], f. [15], g. [7], h. [16], i. [17], j. [18], and k. [19]. (Idea of figure from [12].)

channel crosstalk measurements were reported for erbium-doped fiber amplifiers [25–28], and it was observed that the EDFA is characterized by slow gain dynamics. This is due to the long lifetime (10 to 15 ms) in the upper laser level in erbium. As a result, both SIC and IMD effects are prevented at the bit rates and channel spacing used in fiber-optic communication systems [26]. This was also in good agreement with the theoretical description of the gain dynamics as described in [17], where it was further quantified that interchannel crosstalk due to gain saturation is effectively damped at frequencies above 1 kHz. In addition, it has been stated [26] that the low-frequency component of this effect of intermodulation between channels can be effectively reduced by means of electronic compensation using a feed-forward gain control loop.

Based on these facts, it is tempting to conclude that models including time variations are of no further interest for fiber amplifier applications. However, for some more specific applications (e.g., for studies of pump modulation or signal switching), the gain dynamic models may be useful. This has recently motivated the development of a complete model of transient effects in EDFAs that takes the ASE spectrum into account [18]. This model has been used to illustrate effects such as overshoot in the total forward output power caused by the EDFA saturation by the amplified signal. We will, however, not go into further detail of the gain dynamic models in this description, since their applications are limited to few and special amplifier applications.

After this schematic presentation of models with different complexity, we will in Section 5.1 describe in detail a steady-state model, which is fully sufficient and highly accurate for the vast majority of amplifier applications. This section also includes a review of the methods to determine emission and absorption cross sections, and present the different steps towards a mathematical description of the EDFA. The drawback of the generality of the full model discussed in Section 5.1 is the loss of calculation convenience and some intuition about the amplifier behavior. As described in [12], for example, the notion of optimum fiber length was first posed in simple analytical form [2], and other analytical formulations for amplifier noise [5], spatial gain characteristics [29], and amplifier gain [30, 31] have appeared in the literature. The simplified but analytical models such as [13–15,29,32], therefore, provide a useful comparison to the numerical solutions of the amplifier equations, and we will consequently review a number of these in Section 5.2. Thereafter, we focus on some of the model applications, and, more specifically, Section 5.3 addresses the problem of higher order pump modes, which is especially relevant to 800-nm pumping of fiber amplifiers. Section 5.4 presents a detailed comparison between the model of [16] and experimental data. This is included not only because the experimental verification of any model is of essential importance, but also because it becomes possible through the description in Section 5.4 to discuss the central properties of fiber amplifiers. Finally, Section 5.5 will give a short description of the temperature influence on gain and noise figure of the EDFA.

5.1 THEORY OF THE ERBIUM-DOPED FIBER

For a description of most amplifier properties, a steady-state solution is quite sufficient because of the long lifetime in the upper laser level. However, if a general fiber design is to be considered, it is necessary to choose a model with an accurate representation of the transverse optical mode fields. In cases where we at the same time require accurate information on noise properties and saturation effects, a spectral representation of the ASE is needed. We will in this section present a model that combines the properties of accurate spatial mode representation [7] with the strength of the full spectral ASE description [10]. This presentation will mainly follow the description of [16], because this model complexity has been shown to provide accurate and detailed information necessary for characterizing the EDFA both as a separate component as well as a part of a larger communication system. The present description only mentions the 800-, 980-, and 1480-nm pumping possibilities, but may without further difficulty be expanded for use in other pump bands. The first subject of this section will be the determination of the cross-section spectra that characterizes the energy transitions involved. Second, the local population inversion will be derived, and then the power distribution along the fiber will be found. The mathematical background for the noise figure evaluation will be given, and finally the section includes a discussion of special problems concerning the numerical solution of the established equations.

5.1.1 Determination of Cross Sections

As with all lasers and amplifiers, accurate cross sections are essential to the qualitative understanding of devices as well as the identification of improved active media. However, despite the extensive work on EDFAs, uncertainty surrounded the values of cross sections for a very long time, and only recent results [33, 34] seem to clarify this subject. We will review the principal results of this description, but before we get to this point, it will be natural to go through two other approaches, which have played a central role in the field of erbium-doped fiber amplifiers over the last few years. This section is therefore divided into three steps: we first review the Judd-Ofelt theory, then focus on the Einstein A-B relationship between probabilities for absorption and spontaneous emission, and, finally, we look at the more general McCumber theory.

The electronic structure of rare-earth ions is dominated by electron-electron interaction, with spin-orbit next in importance [33]. These produce the free-ion J multiplets, which are usually labeled by their dominant Russell-Saunders LSJ term, where S is the total spin momentum, L the orbital angular momentum, and J the total angular momentum. A knowledge of the strength of the radiative transitions between different excited LSJ levels is of great importance, when looking for new laser transitions, for assessing new host glasses and in understanding any limitation of device efficiency. A useful semi-empirical technique for the analysis of these rare-earth transitions was developed independently, but almost simultaneously, by Judd [35] and Ofelt [36]. The Judd-Ofelt theory allows estimates to be made of the electric dipole transition strengths between any pair of *LSJ* manifolds.

A Judd-Ofelt analysis of a rare-earth-ion-doped material essentially involves the use of the line strengths of all the observable transitions from the ground state to excited manifolds, which can be determined from spectral absorption measurements (see Section 4.2.2). The determined line strengths are then used to estimate line strengths of transitions between pairs of excited levels. The Judd-Ofelt analysis relies on the assumption that the sublevels within a term are uniformly occupied and that the separation between term values is small compared to the energy separation to the 5d- or 5g-electron configuration admixed by the ligand field. The spectral integrated strength of the transition between a given pair of J multiplets is then determined by three empirical parameters. The resulting transition probability p for absorption or induced emission from an initial state J to a final state J' can be calculated as

$$p = a[(n^2 + 2)^2/n](2J + 1)^{-1}S(J, J') \tag{5.1}$$

Here, parameter a has only a slight host dependence due to the frequency of the transition, and the term containing n is a local field correction for the refractive index n of the host material. $S(J, J')$ is the line strength, which can be written as the sum of three terms:

$$S(J, J') = \Omega_2 U_2 + \Omega_4 U_4 + \Omega_6 U_6 \tag{5.2}$$

U_2, U_4, and U_6 are calculated quantities, for which tabulations are available [37]. In (5.2), Ω_2, Ω_4, and Ω_6 are the Judd-Ofelt parameters, determined by a least-squares fit in the absorption spectra of a rare-earth host combination. It is remarkable that this theory, from only the empirical determination of three parameters, is able to predict transition strength with an accuracy of about 10% to 15% [33]. Judd-Ofelt theory is particularly valuable for providing line strengths of transitions for which direct measurements are difficult. However, it applies only to the integrated transition strengths, and it will consequently not provide any spectral information.

The spectral information of emission and absorption cross sections in an erbium-doped fiber, however, is of central importance for the description of its properties. Therefore, other methods for the determination of transition strengths are necessary. The most straightforward method would of course be to measure the wavelength-dependent emission and absorption cross section, σ_e and σ_a, directly. If no erbium ions are excited, the absorption cross sections can be determined directly from an attenuation measurement according to

$$\sigma_a(\nu) = \frac{\text{att}(\nu)}{10 \log_{10}(e) \cdot 2\pi \int_0^{a_{\text{dot}}} \rho_{\text{Er}}(r) I^{01}(\nu, r) r dr} \tag{5.3}$$

where att(ν) is the erbium-induced attenuation in decibels per meter at the frequency ν, $\rho_{\text{Er}}(r)$ is the erbium concentration, a_{dot} is the erbium dopant radius, and $I^{01}(\nu, r)$ is the normalized LP_{01} mode intensity, which is determined according to the following:

$$2\pi \int_0^{\infty} I^{01}(\nu, r) r dr = 1$$

The emission cross sections can be determined from a gain measurement (see Chapter 4), assuming that all erbium ions are excited:

$$\sigma_e(\nu) = \frac{g(\nu)}{10 \log_{10}(e) \cdot 2\pi \int_0^{a_{\text{dot}}} \rho_{\text{Er}}(r) \cdot I^{01}(\nu, r) r dr} \tag{5.4}$$

where $g(\nu)$ is the gain in decibels per meter at frequency ν.

Determination of the cross sections, using (5.3) and (5.4), implies an accurate knowledge of the overlap integral between the erbium concentration and the LP_{01} mode. Furthermore, it is assumed that only the LP_{01} mode propagates in the fiber, and, more importantly, it might be difficult to fulfill the requirement that all erbium ions are excited. Therefore, it is also interesting that the emission cross sections may be determined from the spectral shape of the fluorescence, $I_e(\nu)$, and the spontaneous emission rate, $A_e = 1/\tau_{21}$. Here, τ_{21} denotes the radiative lifetime. The following equation may be derived from the Fuchtbauer-Ladenburg equations [38]:

$$\sigma_e(\nu) = \frac{A_e \lambda^2}{8\pi n^2} \cdot \frac{I_e(\nu)}{\int I_e(\nu) d\nu} \tag{5.5}$$

where n is the refractive index and λ is an average wavelength.

The absorption cross-section spectrum can also be determined from the spectral shape of the attenuation, $I_a(\nu)$, according to the Fuchtbauer-Ladenburg equations:

$$\sigma_a(\nu) = \frac{A_e\lambda^2}{8\pi n^2}\cdot\frac{2J_2 + 1}{2J_1 + 1}\cdot\frac{I_a(\nu)}{\int I_a(\nu)\,d\nu} \tag{5.6}$$

where J_1 and J_2 are the angular momentum quantum numbers for the ground and the excited state, respectively.

If the absorption cross-section spectrum is evaluated from (5.6), the resulting cross sections will, according to [16], be 15% lower than the values determined from (5.3). The validity of the Fuchtbauer-Ladenburg equations has been questioned by other authors [39,40]. In [40], it is shown that the absorption cross sections, found from (5.6) in Ge/Er preforms, are 10% to 20% too low. This often leads to cross sections determined from (5.6) being multiplied by a correction factor of 1.15 [16].

However, when it is not straightforward to measure both the absorption and stimulated-emission cross sections for a rare-earth material, one is often calculated from the other using the Einstein A and B coefficients for a two-level system [41]. In the general case in which the lower state $^4I_{15/2}$ and the upper state $^4I_{13/2}$ are split into multiple components, the relationship becomes

$$g_1 \int \nu^2\sigma_a(\nu)\,d\nu = g_2 \int \nu^2\sigma_e(\nu)\,d\nu \tag{5.7}$$

where $g_i = 2J_i + 1$ is the degeneracy of level i. By replacing ν^2 with the square of an effective frequency and taking this term outside the integral, (5.7) applies to the Fuchtbauer-Ladenburg relationship [38]. As pointed out in [34], (5.7) is valid for rare-earth ions only if one of two conditions is met. Either all components of the two levels must be equally populated or all the transitions must have the same strength regardless of the components involved. Either of these conditions ensures that the observed oscillator strength is identical to that of the degenerate (unsplit) two-level case. The manifold widths of the $^4I_{15/2}$ and the $^4I_{13/2}$ states of Er^{3+}-doped glasses are typically 300 to 400 cm^{-1} [34], and the first condition is therefore not satisfied at room temperature. Moreover, low-temperature absorption and emission measurements indicate for all glasses that the transition strength is sensitive to the Stark levels involved. Since, therefore, neither of the above requirements is satisfied, it is not surprising, as expressed in [34], that significant deviations from (5.7) have been reported [8,39,40].

As mentioned earlier, one way to overcome the discrepancies is to apply an empirical correction factor [40] to the absorption cross-section spectra. However, a more accurate procedure for the analysis of Er^{3+} cross sections was described in [34], where it was shown that the McCumber theory [42] is ideally suited to rare-earth ions, and it not only provides absolute cross sections, but spectral information as well. In contrast to the former description, the McCumber theory only assumes that the time required to establish a thermal distribution within each manifold is short compared to the lifetime of that manifold. From the consideration of a detailed balance, the absorption and emission cross sections are then related by [42]

$$\sigma_e(\nu) = \sigma_a(\nu) \cdot \exp\left[\frac{(\epsilon - h\nu)}{kT}\right] \tag{5.8}$$

where ϵ is the temperature-dependent excitation energy and $k = 1.381 \times 10^{-23}$ $J \cdot K^{-1}$ is the Boltzmann constant. ϵ can be seen as the net free energy required to excite one Er^{3+} ion from the $^4I_{15/2}$ to the $^4I_{13/2}$ state at temperature T. Since the manifold widths exceed kT, the absorption and emission spectra are offset from each other, from the absorption to lower wavelengths, and from the emission to higher wavelengths (see Figure 3.4). An examination of (5.8) shows that the cross sections are equal only at the crossing frequency, $\nu_c = \epsilon/h$. At frequencies higher than ν_c, the emission cross section is smaller than the absorption cross section, and vice versa for $\nu < \nu_c$. Another important characteristic of (5.8) is that a relative measurement of either cross-section spectrum is sufficient to generate a relative spectrum of the other cross section. This feature is clearly not available from the Einstein relation in (5.7).

We finally bring forward the relationship between the radiative lifetime τ_{21} and the emission cross section spectrum

$$\frac{1}{\tau_{21}} = \frac{8\pi n^2}{c^2} \int \nu^2 \sigma_e(\nu)\, d\nu \tag{5.9}$$

which may be used to scale the emission spectrum when only its spectral dependence is known [34]. In (5.9), c denotes the velocity of light in vacuum. However, it may often be easier to measure the relative absorption spectrum than the emission spectrum, and $\sigma_e(\nu)$ can then be replaced in (5.9) with the use of (5.8). This will require the determination of the parameter ϵ. A method for the evaluation of this parameter has been described in [34], and we will resume the procedure shortly.

The low symmetry of the crystal field lifts all but the Kramers degeneracy of the Er^{3+} ion and splits the ground state into eight components and the excited state into seven components (see Figure 5.2). If the positions of all these Stark components are known, ϵ may be calculated according to [42]

$$\exp\left[\frac{\epsilon}{kT}\right] = \frac{1 + \sum_{j=2}^{8} \exp(-E_{1j}/kT)}{\exp(-E_0/kT)\left[1 + \sum_{j=2}^{7} \exp(-E_{2j}/kT)\right]} \tag{5.10}$$

In this expression, E_0 is the separation between the lowest component of each manifold, and E_{ij} the difference in energy between the jth and the lowest component of level i. The electronic structure required for (5.10) is usually difficult to obtain, and to eliminate this problem, a simple phenomenological procedure has been developed in [34]. Here, it is assumed that the Stark levels for a given manifold are equally spaced; $E_{ij} = (j - 1)E_i$. As a result, the number of parameters in (5.10) is reduced, so only the component spacing for each manifold E_i and the separation between two manifolds E_0 have to be found. In room-temperature measurements, the peaks of the absorption and emission spectra are usually within 5 cm^{-1} (1 nm) of each other. Since low-temperature measurements [43] confirm that this wavelength corresponds to transitions between the lowest component of each manifold, E_0 is accordingly taken as the average of the absorption and emission peaks [34]. The determination of the energy spacing between stark levels is more

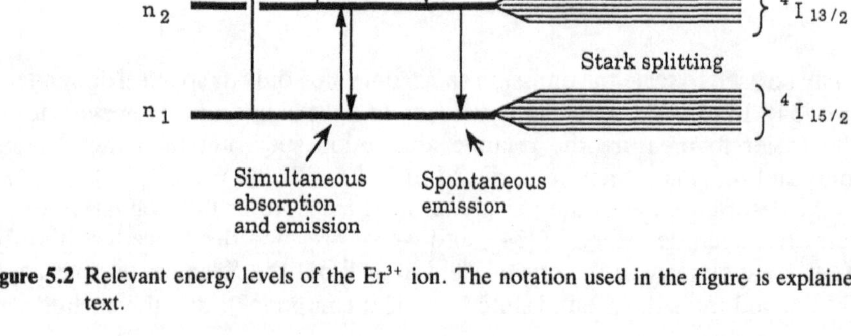

Figure 5.2 Relevant energy levels of the Er^{3+} ion. The notation used in the figure is explained in the text.

difficult, but may be made from low-temperature measurements, where the level-one bandwidth would be the separation between the highest and lowest energy emission lines. Therefore, the low-energy half-width of the room-temperature emission spectrum is used for $7E_1$, since the peak has already been identified with the transition between the lowest components of the two manifolds [34]. In a similar way the high-energy half-width of the absorption spectrum is adopted for the width of the excited-state manifold, $6E_2$. For the half-widths, the separation between the peak and the energy at which the spectrum falls to 5% of the peak was used, which is an uncritical criterion. In [34], it is concluded that this very simple model of the spectra and electronic structure of Er^{3+} in most cases yields results significantly more accurate than those obtained from the Einstein relation.

We may therefore conclude this section by noting that though the Judd-Ofelt theory has its advantage in being able to predict transition strength in cases where measurements are difficult (e.g., ESA cross sections), it provides no spectral information. To obtain this information, we might turn to either the Einstein relation or the McCumber theory. However, it is noteworthy that the Einstein analysis, which for a long time has been the accepted procedure for connecting absorption and stimulated emission, is inaccurate for Er^{3+}-doped glasses. The McCumber theory, on the other hand, has been shown to provide accurate absolute values and spectral information on the cross sections of rare-earth doped glasses. The only drawback is that it requires knowledge of the electronic structure of the dopant ion to provide absolute cross sections. However, the electronic structure may be described by a very simple model, and the theory will still provide highly accurate results.

5.1.2 Evaluation of the Local Population Inversion Factor

We will now turn to the application of the cross section spectra in an accurate steady-state model of the EDFA. To describe the interaction between the Er^{3+} ion and the signal and pump photons, we will use the simplified energy level diagram shown in Figure 5.2. When pumping at 800 or 980 nm, the EDFA acts as a three-level laser system. This means that the Er^{3+} ion is excited from the ground-state level, $^4I_{15/2}$, to a third level (level $^4I_{11/2}$ for 980-nm pumping and level $^4I_{9/2}$ for 800-nm pumping), from which it momentarily decays to the upper laser level, $^4I_{13/2}$. This raises the possibility of a complete inversion in the upper laser level. Besides the upper pump level and the two levels $^4I_{13/2}$ and $^4I_{15/2}$ corresponding to the relevant laser transition, the energy level diagram also includes a fourth level, which takes the pump ESA into account in the case of 800-nm pumping. Also, the nonradiative transition, $4 \rightarrow 3$, is assumed, in accordance with [6], to be so fast that the energy level four is completely depopulated.

When pumping at 1480 nm, the Er^{3+} ion is excited directly into the upper laser level, and, therefore, the EDFA acts as a two-level laser system and the relative inversion in the upper laser level cannot pass a value n_2^{max} set by the emission-to-absorption cross-section ratio, $\epsilon = \sigma_{pe}/\sigma_{pa}$, at 1480 nm. This maximum population can be expressed as $n_2^{max} = 1/(1 + \epsilon)$.

The energy level diagram in Figure 5.2 also illustrates the Stark splitting of the Er^{3+} orbitals (J manifolds), which is caused by the surrounding crystal field. As can be seen in Figure 5.2, the eightfold and sevenfold Stark splittings of the ground state ($^4I_{15/2}$) and upper laser level ($^4I_{13/2}$), respectively, make Er^{3+} glass a rather complicated multilevel system. However, as we shall see in the following description, the effect of this multilevel nature may be expressed by the spectral representation of the emission and absorption cross sections.

The EDFA is described in an (r, ϕ, z) cylindrical coordinate system with the z-axis being the fiber axis. To be able to calculate the local amplification and attenuation of ASE, signal, and pump, it is necessary to look at their local power spectral densities at each spectral component ν. In other words, this may be represented by the mode field intensities in the optical fiber, which, according to the weakly guiding approximation [44], may be expressed by LP_{lp} modes. For a general circular symmetric index profile, these may be calculated using various methods, but a multilayer description was chosen in the present case [45].

Considering a circular symmetry for the EDFA, the steady state population concentrations, $n_1(r, \phi, z)$ and $n_2(r, \phi, z)$ in the ground and excited states, respectively, are evaluated from the transition rates of the pump, signal, and spontaneous emission:

$$n_2(r, \phi, z) = \rho_{Er}(r) \frac{R_{pa}(r, \phi, z) + W_{sa}(r, z)}{R_{pe}(r, \phi, z) + R_{pa}(r, \phi, z) + W_{se}(r, z) + W_{sa}(r, z) + A_e} \tag{5.11}$$

$$n_1(r, \phi, z) = \rho_{Er}(r) - n_2(r, \phi, z)$$

The transition rates are determined as follows. The pump absorption rate, which is given as a summation over the pump modes, can be expressed as follows:

$$R_{pa}(r, \phi, z) = \sigma_{pa} \sum_{lp} I_p^{lp}(r, \phi, z) \frac{1}{h\nu_p} \tag{5.12}$$

where σ_{pa} is the absorption cross section at the pump wavelength. The pump mode intensity is referred to as $I_p^{lp}(r, \phi, z) = P_p^{lp}(z)|E_p^{lp}(r, \phi)|^2$, where $P_p^{lp}(z)$ is the power of the pump mode LP_{lp} at position z in the fiber, h is Planck's constant, and

ν_p is the pump frequency. Note here that all mode fields are normalized according to

$$\int_0^{2\pi} \int_0^b |E(r, \phi)|^2 \, r dr d\phi = 1 \tag{5.13}$$

where b is the outer fiber radius. To illustrate the relative intensity distribution across the erbium-doped fiber, Figure 5.3 shows the LP_{01} signal mode together with the radial variation of the LP_{01} and LP_{11} pump modes for a step-index profile with core radius a = 2.5 μm and NA = 0.18. The signal wavelength is 1535 nm and the pump wavelength is chosen as 800 nm to illustrate the difference between pump and signal modes most clearly.

The pump emission rate is considered only when pumping around 1480 nm and may be expressed, as in (5.12), as:

$$R_{pe}(r, \phi, z) = \sigma_{pe} \sum_{lp} I_p^{lp}(r, \phi, z) \frac{1}{h\nu_p} \tag{5.14}$$

where σ_{pe} is the emission cross section at the pump wavelength ($\approx$1480 nm). For 980-nm pumping, σ_{pe} is zero, and consequently $R_{pe}(r, \phi, z)$ in (5.11) is zero.

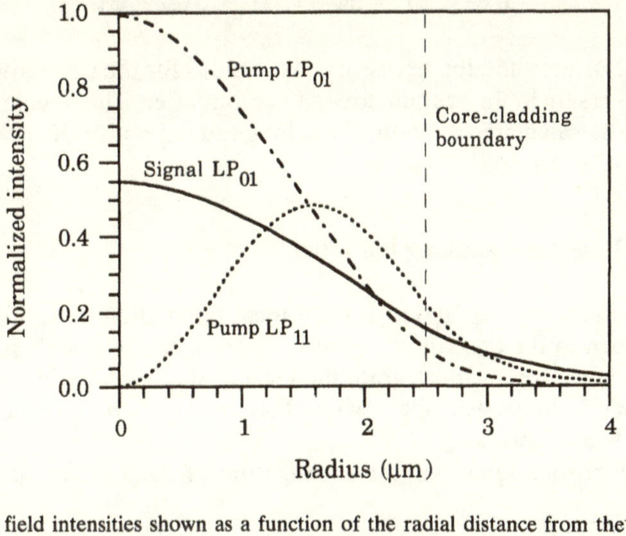

Figure 5.3 Mode field intensities shown as a function of the radial distance from the fiber center for the LP_{01} signal mode and the pump modes LP_{01} and LP_{11}. The signal wavelength is 1535 nm and the pump wavelength is 800 nm. (Data from [7].)

The signal emission and absorption rates, W_{se} and W_{sa}, are calculated from the signal and the amplified spontaneous emission, which are only considered to be excited in the fundamental LP_{01} mode:

$$W_{se}(r, z) = \left[\frac{\sigma_e(\nu_s)}{h\nu_s}P_s(z) + \int_0^\infty \frac{\sigma_e(\nu)}{h\nu}\cdot S_{ASE}(\nu, z)\,d\nu\right]\cdot I_s^{01}(r) \tag{5.15}$$

$$W_{sa}(r, z) = \left[\frac{\sigma_a(\nu_s)}{h\nu_s}P_s(z) + \int_0^\infty \frac{\sigma_a(\nu)}{h\nu}\cdot S_{ASE}(\nu, z)\,d\nu\right]\cdot I_s^{01}(r) \tag{5.16}$$

Here, $\sigma_e(\nu)$ and $\sigma_a(\nu)$ are the emission and absorption cross sections, respectively, at the frequency ν. ν_s is the signal frequency, $S_{ASE}(\nu, z)$ is the amplified spontaneous emission power spectral density at the position z, and $P_s(z)$ is the signal power. Note that both of the ASE polarizations are represented by $S_{ASE}(\nu, z)$. $I_s^{01}(r) = |E_s^{01}(r)|^2$ is the normalized signal mode intensity, and the normalization is made in agreement with (5.13).

The spontaneous emission is amplified in both the forward and backward directions of the fiber axis. This means that the total amplified spontaneous emission spectrum, $S_{ASE}(\nu, z)$ used in (5.15) and (5.16), has to be determined from a forward- as well as backward-traveling amplified spontaneous emission spectrum, $S_{ASE}^+(\nu, z)$ and $S_{ASE}^-(\nu, z)$:

$$S_{ASE}(\nu, z) = S_{ASE}^+(\nu, z) + S_{ASE}^-(\nu, z) \tag{5.17}$$

We have established the necessary expressions for the calculation of the local population inversion in the erbium doped fiber amplifier. The population expressed in (5.11) may be calculated as soon as the longitudinal power distribution of signal pump and ASE is known.

5.1.3 Power Distribution Along the Fiber

After having found an expression for the local population inversion in the fiber, we can now turn to the description of the pump, signal, and ASE passing through the fiber. To do so, we must integrate the local gain or attenuation across the cross section of the erbium-doped fiber, which then provides the guiding equations for the traveling wave powers.

The differential equations for propagation of $S_{ASE}^+(\nu, z)$ and $S_{ASE}^-(\nu, z)$ are given by

$$\frac{dS_{ASE}^-(\nu, z)}{dz} = -2h\nu\gamma_e(\nu, z) - [\gamma_e(\nu, z) - \gamma_a(\nu, z)]\cdot S_{ASE}^-(\nu, z) \tag{5.18}$$

and

$$\frac{dS^{+}_{ASE}(\nu, z)}{dz} = +2h\nu\gamma_e(\nu, z) + [\gamma_e(\nu, z) - \gamma_a(\nu, z)]\cdot S^{+}_{ASE}(\nu, z) \quad (5.19)$$

where the emission, $\gamma_e(\nu, z)$, and absorption, $\gamma_a(\nu, z)$, factors are determined from the respective emission and absorption cross sections, and the overlap integral between the signal mode and the population concentration in the excited and ground states, respectively:

$$\begin{aligned}\gamma_e(\nu, z) &= \sigma_e(\nu)\int_0^{2\pi}\int_0^{a_{dot}} n_2(r, \phi, z)I_s^{01}(r)\,rdrd\phi \\ \gamma_a(\nu, z) &= \sigma_a(\nu)\int_0^{2\pi}\int_0^{a_{dot}} n_1(r, \phi, z)I_s^{01}(r)\,rdrd\phi\end{aligned} \quad (5.20)$$

where a_{dot} is the erbium doping radius. The signal is amplified in the positive direction of the fiber axis according to

$$\frac{dP_s(z)}{dz} = [\gamma_e(\nu_s, z) - \gamma_a(\nu_s, z)]\cdot P_s(z) \quad (5.21)$$

The LP_{lp} pump mode propagates in the fiber with a power variation given as follows:

$$\frac{dP_p^{lp}(z)}{dz} = g_p^{lp}(z)\cdot P_p^{lp}(z) \quad (5.22)$$

where the pump absorption factor $g_p^{lp}(z)$ is introduced as

$$\begin{aligned}g_p^{lp}(z) = &-\mathcal{F}\int_0^{2\pi}\int_0^{a_{dot}} |E_p^{lp}(r, \phi)|^2[(\sigma_{24}(\nu_p) - \sigma_e(\nu_p))\cdot n_2(r, \phi, z)]\,rdrd\phi \\ &-\mathcal{F}\int_0^{2\pi}\int_0^{a_{dot}} |E_p^{lp}(r, \phi)|^2\,\sigma_a(\nu_p)\cdot n_1(r, \phi, z)]\,rdrd\phi\end{aligned} \quad (5.23)$$

In (5.23), the coefficient $\mathcal{F}$ is used to distinguish between the situation in which the pump is co-propagating with the signal ($\mathcal{F} = +1$) and that in which it is counter-propagating ($\mathcal{F} = -1$). We also note that the first term in (5.23) may often be ignored. This is the case when 980-nm pumping is considered, since at this wavelength neither ESA (represented by σ_{24}) nor stimulated pump emission is present. When there is pumping in the 800-nm band, $\sigma_{24}(\nu_p) \neq 0$ and $\sigma_e(\nu_p) = 0$; whereas the opposite (i.e., $\sigma_{24}(\nu_p) = 0$ and $\sigma_e(\nu_p) = 0$) is the case for 1480-nm pumping.

Note that if only pumping in the LP_{01} mode is considered, (5.22) may be expressed by γ_e and γ_a factors calculated from equations similar to (5.20) just using the cross sections at the pump wavelength as well as the pump LP_{01} mode, $I_p^{01}(r)$. For 980-nm pumping, $\gamma_e(\nu_p, z)$ is zero. Note that the background loss is ignored in this description, since it is in most cases a very good approximation of the actual situation of amplification.

Above, we established equations (5.18), (5.19), (5.21), and (5.22), which guide the evolution of pump signal and ASE powers through the erbium-doped fiber amplifier. These equations cannot be solved analytically, and numerical methods must therefore be applied in the general case. In Section 5.1.5, we will return to some of the details of this solution, but at this point we will focus on a more fundamental understanding of the involved elements. Figure 5.4 has been included for this purpose. It shows schematically the involved pump, signal, and ASE powers for a co-propagating pump configuration. It is also indicated that in each cross section of the fiber, the population inversion must be evaluated based on the present field intensities. Finally, the dashed lines are included to indicate that population inversion can only be found in the central part of the fiber core. As the pump power is absorbed along the fiber, the area of population inversion decreases, and eventually the signal power will also start to decrease, since the majority of interacting Er^{3+} ions will be in the ground state.

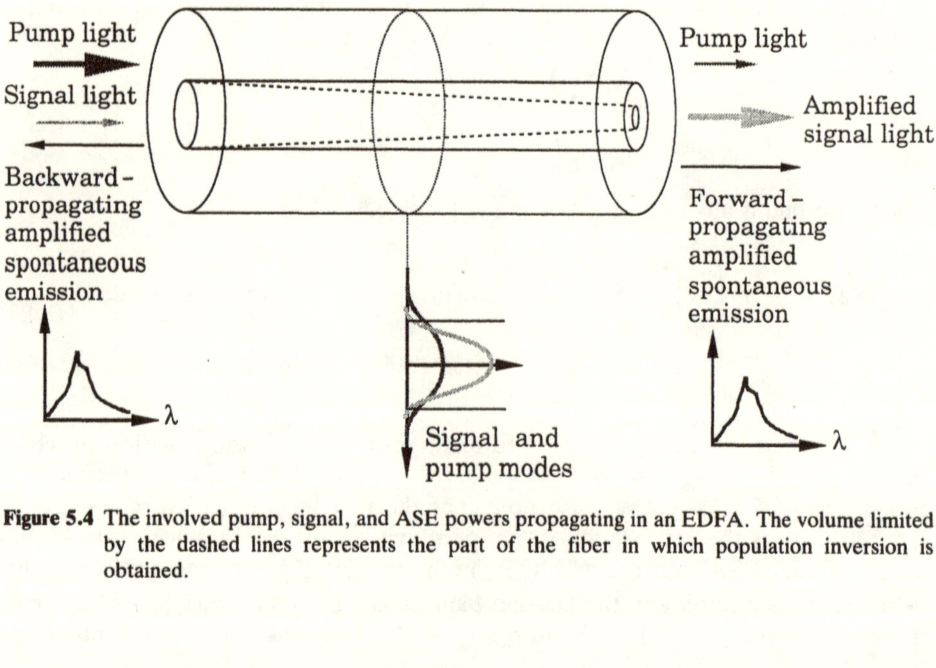

Figure 5.4 The involved pump, signal, and ASE powers propagating in an EDFA. The volume limited by the dashed lines represents the part of the fiber in which population inversion is obtained.

To look more closely at the fundamental properties of gain and ASE generation in an EDFA, calculations have been performed on different lengths of a specific amplifier. In each case, the residual pump power, the output signal power, and the output ASE power are found. This leads to the typical calculation result shown in Figure 5.5 for a step-index fiber (a = 2.5 μm and NA = 0.18) with the full core area erbium-doped in a concentration of $\rho_{Er}(r) = 1.8 \times 10^{24}$ m^{-3}. A signal power of 40 μW is launched into the fiber together with a co-propagating pump power of 100 mW at the pump wavelength of 806 nm. From Figure 5.5, it can be seen that the backward- and forward-propagating ASE are comparable in magnitude for short fiber lengths, whereas the backward ASE becomes largest for longer fiber lengths. The reason for this is that the forward ASE, as with the amplified signal power, starts to suffer from decreasing population inversion along the fiber length, and eventually a gain optimal length is reached. Beyond this length, the amplifier gain will decrease with increasing fiber length, whereas the noise figure (as we will see later) will increase. As long as only a single wavelength application is considered, there is no motivation for using amplifier lengths longer than the maximum gain length as shown in Figure 5.5.

5.1.4 Amplifier Noise Figure

Besides the amplifier gain, the other primary EDFA parameter is the noise figure, which is often denoted by F. For the sake of clarity, the noise figure is defined with respect to the idealized case of a coherent input signal and an infinitesimal receiver bandwidth around the optical frequency of the signal. In this way it becomes possible to separately express the noise properties of the EDFA without specifying the statistical properties of the signal or physical bandwidth limitations of filters (e.g., located in front of the detector).

The noise figure is closely related to the spectral intensity of the forward-propagating ASE as well as to the amplifier gain, and we will follow the description of [16] to derive the accurate expression for the noise figure. To define the noise figure for a fiber amplifier, it is necessary to look at the signal-to-noise ratio (SNR) at both the amplifier input end, SNR_{in}, and the output end, SNR_{out}. We use here the effective electrical signal-to-noise ratios, and the noise figure may be expressed as

$$F = \frac{\text{SNR}_{in}}{\text{SNR}_{out}} \tag{5.24}$$

The significance of (5.24) can be understood from the fact that as the signal propagates through the EDFA, the signal-to-noise ratio deteriorates due to the sto-

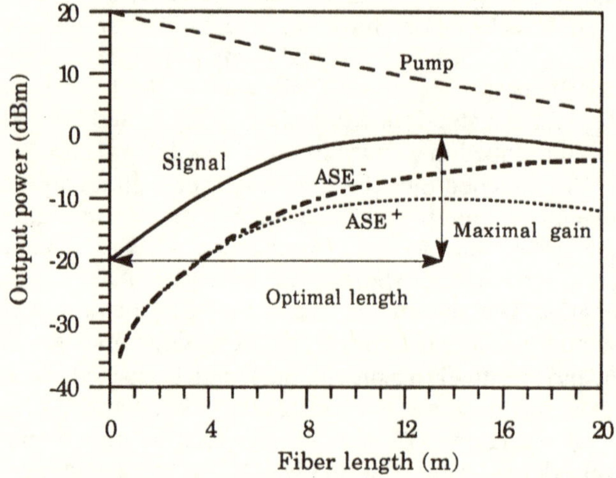

Figure 5.5 Residual power of the LP_{01} pump mode versus the fiber length, together with signal power and co-propagating and counter-propagating ASE power. The fiber is given by $a = 2.5$ μm and NA = 0.18, and the step-index core is fully doped with erbium at a concentration of $\rho_{Er}(r) = 1.8 \times 10^{24}$ m^{-3}. $\lambda_p = 806$ nm and $\lambda_s = 1535$ nm. (Data from [46].)

chastic nature of the interaction between the photons and Er^{3+} ions. In the theoretical limit where an optical filter with an infinitesimal bandwidth around the signal frequency is applied at the output end of the EDFA, it can be shown [47] that the variance V_s of the signal power may be found by solving the equation

$$\frac{dV_s}{dz} = 2 \cdot [\gamma_e(\nu, z) - \gamma_a(\nu, z)] \cdot V_s(z) + [\gamma_e(\nu, z) + \gamma_a(\nu, z)] \cdot P_s(z) \quad (5.25)$$

where the signal power $P_s(z)$ is given as the solution to (5.21).

The noise figure is defined with respect to the idealized case of a coherent input signal ($V_s(0) = P_s(0)$), and if the total amplifier length is denoted by L, (5.24) may be written as

$$F = \frac{(P_s(0))^2/V_s(0)}{(P_s(L))^2/V_s(L)} = P_s(0) \cdot \frac{V_s(L)}{(P_s(L))^2} \quad (5.26)$$

Using (5.21), (5.25,) and (5.26), F can be found from

$$F = \int_0^\infty \frac{\gamma_e(\nu_s, z) + \gamma_a(\nu_s, z)}{P_s(z)} \cdot P_s(0)\,dz + 1 \quad (5.27)$$

To illustrate the close relationship between the noise figure and the spectral density of the forward-propagating ASE, it is convenient to define the quantity $p(z)$:

$$p(z) = h\nu_s \cdot \left(\frac{V_s(z)}{P_s(z)} - 1 \right) \tag{5.28}$$

By using (5.21) and (5.25), a differential equation for $p(z)$ can be obtained:

$$\frac{dp(z)}{dz} = h\nu_s \cdot \frac{V_s(z)}{P_s(z)} \cdot \frac{d}{dz} \left[\log_e \left(\frac{V_s(z)}{P_s(z)} \right) \right] \tag{5.29}$$

which accordingly gives

$$\frac{dp(z)}{dz} = h\nu_s \cdot \frac{V_s(z)}{P_s(z)} \cdot [\gamma_e(\nu, z) - \gamma_a(\nu, z)] \cdot V_s(z) + h\nu_s \cdot [\gamma_e(\nu, z) + \gamma_a(\nu, z)] \tag{5.30}$$

By combining (5.28) and (5.30), we finally find that

$$\frac{dp(z)}{dz} = [\gamma_e(\nu, z) - \gamma_a(\nu, z)] \cdot p(z) + 2h\nu_s \cdot \gamma_e(\nu, z) \tag{5.31}$$

If we now compare (5.31) with (5.19), it can be seen that $p(z)$ and $S^+_{ASE}(\nu, z)$ obey the same differential equation, and since both quantities vanish at $z = 0$, the following identity is derived:.

$$S^+_{ASE}(\nu_s, L) = p(L) = h\nu_s \cdot \left(\frac{V_s(L)}{P_s(L)} - 1 \right) \tag{5.32}$$

which leads (see (5.26)) to the following relationship between the noise figure, the gain, and the forward propagating ASE power spectral density at the signal frequency:

$$F = \frac{1}{G} \cdot \left(\frac{S^+_{ASE}(\nu_s, L)}{h\nu_s} + 1 \right) \tag{5.33}$$

where G is the amplifier gain. The noise figure can accordingly be interpreted as a measure of the spectral density of the ASE around the signal frequency.

The value of F can, according to [48], also be determined from (5.27) by replacing the upper integration limit with the fiber length L. From (5.27), it can be seen that the more rapid the signal increases in the beginning of the EDFA the lower the noise figure becomes, and thereby also, from (5.33), the forward-propagating ASE spectral density $S^+_{ASE}(\nu_s, z)$ decreases. This fundamental importance of providing a rapid gain increase in the input end of the amplifier becomes especially clear if we compare co-propagating and counter-propagating amplifiers with equal pump consumption. As illustrated in Figure 5.6(a), the counter-propagating pump will only be able to provide sufficiently high pump power intensity in the input end of the EDFA as long as very short fiber lengths are used. For longer fiber lengths, or for less ideal radial pump distribution, the noise figure is increased.

Another significant noise property of the EDFA is illustrated in Figure 5.6(b), where the noise figure of fibers with maximum gain length has been determined as a function of the signal input power. For small-signal input power (in this case, below -40 dBm), the noise figure is determined solely by the pump and ASE powers. It is, however, important to note that even though only the co-propagating ASE power spectral density enters the noise figure expression (5.33), the counter-

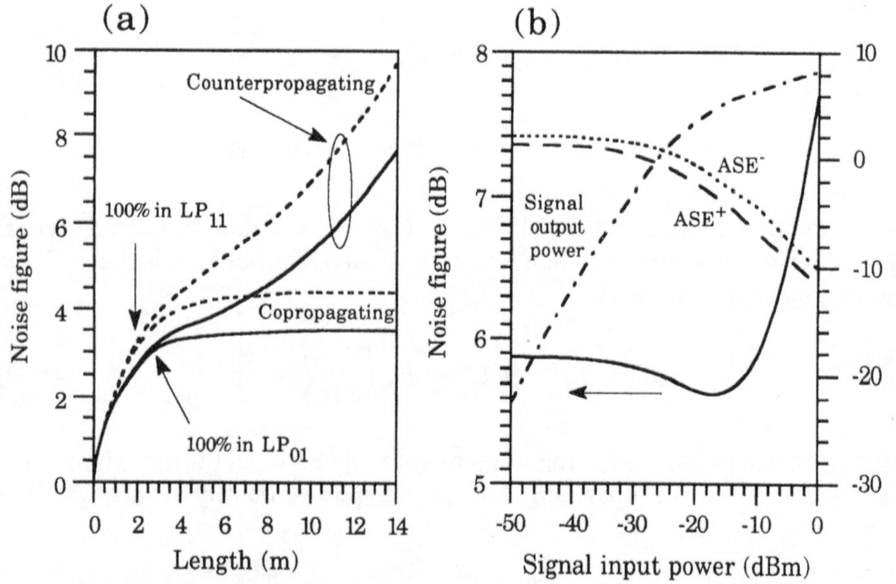

Figure 5.6 (a) Noise figure shown as a function of amplifier length. The signal input power is 10 μW, and 100 mW of pump power at 806 nm is used. The fiber is given by $a = 2.5\ \mu$m and NA $= 0.18$, and the step-index core is fully doped with erbium at a concentration of $\rho(r) = 1.8 \times 10^{24}\ \text{m}^{-3}$. $\lambda_s = 1535$ nm. (Data from [7].) (b) Noise figure, output signal power, and ASE$^+$ and ASE$^-$ powers shown as a function of the signal input power. The fiber has a step-index core ($a = 1.52\ \mu$m, NA $= 0.22$), and the erbium has been confined within a radius $a_{dot} = 0.6 \cdot a$. The pump wavelength is 1480 nm and the maximum gain fiber length is chosen. (Figure from [48].)

propagating ASE also plays a significant role. This happens because the relatively large counter-propagating ASE power lowers the population inversion in the amplifier input end. As the signal input power increases, it will suppress the ASE powers, as can be seen in Figure 5.6(b), and the population inversion in the fiber input end will consequently rise. This results in the lowering of the noise figure, which has a minimum around −16 dBm in this example. For signal input power above this level, it is the signal itself that deteriorates the population inversion, and the noise figure will consequently start to increase with increasing signal input power levels.

5.1.5 Numerical Solution

As mentioned earlier, the differential equations, which were derived in Section 5.1.3, cannot be solved analytically and numerical methods are therefore required. Even though these are basically straightforward, a few comments should be given at this point.

The chief property of the numerical method is that it uses iterative forward and backward integrations through the fiber until a convergence criterion is fulfilled. Such a criterion could typically be to demand that the changes in gain from two successive integrations be within an error limit of 10^{-4} dB. Furthermore, fulfilling this criterion for the gain ensures that the analogous criteria for the ASE and the pump power are also fulfilled. Before we enter into a closer discussion of the use of this convergence criterion, we will describe a typical application for establishing an idea of the number of parameters involved.

In all of the integrations, the co-propagating and counter-propagating ASEs are presented in several hundred (typically 200) frequency slots equally spaced in the wavelength range of λ = 1430 nm to λ = 1630 nm. The resulting spacing of, for example, 1 nm, is found to be necessary when calculating with the spectra from the Ge/Er fiber. These spectra are the most critical because the linewidth of the peaks are narrower than that of the spectra from the Al/Er fiber. As a typical example, in the case of 200 frequency slots, (5.18), (5.19), (5.21), and (5.22) yield a total of 402 coupled differential equations.

In each step of the integration, the overlap integrals involved in (5.20) and (5.23) are solved numerically by representing $I_s^{01}(r)$ and $I_p^{01}(r)$ in a number of equidistant points (typically 40) from r = 0 to $r = a_{\text{dot}}$. $n_2(r, \phi, z)$ and $n_1(r, \phi, z)$ are evaluated from (5.11) to (5.16) in the same 40 points for each step in the fiber.

The first step in the solution of the two boundary problems, which is introduced by the backward ASE, is to perform a forward integration, from z = 0, of the signal, the pump, and the forward ASE, in which the backward ASE is zero. After this, each backward integration is started at $z = L$, with the results $P_s(L)$, $P_p(L)$ and $S_{\text{ASE}}^{+}(\nu, L)$ from the previous forward integration together with the boundary condition $S_{\text{ASE}}^{-}(\nu, L) = 0$ as starting conditions. Analogously, the starting conditions at $z = 0$ for each forward integration are the results of the previous integration as a guess for $S^-_{ASE}(\nu, 0)$ and the boundary conditions, $S^+_{ASE}(\nu, 0) = 0$, $P_p(0) =$ input pump power and $P_s(0) =$ input signal power. Note that this explanation has to be slightly changed in the case of backward pumping. The number of iterations varies from 5 to 30, depending on the actual conditions, type of fiber, fiber length, pump and signal power, and pump and signal wavelength.

This description accounts for many typical calculation examples for EDFAs; however, for high-gain fibers and especially when the fiber length is near the gain optimum, the method can be unstable with the random occurrence of divergence. To be able to perform accurate calculations on amplifiers in more advanced configurations (see Chapter 8), it is essential to stabilize the algorithms, and we will therefore review the involved considerations as described in [49].

The minor but important changes, which in [49] were reported useful for the stability of the algorithm, are solely to integrate ASE, signal, and pump powers in the direction of propagation. This means that forward-propagating powers are found by forward integration, while backward-directed powers are found by backward integration. The fraction of Er^{3+} ions at a given place in the fiber (i.e., the inversion) may then, with the use of (5.11) to (5.16), be calculated simultaneously with the integration from the calculated values of co-directed powers and stored values of contra-directed powers found from a previously applied oppositely directed integration. The implementation of this co-directed integration scheme leads to a substantial decrease in the frequency of divergence failures, but they should not be removed completely. An apparently complete removal of numerical instabilities can, however, be obtained by adding another algorithm. Instead of using the excitation/de-excitation rates $R_n = n_2/n_1$ found directly from the forward- and backward-propagating powers in the nth iteration step, R_n is averaged with R_{n-1} found in the $(n - 1)$th step. The averaging is done as follows:

$$R_n = \alpha \cdot (n_2/n_1) + (1 - \alpha) \cdot R_{n-1} \tag{5.34}$$

where $\alpha = 0.75$, according to [49], gives a safe convergence. This postulate has been based on numerous forward and backward pump configurations, laser situations, and erbium fiber cascades with up to 10 amplifiers, in which situations no problems were detected.

In the final part of this section, we will discuss some typical results, which may be achieved by the use of the described model. We will first look at the spectral dependence of gain and forward-propagating ASE, as shown in Figure 5.7. The results are calculated for a typical Al co-doped EDFA pumped at 980 nm, and the noise figure has been determined from (5.33). Note that the specific fiber parameters have been omitted, since they are of no significance at this point. It is clear from the figure that even though the gain and ASE have a very similar wavelength dependence, the noise figure decreases anyway with increasing wavelength. It

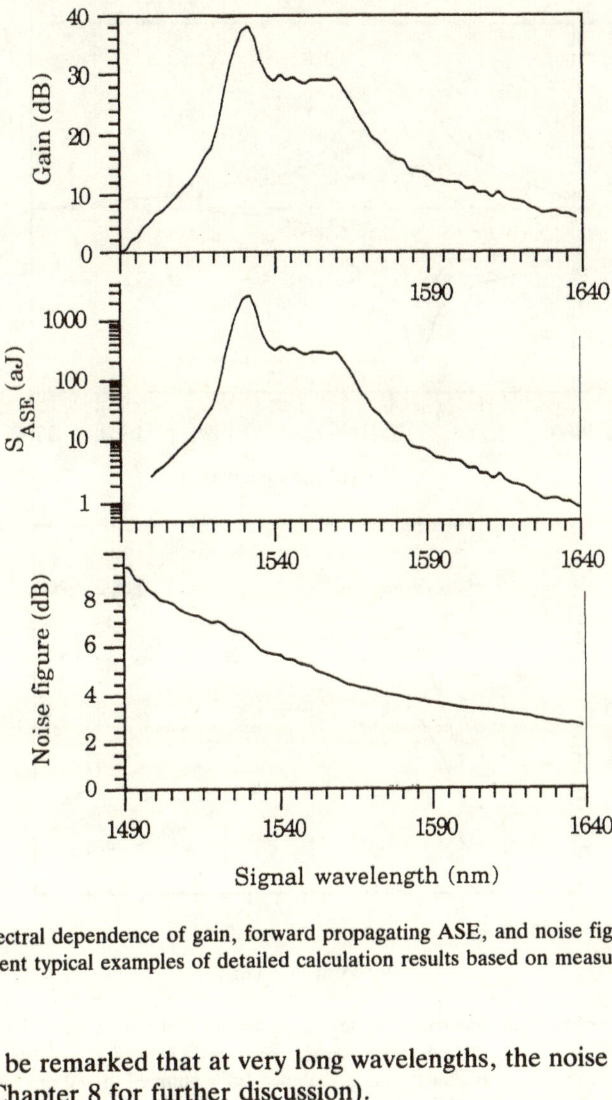

Figure 5.7 Spectral dependence of gain, forward propagating ASE, and noise figure. The curves represent typical examples of detailed calculation results based on measured spectra.

should also be remarked that at very long wavelengths, the noise figure is close to 3 dB (see Chapter 8 for further discussion).

One of the advantages of a reliable numerical model is that it may provide detailed information on primary parameters, which are difficult to measure, and it can thereby help the understanding of the physical limitations of the EDFA. A characteristic example is the evaluation of amplifier noise properties.

Detailed models are, however, not always necessary for the understanding of EDFAs. An example where simpler approaches are valid is the mapping of the spectral dependence of the relative gain coefficient as shown in Figure 5.8(a). This

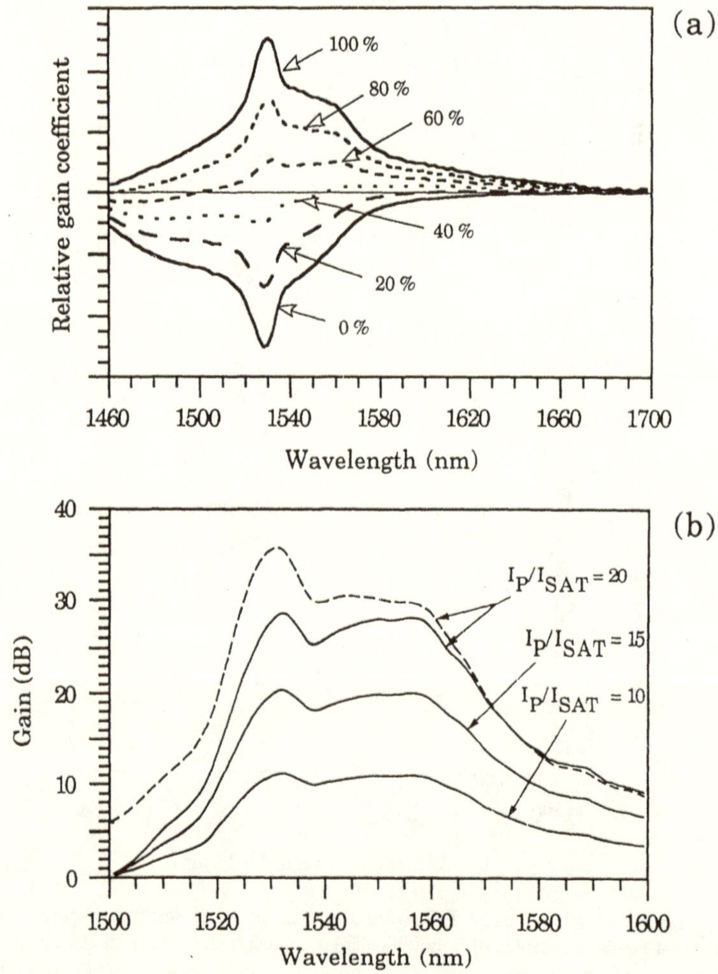

Figure 5.8 (a) Shape of the gain spectrum for various percentages of the Er^{3+} ions in the upper laser manifold. The figure is generated from the measured absorption and emission spectra of Figure 3.4. (b) Gain spectra for an Er^{3+}-doped fiber pumped at 980 nm for various pump intensities. The fiber length has for each pump power been chosen to yield either a flattened gain spectrum (solid lines) or a maximum gain at 1532 nm (dashed line). (Data from [29].)

curve illustrates the wavelengths at which positive gain may be obtained for different percentages of the Er^{3+} ions in the excited state. For example, Figure 5.8(a) demonstrates that for an amplifier section in which only 40% of the erbium ions are excited to the upper laser level, only wavelength components above 1550 nm are amplified, whereas the lower wavelengths show a net attenuation. This also explains why the gain spectrum mean wavelength of an EDFA moves towards longer wavelengths as the fiber length is increased. For long fibers, the pump power will not be sufficient to excite the majority of erbium ions, and, as a consequence, only the longer wavelengths will be amplified.

One direct use of this property has been demonstrated in [29], where the goal was to obtain a flat and broad gain spectrum. This type of spectrum, which is of large potential use in wavelength-division-multiplexed telecommunication systems, may for a given pump power be obtained by adjusting the fiber length. An illustration of this is shown in Figure 5.8(b), and it must be pointed out here that this is an application of a fiber where the fiber length is longer than the gain optimal length discussed in the previous sections. The idea is to use the attenuation of the fiber end to suppress the 1532-nm signals more than those located around 1550 nm. In the figure, I_P is the pump intensity and $I_{SAT} = hc/\sigma_a(\nu_p)\tau\lambda$ indicates the saturation intensity (see Section 9.1.3 for further discussion on similar parameters).

5.2 APPROXIMATIVE METHODS FOR AMPLIFIER ANALYSIS

Models such as the comprehensive approach described in Section 5.1 have the drawback that they require time-consuming computations. This is naturally acceptable in all cases where accurate analysis or amplifier synthesis is required, but may be an obstacle for more intuitive understanding of amplifier properties. A full model approach may be further unnecessary in many applications where only approximate results are needed. For these reasons, a number of approximative methods have been developed over the past years, and we will in this section review some of these. They have been selected to illustrate often applied simplifications.

5.2.1 The Equivalent Bandwidth Method

This method [7] was originally developed to establish a basis for an accurate spatial analysis of EDFAs. In this respect, it formed the foundation of the model of Section 5.1. However, it is very accurate for many applications in which only one signal wavelength is used, and it will only require a fraction of the computation time necessary for the model described in Section 5.1.

The idea behind the equivalent bandwidth method is to describe the ASE evolution with only one single differential equation. This means that (5.18) and (5.19) simplify to the following equation for the total ASE power P_{ASE}:

$$\frac{dP_{ASE}^{\pm}(z)}{dz} = \pm g_s(z) P_{ASE}^{\pm}(z) \pm Bh\nu_s a(z) \tag{5.35}$$

In (5.35), the + and − denote the spontaneous emission that respectively travels in the same or opposite direction of the signal. The local gain coefficient $g_s(z)$ may be expressed as

$$g_s(z) = \int_0^{2\pi}\int_0^{a_{dot}} I_s^{01}(r)[\sigma_e(\nu_s)n_2(r, \phi, z) - \sigma_a(\nu_s)n_1(r, \phi, z)]rdrd\phi \quad (5.36)$$

and the emission factor of the LP_{01} mode at the signal wavelength $a(z)$ can be written as follows:

$$a(z) = \sigma_e(\nu_s)\int_0^{2\pi}\int_0^{a_{dot}} I_s^{01}(r)n_2(r, \phi, z)rdrd\phi \quad (5.37)$$

The central parameter in this simplified description is the equivalent spontaneous emission bandwidth, which is defined as

$$B = \frac{\int_{-\infty}^{+\infty}\sigma_e(\nu)d\nu}{\sigma_e(\nu_s)} \quad (5.38)$$

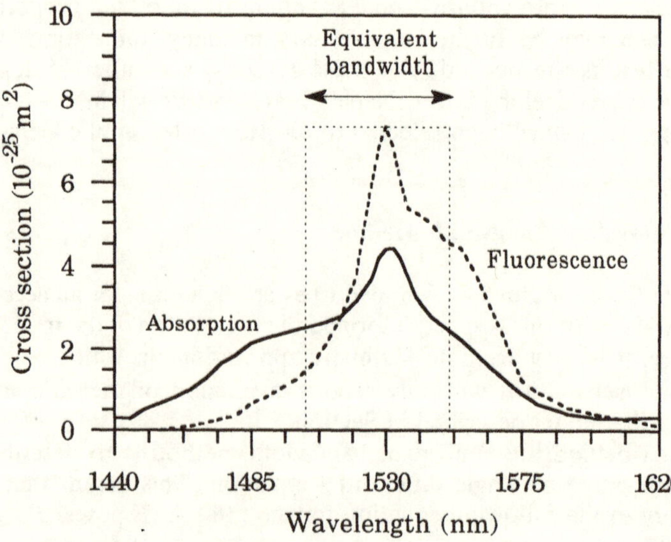

Figure 5.9 Typical absorption and emission spectra for an erbium-doped fiber amplifier with an indication of the equivalent bandwidth.

Figure 5.9 illustrates the equivalent bandwidth, which may be typically determined on the order of magnitude of $B = 7.0$ THz [7]. When B is determined from (5.38), the fact that the noise bandwidth changes with the population inversion has obviously not been taken into account. Note, however, that instead of using (5.38), a better accuracy might be obtained if B is minimized according to a priori values determined by the amplifier operation point [12].

5.2.2 Approximative Formulas for Gain and Optimal Length

To achieve a more significant simplification of the EDFA model, it is necessary to further limit the generality of the amplifier description. One way to accomplish this is to describe only medium gain amplifiers in which gain saturation by ASE can be ignored [15, 30]. A recent publication [15] describes the results of these simplifications, for which the first step is to write the local population of the upper laser level as in (5.11), except with the significant simplification that the ASE powers are ignored in (5.15) and (5.16). The transverse spatial integration of the mode fields is then performed under the assumption that the overlap integrals between pump modes and erbium dopant profile are power-independent. This is, according to [12], not the case if a more detailed description is applied; but in the case with sufficiently high pump powers or for confined erbium doping, they may be set constant. The primary result of [15] may be expressed in the following equation for the calculation of the amplifier gain G:

$$\frac{P_p^{in}}{h\nu_p}\left(1 - \left(\frac{G}{G_{max}}\right)^{\delta}\right) = \frac{P_s^{sat}}{h\nu_s}(\alpha_s L + \ln(G)) + \frac{P_s^{in}}{h\nu_s}(G - 1) \tag{5.39}$$

In (5.39), P_p^{in} and P_s^{in} are the input powers of pump and signal, respectively. The signal saturation power is defined as

$$P_s^{sat} = I_s^{sat}\frac{\int \rho_{Er}(r)\,dA \int |E_s(r,\phi)|^2 dA}{\int \rho_{Er}(r)|E_s(r,\phi)|^2 dA} \tag{5.40}$$

where saturation intensity is determined from $I_s^{sat} = h\nu_s/((\sigma_e(\nu_s) + \sigma_a(\nu_s))\tau)$, and A is the fiber cross section. The other parameters in (5.40) are defined in Section 5.1. In (5.39), furthermore, we need to define the length L of the erbium-doped fiber and the maximum gain

$$G_{\max} = \exp\left\{\left(\frac{\alpha_p}{\delta} - \alpha_s\right)L\right\} \tag{5.41}$$

We also use here the small-signal absorption coefficients

$$\alpha_k = \sigma_a(\nu_k)\frac{\int \rho_{\mathrm{Er}}(r)|E_s(r, \phi)|^2\, rdrd\phi}{\int |E_s(r, \phi)|^2\, rdrd\phi} \qquad k = s, p \tag{5.42}$$

and the ratio of saturation powers

$$\delta = \frac{P_s^{\mathrm{sat}}/\mathrm{h}\nu_s}{P_p^{\mathrm{sat}}/\mathrm{h}\nu_p} \tag{5.43}$$

where the pump saturation power is defined by replacing the appropriate parameters in (5.40).

After having defined the parameters of (5.39), we note that the term on the left-hand side represents the number of absorbed pump photons per unit time. The terms on the right-hand side correspond to the rate of spontaneously emitted photons and to the number of photons added to the signal per unit time, respectively.

To illustrate the applicability of the simplified model, it was used in [15] to determine the gain of a step-index fiber (a = 1.4 μm, NA = 0.26) with the core fully erbium-doped in a concentration of $\rho_{\mathrm{Er}} = 1.5 \cdot 10^{24}\ \mathrm{m}^{-3}$. A fiber with an erbium dopant radius $a_{\mathrm{dot}} = 1.7 \cdot a$ was also examined. As illustrated in Figure 5.10, the results were compared with calculations in which a full numerical integration of differential equations over the fiber cross section and fiber length is applied (see [50]). The specific parameter values entered into the calculations were as follows; λ_p = 980 nm, λ_s = 1550 nm, $\sigma_a(\nu_p) = 2 \cdot 10^{-25}\ \mathrm{m}^2$, $\sigma_e(\nu_p) = 0$, $\sigma_a(\nu_s) = 2 \cdot 10^{-25}\ \mathrm{m}^2$, $\sigma_e(\nu_s) = 3 \cdot 10^{-25}\ \mathrm{m}^2$, and τ = 10 ms.

From Figure 5.10 we find that the simplified gain expression (5.39) gives correct values for the gain. This is provided that either the doping radius does not exceed the pump mode field radius or that the powers (pump and/or signal) are sufficiently large. For $a_{\mathrm{dot}} = a$, the calculated gain difference is less than 0.5 dB for the small-signal gain, but, not surprisingly, the discrepancy increases for larger doping diameters. For realistic fibers, however, the method of [15] seems reliable for medium gain amplifiers (G about 20 dB) or for high signal input powers larger than −20 dBm.

A good estimate of the amplifier gain as described above is naturally of importance to the constructor of an EDFA. The first question that may come to

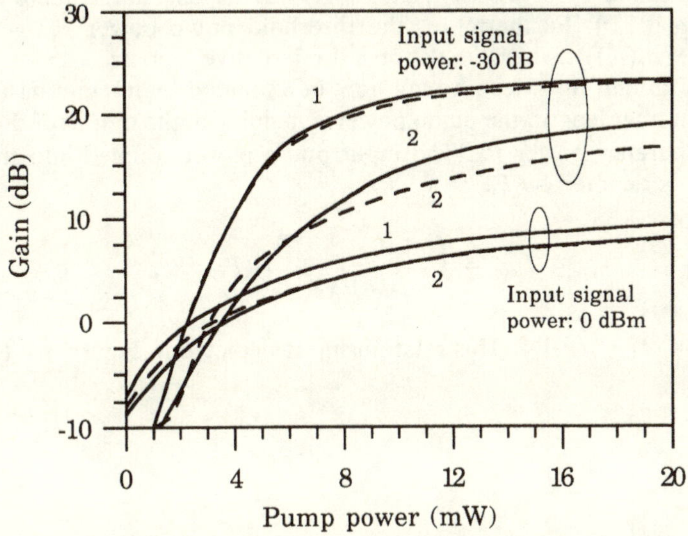

Figure 5.10 Gain at 1550 nm for 980-nm pumping calculated by the use of (5.39) (solid lines) and a full model [50] (dashed lines). Signal input powers are -30 dBm and 0 dBm. "1" refers to an erbium doping radius $a_{dot} = a$ and $L = 30$ m, whereas "2" indicates a fiber with the parameters $a_{dot} = 1.7 \cdot a$ and $L = 15$m. (Data from [15].)

mind is, however, not so much the gain as how to determine the gain optimal fiber length. Therefore, a discussion of simplified methods for length determination, such as those described in [13,14,51], is also called for. Therefore, we will first note that the simplification of this theory is obtained by considering only the pump power (i.e., ignoring both signal and ASE powers) when calculating the population inversion. However, in contrast to the former, the spatial mode description is included through a Gaussian approximation of the pump mode; that is, $|E_p^{01}(r)|^2 = (\pi\omega_p^2)^{-1} \exp(-r^2/\omega_p^2)$. By using this, the following relationship can be derived between the normalized pump power $\kappa = P_p(z)/P_{th}$ and the fiber length [52]:

$$[f_K(\kappa)]^{-1} d\kappa = dz \tag{5.44}$$

where the function

$$f_K(\kappa) = \rho_{Er}\sigma_a(\nu_p)\left(1 - \frac{\sigma_{ESA}(\nu_p)}{\sigma_a(\nu_p) - \sigma_e(\nu_p)} \cdot \ln\left(\frac{1 + \kappa}{1 + \kappa \exp(-\tau_p)}\right)\right. \\ \left. + \rho_{Er}\sigma_a(\nu_p)\frac{\sigma_{ESA}(\nu_p)}{\sigma_a(\nu_p) - \sigma_e(\nu_p)} \cdot \kappa \cdot (1 - \exp(-\tau_p))\right. \tag{5.45}$$

Here the parameter $\tau_p = (a_{dot}/\omega_p)^2$, and $\sigma_{ESA}(\nu_p)$ is the ESA at the pump wavelength (see Section 3.1.3). Furthermore, the threshold pump power $P_{th} = (\pi\omega_p^2 h\nu_p)/(\tau_{21}(\sigma_a(\nu_p) + \sigma_e(\nu_p)))$, where τ_{21} denotes the radiative lifetime.

The optimum fiber length may now be obtained by integrating (5.44). For the optimum fiber length, the pump power remaining at the output of the amplifier is the transparency power P_{tr}. The input pump power coupled into the erbium-doped fiber is denoted by $P_{in} = \kappa_{in} \cdot P_{th}$.

$$L_{opt} = \int_{\kappa_{tr}}^{\kappa_{in}} \frac{d\kappa}{f_K(\kappa)} = \int_{P_{tr}}^{P_{in}} f(P_p)\,dP_p \tag{5.46}$$

with $f(P_p) = [P_{th} \cdot f_K(\kappa)]^{-1}$. This relationship is depicted in Figure 5.11(a).

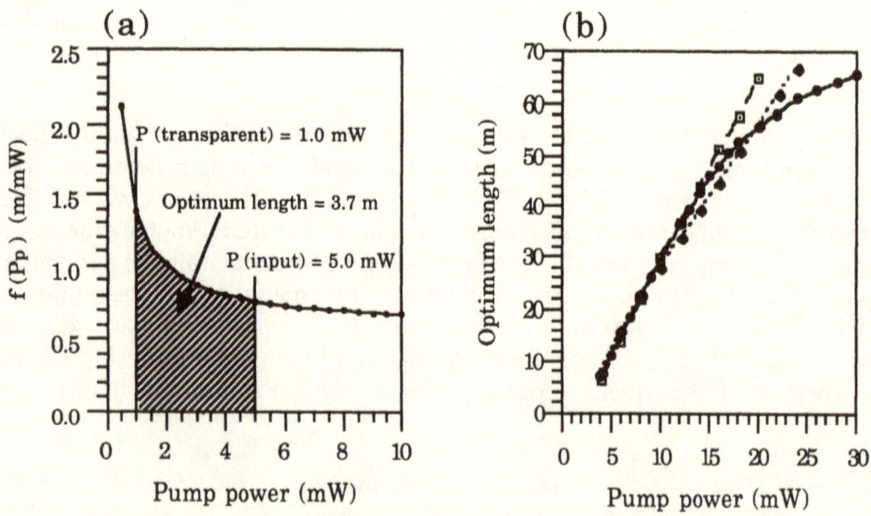

Figure 5.11 (a) Calculation of optimum fiber length by integration of the function $f(P_p) = [P_{th} \cdot f_K(\kappa)]^{-1}$ over pump power P_p. (b) Calculation of optimum fiber length in the presence of pump-induced depletion of metastable level. The fiber parameters are $a = 3.0\ \mu m$, NA = 0.13, and the core is erbium-doped at a concentration of 100 ppm within the radius $a_{dot} = 0.707 \cdot a$. Solid line: full model calculation. Long dashes and square markers: method of [13] including stimulated emission at the pump wavelength, 1480 nm. Short dashes: method of [13] excluding stimulated pump emission. (Figures from [13].)

The transparency power $P_{tr} = \kappa_{tr} \cdot P_{th}$ can be obtained from the condition that the gain coefficient [6, 50] $\gamma(z) = 0$. This yields in the Gaussian approximation

$$\int_0^{\tau_s} e^{-y} \frac{\left[\frac{\sigma_a(\nu_p)}{(\sigma_a(\nu_p) - \sigma_e(\nu_p))} + \left(\frac{\sigma_a(\nu_p)}{(\sigma_a(\nu_p) - \sigma_e(\nu_p))} - 1\right) \cdot \Gamma\right] \cdot \kappa_{tr} \cdot e^{-qy} - \Gamma}{\kappa_{tr} \cdot e^{-qy} + \Gamma} dy = 0 \tag{5.47}$$

where $\Gamma = (\sigma_a(\nu_s)/\sigma_e(\nu_s))$, $q = (\omega_s^2/\omega_p^2)$, $\tau_s = (a_{dot}/\omega_s)^2$, and ω_s is the fiber spot size at the signal wavelength calculated from the Gaussian approximation.

In the small-signal regime, the method gives good agreement with results obtained by a full model such as that described in Section 5.1. This is illustrated for 1480-nm pumping in Figure 5.11(b). At large pump powers, the growing signal eventually saturates the gain of the amplifier. This reduces the optimal length of the EDFA from the small-signal gain value [51], which consequently results in a discrepancy between the results obtained from (5.46) and those of a full model. For the example in Figure 5.11(b), such saturation occurs above a signal power of about 250 μW (at $P_p = 15$ mW). This compares well [52] with a predicted saturation power (5.40), which is defined in terms of reduced population inversion at a point in the fiber. Note, however, that the specific spot-size definition may vary the estimate of (5.40) significantly. Figure 5.11(b) also compares results of (5.46) in the cases where the stimulated pump emission has been included and omitted, respectively. The effect of the pump induced stimulated emission is a raising of the transparency power and, in general, an increased optimum fiber length.

5.3 THE INFLUENCE OF HIGHER ORDER PUMP MODES

The research in active fibers during the last few years has been mainly directed towards 980- and 1480-nm-pumped amplifiers. The reason is that these pump bands do not suffer from any significant ESA (see Chapter 3). However, the reliability and cost advantage of pump sources in the 800-nm wavelength band have, together with the development of more efficient active fibers, renewed the interest for this pump band. This has led to a demonstration [53] of an 820-nm pumped amplifier with a gain of 24 dB using a pump power of no more than 50 mW.

The use of 800-nm pump sources, however, makes it of special importance to consider the mode nature of the light propagating in the EDFA, since the fiber might be dual-moded at this wavelength. We will, therefore, in this section analyze what influence it has on the amplification properties of the EDFA if all the pump power is not only concentrated in the LP_{01} mode but is also launched into the LP_{11} mode.

This analysis will follow the description of [7, 46], which include an accurate mode representation. However, for the sake of simplicity, the equivalent bandwidth method (described in Section 5.2.1) is used. Furthermore, only the signal wavelength of 1535 nm is considered, and the pump wavelength is fixed at 806 nm in the following examples. This simplification agrees with the theoretical treatment of [2,5], and is not considered to affect the generality of the results. To calculate the local amplification and attenuation of signal and pump, it is necessary to know the mode intensities, which are described by the shape of the LP_{01} and LP_{11} modes that may be calculated for a general circular symmetric index profile [45]. In the discussion of the results described in this section, we will, however, only deal with step-index fibers. More specifically, the analysis is limited to a single erbium-doped fiber design, which has already been presented in [6]; namely, a step-index fiber given by the core radius $a = 2.5$ μm and NA $= 0.18$. The total core area is homogeneously doped with erbium in a concentration of $\rho_{Er}(r) = 1.8 \cdot 10^{24}$ m^{-3}.

The absorption and emission cross-section values corresponding to the signal wavelength ($\lambda = 1535$ nm) are $\sigma_e(\nu_s) = \sigma_a(\nu_s) = 7.0 \cdot 10^{-25}$ m^2 [40]. The pump absorption cross section is $\sigma_a(\nu_p) = 6.8 \cdot 10^{-26}$ m^2, and the ESA appears through the absorption cross-section value $\sigma_{24} = 2\sigma_a(\nu_p)$. Furthermore, we have used $\tau_{21} = 10.6$ ms as the spontaneous emission lifetime, and an equivalent bandwidth $B = 7.0$ THz was applied.

When we want to investigate the influence of the power distribution of the pump modes, it is naturally the basic EDFA characteristics such as gain and noise figure that are relevant. To obtain a detailed picture, we have in Figure 5.12, furthermore, shown the maximum gain as contour curves for pump powers of 50 mW and 100 mW, respectively. Calculations have been made for different signal powers between -30 and 0 dBm, and the pump mode distributions at the input end of the fiber have been varied. In all cases, the total pump power is assumed to be guided in the LP_{01} and the LP_{11} modes, and the fiber length has been optimized with respect to maximum gain.

We first note that for a given signal power in the low end of the considered interval, the largest gain will not be found for the situation in which all the pump power is launched into the LP_{01} mode. The curves show that the largest gain is found when 20% to 25% of the pump power excites the LP_{11} mode. The explanation is that pump ESA makes pump power less and less useful as the inversion level increases in the center of the fiber core. Thus, despite the poorer overlaps, pump power near the core-cladding interface (i.e., propagating in the LP_{11} mode) may still give more of a gain improvement than simply increasing the on-axis intensity. Therefore, by launching a part of the pump power into this higher order mode, better use is made of the erbium at the core-cladding boundary. By further decreasing the LP_{01}-to-LP_{11} ratio of the power in the pump modes, the maximum gain for a given signal power decreases slowly. For instance, if we look at a signal power of -20 dBm for a power of 100 mW, a maximum gain decrease of more than 1

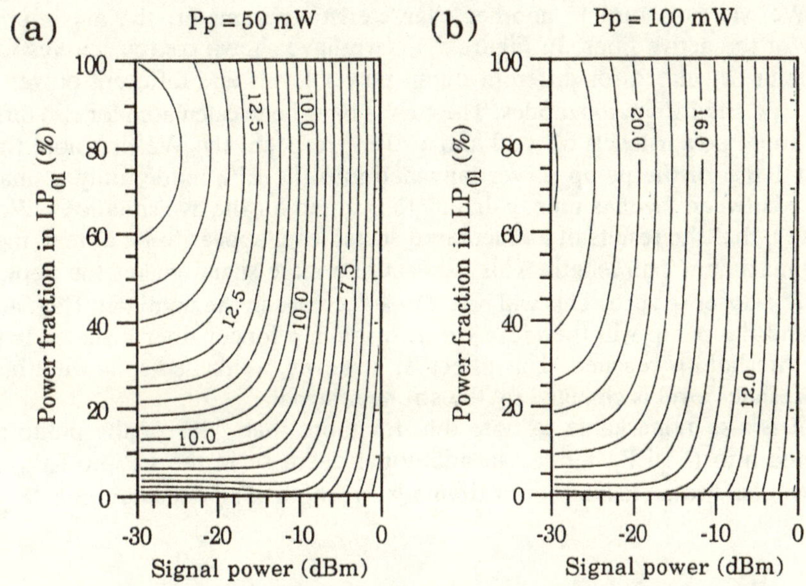

Figure 5.12 Contour curves of maximum gain shown for variable ratios of the LP_{01} and LP_{11} pump mode excitation and for different input signal powers. The parameters describing the fiber and erbium ion model are given in the text. (a) Total pump power at the input end of the fiber is 50 mW. (b) The pump power with the value of P_p = 100 mW. (From [46].)

dB will first appear when less than 25% of the total pump power is launched into the LP_{01} mode. For P_p = 50 mW and P_s = −20 dBm, the 1-dB gain decrease limit changes, so we need to launch at least 40% of the pump power into the LP_{01} mode, but it is still a very nonrestrictive requirement.

At high signal power levels, the influence of the pump mode distribution on the maximum gain becomes very weak. For P_p = 100 mW and P_s = 1 mW, the maximum gain will only decrease by 1.5 dB (from a value of 9 dB) when the LP_{01} pump mode is totally replaced by the LP_{11} mode. The reason is that, due to the high signal power, the population inversion all over a cross section of the active fiber is now brought closer to the situation in which the population inversion equals 1/2. The different pump mode distributions have, as a result, a smaller influence on the population inversion. If we finally compare the situation in which the pump power P_p = 50 mW with the situation in which P_p = 100 mW, it is found that if only a certain gain decrease can be accepted, then a higher pump power will allow a larger fraction of the total pump power to be launched into the LP_{11} mode. This is not surprising, since a larger pump power will bring the overall population inversion closer to unity.

We will now turn to another characteristic parameter: the maximum gain length of the active fiber. In Figure 5.13, we have shown contour curves for this parameter calculated for different pump power levels and different power ratios of the LP_{01} and LP_{11} pump modes. The curves have been calculated for two different signal input power levels of −20 and 1 dBm, respectively. We first note that for almost 100% of the pump power launched into the LP_{01} mode, only a small difference between the maximum gain length is found for the two situations. We find, however, that the results of an increased signal level appear to be a small increase in the maximum gain length. This is because a large signal makes the population inversion closer to zero. This will result in a decrease in the dominant ESA absorption, and the pump will therefore travel through a longer fiber distance before a given threshold is reached. This effect is, however, weakened somewhat because the threshold level is changed by the stronger signal.

It is also remarkable to note that for more than 50% of the pump power launched into the LP_{11} mode, an additional decrease in the LP_{01}-to-LP_{11} power ratio will lead to an increased maximum gain length. The explanation is that, due

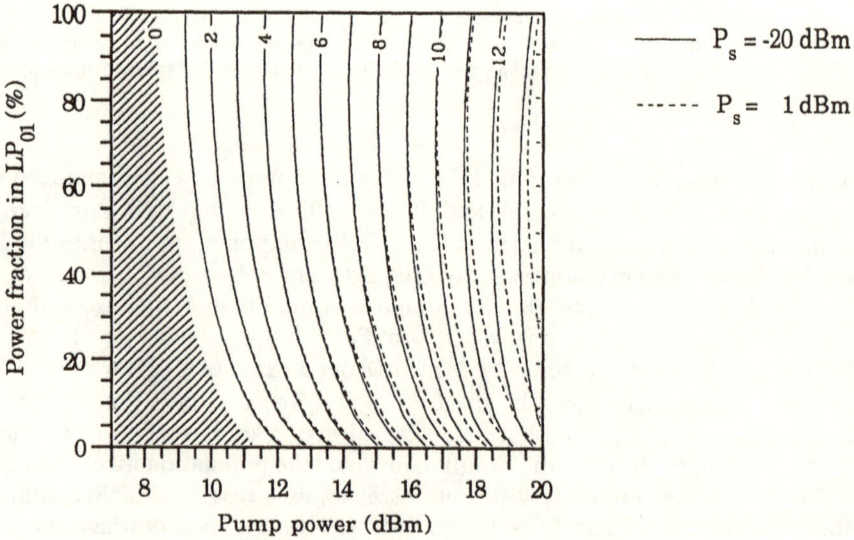

Figure 5.13 Contour curves of optimal length shown for different ratios of the LP_{01} and LP_{11} pump mode excitation and different values of the total pump power. The two curve sets correspond to respective input signal levels of P_s = 10 μW and P_s = 1 mW, and the parameters describing the fiber are the same as those used for finding the results of Figure 5.12. (Figure from [46].)

to a smaller relative power in the erbium-doped area, the LP_{11} mode will be attenuated at a slightly slower rate than the LP_{01} mode. This means that it may travel longer than the LP_{01} mode before their almost common threshold value is reached. This effect is strongest for large-signal levels, where the threshold levels of the two pump modes are closest to each other. Note that, in this discussion, for a pump power close to $P_p = 100$ mW, the maximum gain length for different pump distributions does not deviate more than 1 m (below 10% of the total length) as long as at least 25% of the pump is launched into the LP_{01} mode.

We will now turn to the noise properties of the EDFA. To be able to show the influence of the pump mode distribution on the noise figure in detail, we have in Figure 5.14 shown contour curves for the noise figure of co-propagating signal and pump. For pump powers of 50 mW and 100 mW, respectively, the input signal power and the LP_{01}-to-LP_{11} power ratio have been varied, and in each case the maximum gain length of the fiber amplifier has been used. As we found for the maximum gain shown in Figure 5.12, the influence of the pump mode distribution is very weak for large-signal powers. Only if less than 40% of the total pump power is launched into the fundamental LP_{01} mode is the noise figure deterioration remarkable for small input signal powers below −20 dBm.

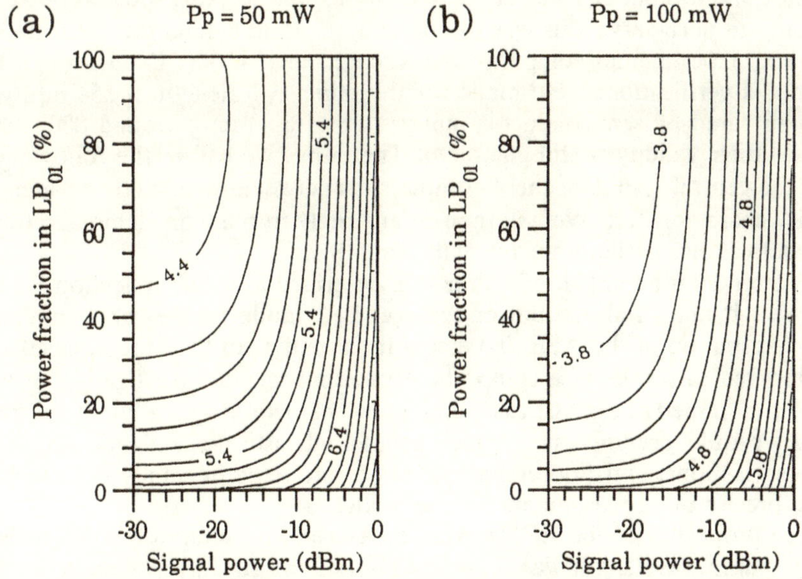

Figure 5.14 Contour curves of noise figure in a co-propagating signal and pump configuration. The results are calculated for different power fractions in the LP_{01} pump mode and for a variable signal power. (Figure from [46].) (a) The total pump power at the input end of the fiber is 50 mW. (b) The pump power is given the value of $P_p = 100$ mW.

It should finally be noted that the excitation of higher order pump modes such as the LP_{11} mode in practical applications is most likely to appear due to nonideal splices (with transversal offset). This will only result in a minor percentage of the total power in the LP_{11} mode. We may therefore conclude, considering the results presented in this section, that the pump mode excitation is not likely to result in any significant problems for EDFAs.

5.4 EXPERIMENTAL VERIFICATION OF MODELS

The early theoretical work on erbium-doped fiber amplifiers was based on the simplification of assumptions in order to achieve more interpretable and understandable results. Another reason was that it was still unclear what was actually required of a fiber amplifier model. Simplifications such as small-signal power and small ASE power were applied, and only step profiles of the index and erbium concentration were considered [8, 25]. Most of these assumptions do not strictly apply to real fibers with high gain and output power. Therefore, adjustable parameters are required to obtain agreement with experimentation. Thus, for applications where quantitative predictions are important, full scale calculations based on realistic index and erbium concentration profiles, as well as ASE spectral power distribution, are necessary. This was the basic motivation behind the accurate theoretical and experimental comparisons presented in [54, 55]. Even though these early model verifications were made for the pump wavelength of 654 nm, which is now outdated by semiconductor pump sources operating around 800, 980, or 1480 nm, their validity is still complete. Therefore, we will in the following text review the central results of these comparisons between calculated and measured amplifier characteristics. We will then refer briefly to the later demonstrations of the model validity in the other relevant pump bands.

The model used in [54, 55] is the one described in detail in Section 5.1. This means that it takes full account of the real LP_{01} mode profiles and the spectral power distribution of the ASE. To assess the accuracy of the model, calculations are compared to results of a comprehensive experimental investigation of a particular Er/Al-doped fiber. All basic parameters characterizing the fiber (i.e., refractive index profile, erbium concentration profile, intrinsic loss, emission and absorption cross sections, and fluorescence lifetime) are experimentally determined to provide precise input parameters for the model.

The properties of the EDFA were measured in a setup as shown in Figure 4.6. The pump wavelength was λ_p = 654 nm and the signal wavelength was λ_s = 1530 nm. The erbium-doped fiber was prepared by the solution doping technique (see Section 2.1.4), with aluminum as the index-raising element. The fiber had an outer diameter of 75 μm. The dopant concentration profiles of aluminum and erbium were measured as described in Section 4.1.1 to 4.1.2. The only preprocessing of data was as follows. The shape of the measured index and erbium profiles were found to be identical and very well approximated by an α-profile $(C(1 - (r/a)^\alpha)$ with $\alpha = 3$ and a core radius of $a = 2.7$ μm. Consequently, this parametrization was used as a convenient representation of the data in the calculations. The absolute scale was measured to be $C = 10^{-2}$ for the index difference and $C = 2.0 \cdot 10^{24}$ m^{-3} for the erbium concentration, the latter value having an estimated uncertainty of 10% due to calibration uncertainties. The calculated LP_{11} cutoff wavelength is 947 nm, close to the measured value of 960 nm.

The attenuation spectrum, the fluorescence spectrum, and the lifetime were measured by standard techniques (see Section 4.2). Emission and absorption cross sections are calculated (as described in Section 5.1.1) from the measured fluorescence and attenuation spectra. This was done using a measured fluorescence lifetime of 10.6 ms and including an empirical correction factor of 1.15 for the absorption [40] necessary, as described, for the Einstein relations used on Er^{3+}. As an example, the emission and absorption cross sections were found to be $5.0 \cdot 10^{-25}$ m^2 at the signal wavelength. The resulting cross section ratio of $\sigma_a/\sigma_e = 1.0$ is supported by the close agreement between the measured asymptotic gain for heavily pumped fibers and the measured (unpumped) fiber attenuation of 2.14 dB/m at 1530 nm, in contrast to the results of [39]. Based on the fiber attenuation and the absorption cross section, a maximum value for the erbium concentration of $\rho_{Er} = 2.1 \cdot 10^{24}$ m^{-3} was calculated, in very good agreement with the measured value. The calculated result was adopted, since it was considered more accurate.

From the knowledge of erbium concentration and the attenuation at the pump wavelength, $\lambda_p = 654$ nm, the absorption cross section is $\sigma_a(\nu_p) = 3.2 \cdot 10^{25}$ m^2. Furthermore, from the attenuation spectrum, the fiber was observed to have an intrinsic background loss of 0.3 dB/m. The ESA cross section, $\sigma_{ESA} = \sigma_{24}$, was extracted from the pump attenuation in a heavily pumped 5m-long fiber corrected for the intrinsic attenuation. An upper limit of $0.06 \cdot 10^{-25}$ m^2 was extracted. In the calculations, $\sigma_{ESA} = 0$.

Visual inspection of the output pump far field indicated that only the LP_{01} pump mode was excited. The intensity in the only other allowed mode LP_{11} was consequently set to zero. Still, trial calculations showed that even 25% of the pump power in this mode will influence the gain by less than 0.5 dB.

The model predictions are compared with the experimental results in Figure 5.15(a) for the small-signal gain, and in Figure 5.15(b) for the residual pump power. As observed, the agreement is excellent over more than four orders of magnitude of the gain. The measured maximum gain of 39.6 dB was reached with a fiber length of 30 m and a launched pump power of 115 mW. The residual pump power tends to be underestimated by the calculations on fiber lengths beyond 20 m. This is attributed to minor inaccuracies in the input parameters.

Further, excellent agreement between the measured and the calculated forward (a) and backward (b) ASE powers is observed in Figure 5.16. Equally good

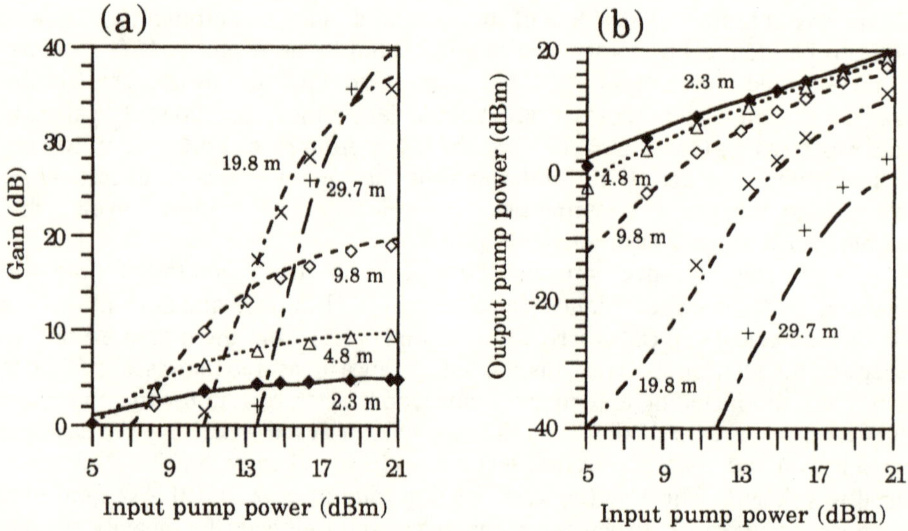

Figure 5.15 (a) Comparison of measured (symbols) and calculated gains for different lengths of fiber as a function of launched pump power. (b) Comparison of measured (symbols) and calculated output pump power as a function of fiber length and launched pump power. (Figures from [54].)

agreement is observed between both the measured and the calculated forward and backward ASE spectra, as illustrated in Figure 5.17. This demonstrates that straightforward calculations of noise characteristics are allowed at arbitrary signal wavelengths.

The present comparison verifies that the model described in Section 5.1 is an accurate and general tool in the analysis of fiber optical amplifiers, because it does not have to incorporate any adjustable parameters to match the experimental results. The demonstrated strength of the model is due to the proper treatment of the guided-light and erbium concentration profiles in combination with the application of correct spectral shapes of the emission and absorption cross sections. Finally, it should be noted that the present model was later tested in other pump bands, and, in both the 800- and 980-nm bands [56], the agreement between the calculated and the measured results was very good. Also, for 1480-nm pumping, agreement between the measured and the calculated results is established, which are also in agreement with the qualitative and quantitative results of [10].

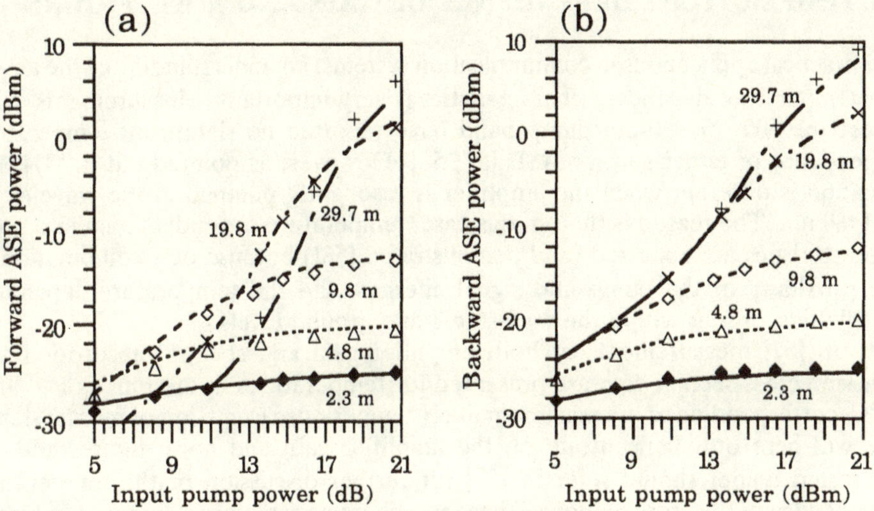

Figure 5.16 Comparison of measured (symbols) and calculated ASE power for different fiber lengths and launched pump power. (Figures from [54].) (a) Forward propagating ASE. (b) Backward propagating ASE.

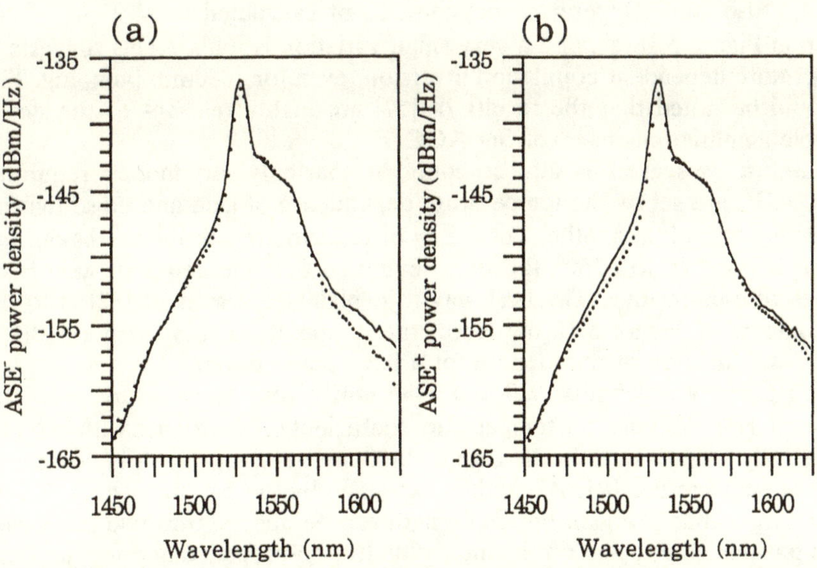

Figure 5.17 Comparison of measured (dotted) and calculated ASE spectra for a fiber length of 10.0m and a launched pump power of 44 mW at 654 nm. (Data from [55].) (a) Backward propagating ASE. (b) Forward propagating ASE.

5.5 TEMPERATURE DEPENDENCE OF GAIN AND NOISE FIGURE

For practical applications in communication systems, an understanding of the amplifier temperature-dependent characteristics is very important. Measurements using either the 600- or 980-nm pump band have reported no significant temperature dependence of either gain or ASE [57, 58]. However, as pointed out in [57], the situation is different when the amplifier is resonantly pumped at the wavelength of 1480 nm. The reason is that, in this case, temperature dependent gain and noise characteristics are expected (and demonstrated [58]) because of a combination of the proximity of the pump and signal energies and the temperature-dependent distribution of ions within the metastable and ground levels.

In [57] measurements of both amplifier gain and spectral absorption and emission cross-section data are presented for temperatures in the range of −40 to 60°C, corresponding to the maximum likely temperature excursion of most systems. We will here only concentrate on the amplifier gain and noise figure, and the interested reader should refer to [57] for further discussion of the temperature dependence of the cross sections. These two characteristics are illustrated in Figure 5.18(a), where we find that the gain change is 1.5 dB within the considered temperature range. This small change in gain with temperature can be readily compensated for with *automatic gain control* (AGC), confirming the practicality of the EDFA. Also, the temperature dependence of calculated noise figure values is shown in Figure 5.18(a), and a very small variation is found. This suggests a low temperature dependent population inversion, even for 1480-nm pumping. Finally, it should be noted that the results of [57] are highly relevant to the design of multiple amplifier chains involving AGC.

Before we leave this subject, confident that only very modest requirements to the AGC are set by the temperature dependence of gain and noise figure, it is interesting to look at another possibility of reducing this problem. As outlined in [52], it is possible to adjust the fiber length so that the amplifier gain becomes temperature-insensitive. The best way to explain the results of [52] is to discuss the contents of Figure 5.18(b). Here, the temperature coefficient (in dB/°C) is shown as a function of fiber length for a fixed pump power. It is noteworthy that at both pump wavelengths, 980 and 1480 nm, a zero temperature coefficient is observed. Note also that the temperature coefficient in accordance with [58] reaches numerically larger values in the case of 1480-nm pumping. The phenomenon of a zero temperature coefficient length is explained by the fact that an EDFA with a length larger than the gain optimal length can be divided into two sections. One is the positive gain region on the input side (for co-propagating pump and signal), and the other is the negative gain region on the output side. When the fiber temperature increases, the absorption and emission cross section increases. The gain in the positive gain region then increases monotonically as the temperature decreases. For the negative gain region, the absorption loss increases for decreasing

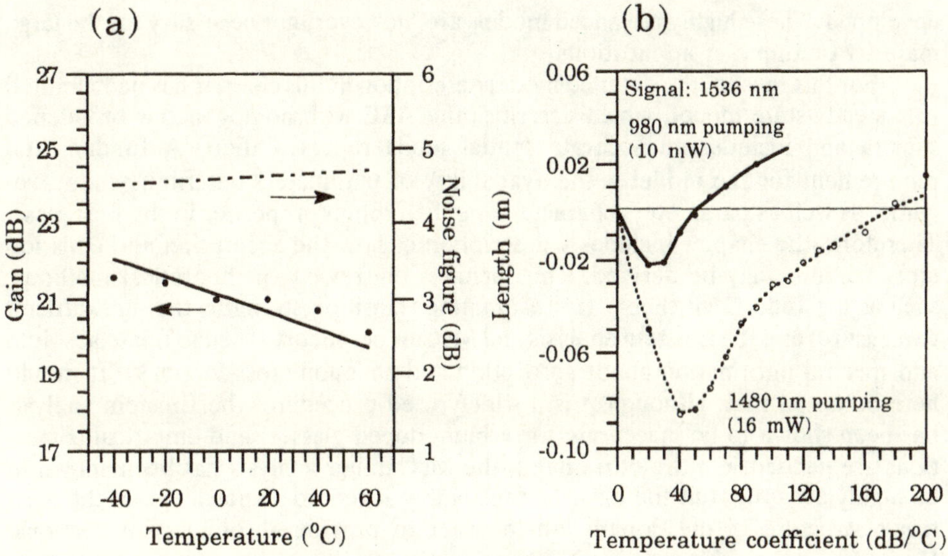

Figure 5.18 (a) Temperature dependence of gain and noise figure of an EDFA. The points are measured values and the lines are calculated. (Data from [57].) (b) Temperature coefficient versus the fiber length. (Data from [52].)

temperature, and this may outbalance the gain increase in the positive gain region. In this case, however, the amplifier gain is significantly reduced, and AGC may therefore be preferred in most cases.

5.6 SUMMARY

In this chapter, we have reviewed and compared the overall properties of models for the erbium-doped fiber amplifier as they developed from simple plane wave descriptions to full steady-state models accurately describing all relevant parameters of the EDFA. Important steps towards these complete models were the introduction of an accurate spatial mode field representation and the representation and spectral resolution of the ASE. Whereas the accurate spatial mode representation is important for waveguide design and optimization, the inclusion of ASE is necessary for describing saturation and noise properties of the amplifiers. In the description of the ASE spectrum, careful examinations have shown that for most applications it is sufficiently accurate to assume homogeneously broadened spectra. However, as discussed in the beginning of the chapter, recent advanced models further include inhomogeneous broadening, and models that explain time variations have been developed. These highly advanced models are, however, not necessary for the large majority of amplifier applications.

For this reason, the detailed model description in this chapter has been limited to a steady-state model, which describes the ASE with homogeneous broadened spectra and includes an accurate spatial mode representation. A fundamental requirement for the model is the availability of parameters describing the waveguide, as well as parameters characterizing the erbium properties in the host glass. Therefore, the chapter includes a description of how the absorption and emission cross sections may be derived. This includes the review of theoretical methods, such as the Judd-Ofelt theory (for estimating transition strengths that are difficult to measure) and the Einstein analysis and McCumber theory (used to relate absolute and spectral information about absorption and emission cross sections). It should here be noted that, although it is a widely used procedure, the Einstein analysis has been shown to be inaccurate for erbium-doped glasses, and empirical corrections are needed. On the other hand, the McCumber analysis has been shown to be highly accurate, but the disadvantage is that it requires knowledge of the electronic structure of the dopant ion in order to provide absolute cross sections. However, after significant simplifications, it still provides results that are as accurate as the empirically corrected ones obtained from the Einstein relations.

The resulting cross sections may then be applied to the model, which in the present formulation was established by first evaluating the local population inversion in the erbium-doped fiber. Afterwards, differential equations for the description of signal, pump, and ASE power distribution along the fiber are derived, and accurate equations for calculating amplifier noise figures are described. The propagation equations cannot be solved analytically in the general case, but direct numerical integration becomes necessary. Therefore, the final part of the model presentation discusses some of the numerical problems (and their solutions) involved in the calculation of the amplifier properties.

An approximative method for determining the gain and noise properties of the EDFA is to replace the spectral ASE representation with an equivalent bandwidth. The number of involved coupled differential equations are thereby reduced from typically several hundred to four or five, depending on the pump configuration.

Analytical methods for modeling the amplifier gain and optimal fiber length have also been described. These methods differ from the full model in the respect that they include a number of further simplifications. For example, a typical approximation is to neglect the ASE powers and to consider only small-signal operation of the EDFA, whereby the population inversion may be determined solely from the pump power intensity. This naturally limits the range within which the signal and pump powers may be varied, but the number of required input parameters are reduced to a minimum. We may therefore conclude that, even though the simplified methods are not suitable for accurate amplifier optimization, they are still important from the point of view that they rapidly provide results. This can help the intuitive understanding of the amplifier properties, as well as the analytical methods, may become a helpful tool in the laboratory. Finally, the analytical models may be used to describe the limiting performance of the amplifier.

After describing amplifier models, we turned to their applications, and a large-signal model with full spatial resolution was used to investigate the pump mode dependence on characteristics of the EDFA. This is relevant for pumping at 800 nm, where the fiber can be dual-moded. It was demonstrated for small-signal applications that only minor gain variations (within 1 dB) were seen when less than 25% of the total pump power was launched into the LP_{11} mode. For large-signal applications, the mode distribution of the pump power is of even less importance, and the existence of higher order pump modes is therefore normally not considered a significant problem.

To discuss further the parameters necessary to characterize an EDFA, and in addition to this establish a confidence in the earlier described model, detailed comparisons between calculations and experiments were presented. A very good agreement was found through these comparisons, and since no adjustable parameters were used to match the experimental results, the validity of the model was confirmed.

A short review of central results on temperature dependence of EDFAs in the temperature range of −40 to 60°C showed a very small sensitivity for gain and noise figure. For 1480-nm pumping, which is the most sensitive, temperature gain coefficients between 0 and −0.1 dB/°C have been demonstrated. This indicates that AGC will have no difficulty in solving the problem of temperature sensitivity. Another solution is to choose a fiber length, which is significantly longer than the gain optimal length. For a given pump power, such a length exists for which the temperature dependence is fully eliminated within the considered temperature range. However, it must be kept in mind that the noise properties of such erbium-doped fibers are very poor. Therefore, the AGC solution is preferable for most applications.

REFERENCES

[1] Mears, R. J., L. Reekie, I. M. Jauncey, and D. N. Payne, "Low Noise Erbium-Doped Fibre Amplifier Operating at 1.54 μm," *IEE Electronics Letters*, Vol. 23, No. 19, 1987, pp. 1026–1028.

[2] Desurvire, E., J. R. Simpson, and P. C. Becker, "High-Gain Erbium-Doped Traveling-Wave Fibre Amplifier," *Optics Letters*, Vol. 12, No. 11, 1987, pp. 888–890.

[3] Giles, C. R., E. Desurvire, J. R. Talman, J. R. Simpson, and P. C. Becker, "2-Gbit/s Signal Amplification at $\lambda = 1.53$ μm in an Erbium-Doped Single-Mode Fiber Amplifier," *IEEE Journal of Lightwave Technology*, Vol. 7, No. 4, 1989, pp. 651–656.

[4] Hagimoto, K., et al., "A 212 km Non-Repeatered Transmission Experiment at 1.8 Gb/s Using LD Pumped Er^{3+}Doped Fiber Amplifiers in an IM/Direct-Detection Repeater System," *Optical Fiber Communications OFC '89*, U.S., 1989, postdeadline paper PD15.

[5] Olshansky, R., "Noise Figure for Erbium-Doped Optical Fibre Amplifiers," *IEE Electronics Letters*, Vol. 24, No. 22, 1988, pp. 1363–1365.

[6] Armitage, J. R., "Three-Level Fiber Laser Amplifier: A Theoretical Model," *Applied Optics*, Vol. 27, 1988, pp. 4831–4836.

[7] Bjarklev, A., S. L. Hansen, and J. H. Povlsen, "Large Signal Modelling of an Erbium Doped Fiber Amplifier," *Proceedings of Society of Photo-Optical Instrumentation Engineers Conference, OE/Fibers '89*, part: Fiber Laser Sources and Amplifiers, Boston, 6–8 Sept. 1989, SPIE Vol. 1171, pp. 118–129.

[8] Morkel, P. R., and R. I. Laming, "Theoretical Modeling of Erbium-Doped Fiber Amplifiers With Excited-State Absorption," *Optics Letters*, Vol. 14, 1989, pp. 1062–1064.

[9] Giles, C. R., and E. Desurvire, "Propagation of Signal and Noise in Concatenated Erbium-Doped Fiber Optical Amplifiers," *IEEE Journal of Lightwave Technology*, Vol. 9, No. 2, 1991, pp. 147–154.

[10] Desurvire, E., and J. R. Simpson, "Amplification of Spontaneous Emission in Erbium-Doped Single Mode Fibers," *IEEE Journal of Lightwave Technology*, Vol. 7, No. 5, 1989, pp. 835–845.

[11] MacFarlane, R. M., and R. M. Shelby, "Homogeneous Line Broadening of Optical Transitions of Ions and Molecules in Glasses," *Journal of Luminescence*, Vol. 36, 1987, p. 187.

[12] Giles, C. R., and E. Desurvire, "Modeling Erbium-Doped Fiber Amplifiers," *IEEE Journal of Lightwave Technology*, Vol. 9, No. 2, 1991, invited paper, pp. 271–283.

[13] Rühl, F. F., "Calculation of Optimum Fibre Lengths for EDFAs at Arbitrary Pump Wavelengths," *IEE Electronics Letters*, Vol. 27, No. 16, 1991, pp. 1443–1445.

[14] Rühl, F. F., "Accurate Analytical Formulas for Gain-Optimized EDFAs," *IEE Electronics Letters*, Vol. 28, No. 3, 1992, pp. 312–313.

[15] Pfeiffer, Th., and H. Bülow, "Analytical Gain Equation for Erbium-Doped Fiber Amplifiers Including Mode Field Profiles and Dopant Distribution," *IEEE Photonics Technology Letters*, Vol. 4, No. 5, 1992, pp. 449–451.

[16] Pedersen, B., A. Bjarklev, J. H. Povlsen, K. Dybdal, and C. C. Larsen, "The Design of Erbium-Doped Fiber Amplifiers," *IEEE Journal of Lightwave Technology*, Vol. 9, No. 9, 1991, pp. 1105–1112.

[17] Desurvire, E., C. R. Giles, and J. R. Simpson, "Gain Dynamics of Erbium-Doped Fiber Amplifiers," *Proceedings of Society of Photo-Optical Instrumentation Engineers Conference, OE/Fibers '89*, part: Fiber Laser Sources and Amplifiers, Boston, 6–8 Sept. 1989, SPIE Vol. 1171, pp. 103–118.

[18] Maigron, Y., and J. F. Marcerou, "Model of Gain Dynamics in EDFAs," *Techn. Dig. of Optical Amplifiers and Their Applications OAA '92*, Santa Fe, 24–26 June 1992, paper FD3, pp. 210–213.

[19] Desurvire, E., J. W. Sulhoff, J. L. Zyskind, and J. R. Simpson, "Spectral Dependence of Gain Saturation and Effect of Inhomogeneous Broadening in Erbium-Doped Aluminosilicate Fiber Amplifiers," *Techn. Dig. of Optical Amplifiers and Their Applications OAA '90*, Monterey, 6–8 Aug. 1990, post deadline paper PD9.

[20] Grosskopf, G., R. Ludwig, and H. G. Weber, "Crosstalk in Optical Amplifiers for Two-Channel Transmission," *IEE Electronics Letters*, Vol. 22, No. 17, 1986, pp. 900–902.

[21] Mukai, T., K. Inoue, and T. Saitoh, "Signal Gain Saturation in Two-Channel Common Amplification Using a 1.5 m InGaAsP Travelling-Wave Laser Amplifier," *IEE Electronics Letters*, Vol. 23, No. 8, 1987, pp. 396–397.

[22] Agrawal, G. P., "Amplifier-Induced Crosstalk in Multichannel Coherent Lightwave Systems," *IEE Electronics Letters*, Vol. 23, No. 22, 1987, pp. 1175–1177.

[23] Jopson, R. M., T. E. Darcie, K. T. Gayliard, R. T. Ku, R. E. Tench, T. C. Rice, and N. A. Olsson, "Measurement of Carrier-Density Mediated Intermodulation Distortion in an Optical Amplifier," *IEE Electronics Letters*, Vol. 23, No. 25, 1987, pp. 1394–1395.

[24] Inoue, K., T. Mukai, and T. Saitoh, "Nearly Degenerate Four Wave Mixing in a Traveling-Wave Semiconductor Laser Amplifier," *Applied Physics Letters*, Vol. 51, No. 14, 1987, p. 1051.

[25] Desurvire, E., C. R. Giles, and J. R. Simpson, "Gain Saturation Effects in High-Speed, Multichannel Erbium-Doped Fiber Amplifiers at $\lambda = 1.53\ \mu m$," *IEEE Journal of Lightwave Technology*, Vol. 7, No. 12, 1989, pp. 2095–2104.

[26] Giles, C. R., E. Desurvire, and J. R. Simpson, "Transient Gain and Crosstalk Effects in Erbium-Doped Fiber-Amplifiers," *Optics Letters*, Vol. 14, No. 16, 1989, pp. 880–882.

[27] Pettitt, M. J., A. Hadjifotiou, and R. A. Baker, "Crosstalk in Er-Doped Fiber Amplifiers," *IEE Electronics Letters*, Vol. 25, No. 6, 1989, pp. 416–417.

[28] Laming, R. I., L. Reekie, P. R. Morkel, and D. N. Payne, "Multichannel Crosstalk and Pump Noise Characterization of Er^{3+}-Doped Fibre Amplifier Pumped at 980 nm," *IEE Electronics Letters*, Vol. 25, No. 7, 1989, pp. 455–456.

[29] Armitage, J. R., "Spectral Dependence of Small-Signal Gain Around 1.5 μm in Erbium Doped Silica Fiber Amplifiers," *IEEE Journal of Quantum Electronics*, Vol. 26, No. 3, 1990, pp. 423–425.

[30] Saleh, A. A. M., R. M. Jopson, J. D. Evankow, and J. Aspell, "Modeling of Gain in Erbium-Doped Fiber Amplifiers," *IEEE Photonics Technology Letters*, Vol. 2, No. 10, 1990, pp. 714–717.

[31] Peroni, M., and M. Tamburrini, "Gain in Erbium-Doped Fiber Amplifiers: A Simple Analytical Solution to the Rate Equations," *Optics Letters*, Vol. 15, No. 15, 1990, pp. 842–844.

[32] Kikuchi, K., "Generalised Formula for Optical-Amplifier Noise and Its Application to Erbium-Doped Fibre Amplifiers," *IEE Electronics Letters*, Vol. 26, No. 22, 1990, pp. 1851–1853.

[33] Miniscalco, W. J., "Erbium-Doped Glasses for Fiber Amplifiers at 1500 nm," *IEEE Journal of Lightwave Technology*, Vol. 9, No. 2, 1991, invited paper, pp. 234–250.

[34] Miniscalco, W. J., and R. S. Quimby, "General Procedure for the Analysis of Er^{3+} Cross Sections," *Optics Letters*, Vol. 16, No. 4, 1991, 258–260.

[35] Judd, B. R., "Optical Absorption Intensities of Rare-Earth Ions," *Physical Review*, Vol. 127, 1962, pp. 750–761.

[36] Ofelt, G. S., "Intensities of Crystal Spectra of Rare-Earth Ions," *Journal of Chemical Physics*, Vol. 37, 1962, pp. 511–520.

[37] Nielson, C. W., and G. F. Koster, *Spectroscopy Coefficients for p^n, d^n and f^n Configurations*, Cambridge, MA: MIT Press, 1964.

[38] Sandoe, J. N., P. H. Sarkies, and S. Parke, "Variation of Er^{+3} Cross Section for Stimulated Emission With Glass Composition," *Journal of Physics D*, Vol. 5, 1972, pp. 1788–1799.

[39] Barnes, W. L., R. I. Laming, P. R. Morkel, and E. J. Tarbox, "Absorption-Emission Cross-Section Ratio for Er^{3+} Doped Fibers at 1.5 μm," *Proceedings of Conference on Lasers and Electro-Optics CLEO '90*, Anaheim, 1990, paper JTUA3, pp. 50.

[40] Dybdal, K., N. Bjerre, J. E. Pedersen, and C. C. Larsen, "Spectroscopic Properties of Er-Doped Silica Fibers and Preforms," *Proceedings of Society of Photo-Optical Instrumentation Engineers Conference, OE/Fibers '89*, part: Fiber Laser Sources and Amplifiers, Boston, 6–8 Sept. 1989, SPIE Vol. 1171, pp. 209–218.

[41] DiBartolo, B., *Optical Interactions in Solids*, New York: Wiley, 1968, pp. 404–411.

[42] McCumber, D. E., *Physical Review*, Vol. 134, No. A299, 1964.

[43] Desurvire, E., and J. R. Simpson, "Evaluation of $^4I_{15/2}$ and $^4I_{13/2}$ Stark-Level Energies in Erbium-Doped Aluminosilicate Glass Fibers," *Optics Letters*, Vol. 15, No. 10, 1990, pp. 547–549.

[44] Gloge, D., "Weakly Guiding Fibers," *Applied Optics*, Vol. 10, 1971, pp. 2252–2258.

[45] Danielsen, P. L., "Analytical Expressions for Group Delay in the Far Field From an Optical Fibre Having an Arbitrary Index Profile," *IEEE Journal of Quantum Electronics*, Vol. 19, 1981, pp. 850–853.

[46] Bjarklev, A., J. H. Povlsen, B. Pedersen, and K. Rottwitt, "The Influence of Nonfundamental Pump Modes in Erbium-Doped Fiber Amplifiers," *Fiber and Integrated Optics*, Vol. 10, 1991, pp. 11–21.

[47] Shimota, K., H. Takahasi, and C. H. Townes, "Fluctuations in Amplification of Quanta With Application to Maser Amplifiers," *Journal of the Physical Society of Japan*, Vol. 12, No. 6, 1957, pp. 686–700.

[48] Vendeltorp-Pommer, H., B. Pedersen, A. Bjarklev, and J. H. Povlsen, "Noise and Gain Performance for an Er^{3+} Doped Fibre Amplifier Pumped at 980 nm or 1480 nm," *Proceedings of Society of Photo-Optical Instrumentation Engineers Conference, OE/Fibers '90*, part: Fiber Laser Sources and Amplifiers II, San Jose, 18–19 Sept. 1990, SPIE Vol. 1373, pp. 254–265.

[49] Povlsen, J. H., A. Bjarklev, O. Lumholt, H. Vendeltorp-Pommer, and K. Rottwitt, "Optimizing Gain and Noise Performance of EDFAs With Insertion of a Filter or an Isolator," *Proceedings of Society of Photo-Optical Instrumentation Engineers Conference, OE/Fibers '91*, part: Fiber Laser Sources and Amplifiers III, Boston, 5–6 Sept. 1991, SPIE Vol. 1581, pp. 107–113.

[50] Desurvire, E., J. L. Zyskind, and C. R. Giles, "Design Optimization for Efficient Erbium-Doped Fibre Amplifiers," *IEEE Journal of Lightwave Technology*, Vol. 8, No. 11, 1990, pp. 1730–1741.

[51] Rühl, F. F., "Prediction of Optimum Fibre Lengths for Erbium Doped Fibre Amplifiers," *IEE Electronics Letters*, Vol. 27, No. 9, 1991, pp. 769–770.

[52] Yamada, M., M. Shimuzu, M. Okayasu, and M. Horiguchi, "Temperature Insensitive Er^{3+}-Doped Optical Fibre Amplifiers," *IEE Electronics Letters*, Vol. 26, No. 20, 1990, pp. 1649–1650.

[53] Horiguchi, M., M. Shimuzu, K. Yamada, H. Yoskino, and H. Hanafusa, "Highly Efficient Optical Fiber Amplifier Pumped by a 0.8 μm Band Laser Diode," *IEE Electronics Letters*, Vol. 26, No. 21, 1990, pp. 1758–1759.

[54] Pedersen, B., K. Dybdal, C. Dam-Hansen, A. Bjarklev, J. H. Povlsen, H. Vendeltorp-Pommer, and C. C. Larsen, "Detailed Theoretical and Experimental Investigation of High-Gain Erbium-Doped Fiber Amplifier," *IEEE Photonics Technology Letters*, Vol. 2, No. 12, 1990, pp. 863–865.

[55] Dybdal, K., A. Bjarklev, C. Dam-Hansen, C. C. Larsen, B. Pedersen, J. H. Povlsen, and H. Vendeltorp-Pommer, "Detailed Experimental and Theoretical Investigation of High-Gain Er-Doped Fibre Amplifier," *Proc. of 16th European Conference on Optical Communication ECOC '90*, Amsterdam, Sept. 16–20, 1990, pp. 443–446.

[56] Pedersen, B., S. Zemon, and W. J. Miniscalco, "Evaluation of the 800 nm Pump Band for Erbium-Doped Fiber Amplifiers," *IEEE Journal of Lightwave Technology*, Vol. 10, No. 8, 1992, pp. 1041–1049

[57] Suyama, M., R. I. Laming, and D. Payne, "Temperature Dependent Gain and Noise Characteristics of a 1480 nm-Pumped Erbium-Doped Fibre Amplifier," *IEE Electronics Letters*, Vol. 26, No. 21, 1990, pp. 1756–1757.

[58] Kagi, N., A. Oyobe, and K. Nakamura, "Temperature Dependence of the Gain in an Erbium-Doped Fiber," *Proceedings of Conference on Lasers and Electro-Optics CLEO '90*, Anaheim, 1990, paper CPDP36.

Chapter 6
Optimization of the Erbium-Doped Fiber

The remarkably rapid development of EDFAs from the first laboratory demonstrations to field trial experiments has been reinforced by the results of the accurate optimization of the fibers. The significance of the optimization is founded on the fact that the resulting optimal waveguide designs of erbium-doped fibers are very different from that of the standard 1300-nm or dispersion shifted 1550-nm single-mode fiber. Moreover, the relationship between optimal erbium-doped fiber design and the choice of pump wavelength complicates the design requirements considerably compared to standard fiber design. The glass host material will, in addition, also affect the fiber properties significantly.

As will become clear from the presentation in this chapter, a point of reference is taken in a short description of the different working conditions of erbium-doped fibers. This defines general characteristics and operation differences, depending on the amplifier applications, and it naturally leads to the preceding choice of characterization parameters relevant to the definition of optimization criteria.

The fundamental requirements for the optimization of the small-signal operation of fiber amplifiers are given, and optimal designs of in-line amplifiers and preamplifiers are presented, depending on whether the pump wavelength is 1480, 980, or 800 nm. Also, the influence of the index-raising co-dopant material is described. From then on we focus on the best possible design of power booster amplifiers, and finally there is a discussion on how different choices of refractive index profiles affect the properties of fiber amplifiers for 1550-nm operation.

6.1 WORKING CONDITIONS OF FIBER AMPLIFIERS

Before we go into the details of the optimization of EDFAs, we will, with reference to [1], discuss some general and fundamental properties of the fiber amplifier. The discussion is closely related to the characteristic illustration in Figure 6.1. Here,

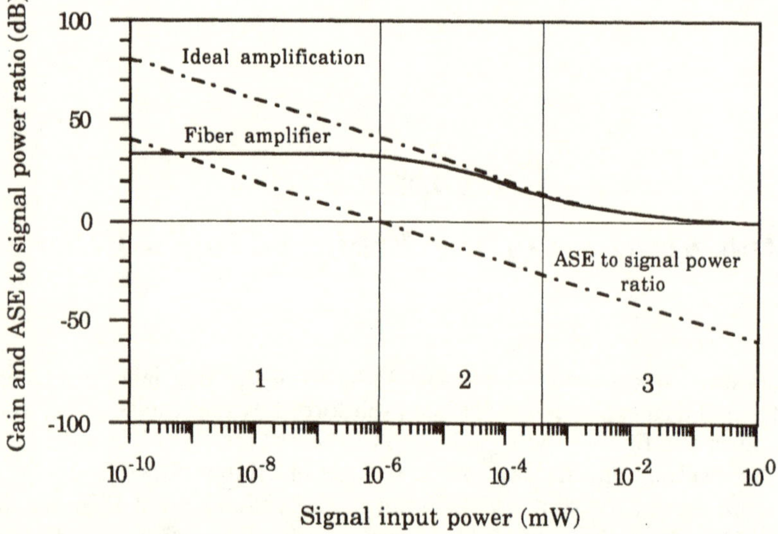

Figure 6.1 Gain- and ASE-to-signal ratio for fiber pumped with 20 mW at 1480 nm [1].

the solid curve shows the gain at the fiber length L_{opt} (defined as the length giving the highest gain for a given signal and pump power). Note that, even though the calculations, which form the basis of Figure 6.1, are performed for the 1480-nm pump wavelength and a signal wavelength of 1530 nm, this does not affect the generality of the results. A similar behavior will be found for any of the possible pump wavelengths.

At small-signal input power levels (region 1 in Figure 6.1), the gain is independent of the signal input power. In this region of operation, the quantum efficiency of the pump power is very poor, with pump photons mainly being converted to useless photons of backward- and forward-traveling ASE.

By increasing the signal input power to higher levels (region 2 in Figure 6.1), the decreasing curve of "ideal amplification" (defined by unity quantum efficiency of pump photons) comes closer to the signal gain. This means that the gain curve starts rather abruptly to fall off following the curve of "ideal amplification." A number of advantages in operating the amplifier in region 2 are:

- Pump efficiency close to unity;
- Signal output power higher than power of the ASE;
- Decreasing gain for increasing signal input power (AGC);
- The noise figure of the amplifier exhibits its minimum [2].

By operating the amplifier in region 2 instead of region 1, the need for spectral filtering of the output signal becomes less significant. Furthermore, the property of partly self-adjusting output signal power makes possible gain control of cascade coupled amplifiers less important than in region 1.

The extreme signal input power case is covered by region 3, where the power of the signal has reached such magnitudes that it dominates the population inversion along the fiber by establishing a counterbalance between stimulated emission and absorption. Operation of the fiber amplifier in this region cannot be recommended, since the gain is weak with the random emission and absorption processes along the fiber act merely as noise sources on the signal.

The shape of the gain curve is similar for arbitrary fiber designs. The location of the intersection between regions 1 and 2 may, however, vary with the fiber design. In a gain optimization of the fiber design, the task is to move this intersection to as small an input power level as possible or, equivalently, to maximize the small-signal gain.

6.2 PARAMETERS FOR OPTIMIZATION OF ERBIUM-DOPED FIBERS

Due to the large number of parameters involved in the accurate description of fiber amplifiers and the resulting complexity of the model, it is important to focus on the figures of merit for the optimization of erbium-doped fibers. In this section, we will discuss how to select such key parameters of the fiber by considering the different applications of fiber amplifiers. We will focus first on parameters relevant to amplifiers operating in the small signal regime [1]. Then the question of minimal pump power consumption is addressed [3,4] and parameters for the optimization of power booster amplifiers are considered [5].

6.2.1 Small-Signal Amplifiers

As discussed in the previous section, the major task in optimizing fiber amplifiers is to maximize the small-signal gain. It is obvious here that the required pump power is an important parameter for the fiber amplifier. To be able to consider optimization of gain related to the pump power, we first look at the calculation results shown in Figure 6.2. The results represent the case of 1480-nm pumping with pump power values of 5, 10, 20, and 40 mW and erbium fiber designs within a design area given by an NA interval [0.1; 0.2]. For the more than 500 gain calculations, the normalized frequency is further varied within the interval [0.8; 2.4] (defined at the pump wavelength), and the erbium-dopant-radius-to-core index ratio is varied within the interval [0.2; 1.0]. In Figure 6.2, the small-signal gain is shown against the threshold pump power defined as the lowest pump power needed to reach transparency in an infinitesimal signal in an infinitesimally short erbium-

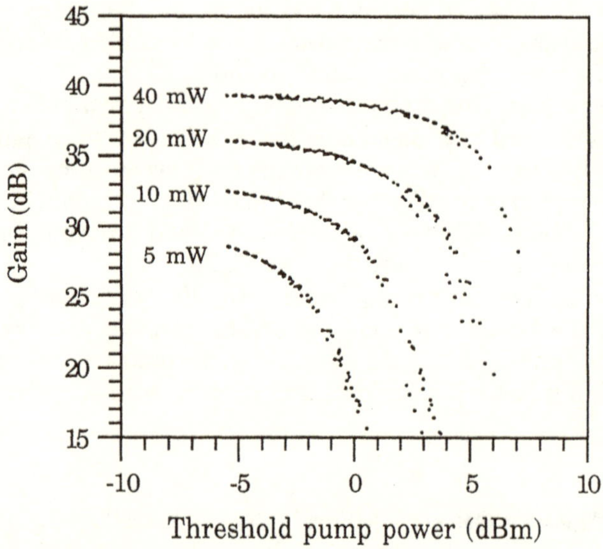

Figure 6.2 Gain at L_{opt} against threshold pump power for infinitesimal fiber length shown for different pump power values [1].

doped fiber. All gain values are presented for maximum gain lengths, L_{opt}, of the erbium-doped fiber.

The threshold pump power P_{th} depends only on the design parameters of the fiber and the absorption and emission cross sections of the Er^{3+} ions. Compared to the gain at L_{opt}, the threshold pump power has the advantage of being a fast calculable quantity because optimization of fiber length and iterative integrations of differential equations are avoided. From Figure 6.2, we conclude that the small-signal gain decreases with increasing threshold pump power. This means that gain optimization is independent of the actual pump power and can be performed by minimizing the threshold pump power. We will return to the application of this simplified strategy for fiber optimization in Section 6.3.1.

When the EDFA is used as an in-line amplifier or preamplifier, the signal input power is so low that the amplifier operates in the unsaturated region [1]. As seen above, the design goal, therefore, is to maximize the gain for a given pump power. We can also regard the optimization situation more from the point of view of the designer, who has a specific gain requirement for the amplifier which is to be developed. This means that we should rather formulate the task to minimize the pump power needed for a given gain. The latter design criterion is posed because a decrease in pump power will lead to an increase in the lifetime of the pump laser diodes. In order to characterize a specific EDFA's performance with respect to this design criterion, the gain coefficient is introduced (see Figure 6.3). This is defined for a given waveguide geometry and pumping scheme as the maximum ratio between small-signal gain in decibels and launched pump power in milliwatts.

In the optimizations performed in Section 6.3, the gain coefficient is found by iteratively changing the pump power, finding the fiber length that gives maximum small-signal gain, and plotting the gain-to-pump power ratio until the maximum ratio is reached.

It is important to note that the gain coefficient defined in Figure 6.3 only represents another optimization parameter, which, for the subject of design optimization, is equally valuable as the threshold pump power presented above. This means that there is no principal difference between small signal optimization of fiber amplifiers, whether it is performed either with respect to threshold pump power or with respect to the gain coefficient. The gain coefficient however has the advantage of providing more directly understandable information about the properties of a given erbium-doped fiber.

At this point, it becomes relevant to elaborate on the advantages of gain optimization a little further. Investigations of the pump power dependence of signal gain have shown that the signal gain improvement due to optimizing the fiber parameters decreases as the pump power is increased. Therefore, optimizing the fiber parameters of the fiber amplifier is most effective at low pump power levels.

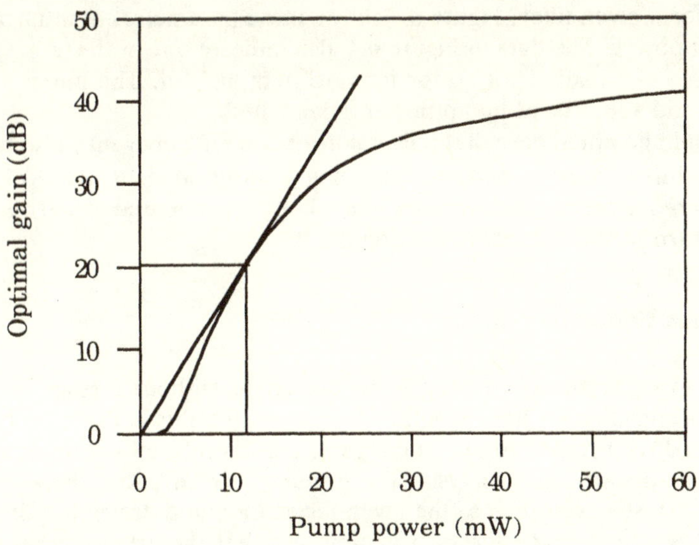

Figure 6.3 Gain calculated at L_{opt} as a function of pump power. The slope of the dashed line indicates the gain coefficient characteristic for the fiber design.

6.2.2 Large-Signal Amplifiers

The large-signal powers desired for power amplifier applications often require that two or more diode laser pumps be included in the amplifier package. The cost and complexity of such packages increase, therefore, for 980- and 1480-nm pumping, and the reliability decreases as the number of pump diode lasers increases. Therefore, it is very desirable that the pump-to-signal conversion efficiency in the fiber be made as large as possible. As a figure of merit for power amplifiers, the *quantum conversion efficiency* (QCE) is obvious. QCE is defined as the increase in the number of signal photons as a result of amplification divided by the number of launched pump photons.

The QCE can therefore be expressed as

$$\mathrm{QCE} = \frac{\lambda_s}{\lambda_p} \cdot \frac{(P_s^{\mathrm{out}} - P_s^{\mathrm{in}})}{P_p^{\mathrm{in}}} \tag{6.1}$$

where P_s^{in} is the signal input power, P_s^{out} is the signal output power, and P_p^{in} is the input pump power. λ_s and λ_p are the signal and pump wavelengths, respectively. When optimizing power booster amplifiers, the goal is to convert as many pump photons as possible into amplified signal photons. This may be equivalently expressed as maximizing the QCE of the fiber amplifier.

In order to illustrate the relationship between the QCE and the gain characteristics for a given fiber, Figure 6.4 shows these parameters as a function of the signal input power. The data in Figure 6.4 also indicate one of the largest observed QCE values (89%) yet reported for forward pumping [6]. The pump wavelength is 980 nm, and 138 mW of pump power is launched.

It should be noted here that power conversion efficiency may also be defined. In terms of this efficiency, 1480-nm pumping is more efficient, due to the smaller energy difference between pump and signal photons, compared to 980 nm pumping. We will return to this subject in Section 8.4.2.

6.2.3 Erbium Concentration

In the following sections, all results are presented without further indication of erbium concentration or fiber length. The reason for this is that all fiber lengths are optimized with regard to maximum gain, and therefore the erbium concentration will not appear in the analysis. The erbium concentration, therefore, may in all cases be chosen according to the given fabrication and design requirements. In saying this, we have made the tacit assumption that the erbium concentration is low enough to avoid the fluorescence quenching effect. This effect, which is due to cooperative upconversion [7], shows as a decrease of the fluorescence lifetime,

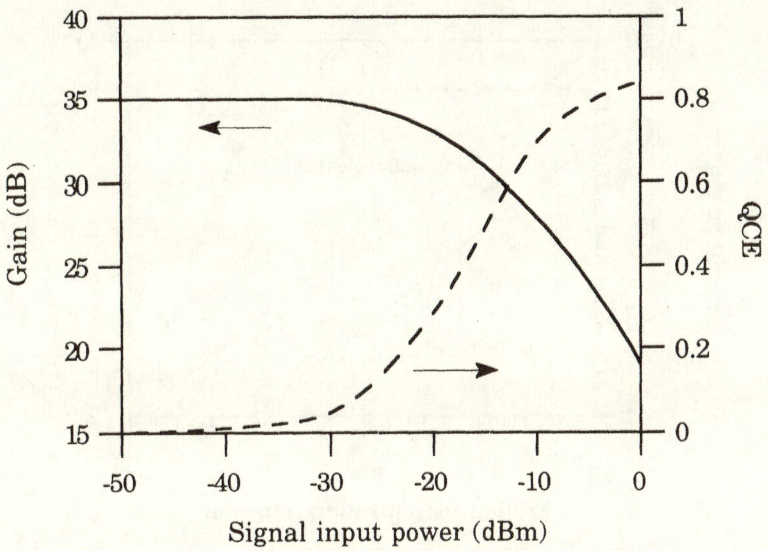

Figure 6.4 Signal gain and QCE versus signal input power (in dBm). 138-mW pump is launched, and the signal wavelength is 1551 nm. (Data from [6].)

and according to [8] it appears when the concentration is increased beyond $\rho_0 = 10^{25}\ m^{-3}$.

To quantify the fluorescence quenching effect and at the same time point out the problem of compatibility between standard and dispersion-shifted single-mode fibers and optimum erbium doped fibers, the description of [8] becomes valuable. This reference assumes that the correspondence between fluorescence lifetime and erbium concentration is typically given as illustrated in Figure 6.5 for all types of silica glasses. The fiber has a core radius $a = 3\ \mu m$ and a refractive index difference $\Delta n = 0.01$ (NA = 0.17), and the case of confinement down to 0.25 is presented, together with the situation where the full core area is doped.

For practical amplifier applications, the value $\rho_0 = 10^{25}\ m^{-3}$ is considered to be an upper limit for the erbium concentration, and for smaller concentrations ρ_0', the optimal length $L_{opt}(\rho_0')$ is 21given by the simple relation [8]

$$L_{opt}(\rho_0') = \frac{\rho_0}{\rho_0'} \cdot L_{opt}(\rho_0) \tag{6.2}$$

because the gain (in decibels) of the EDFA according to the description in Section 5.1 is proportional to the Er^{3+} concentration. The straight lines describing $L_{opt}(\rho)$

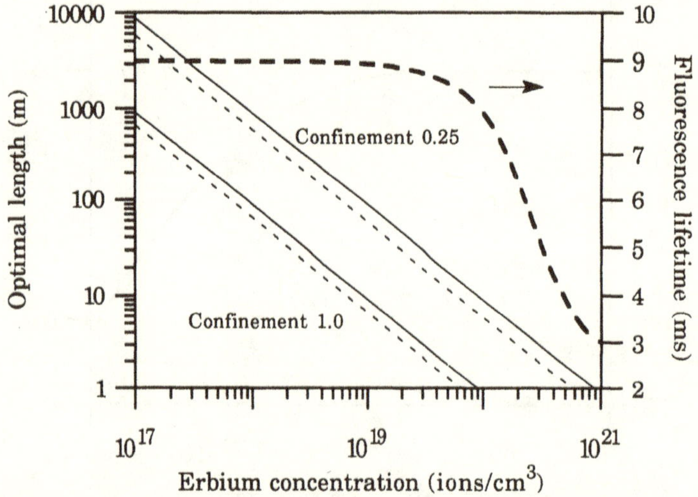

Figure 6.5 Optimal length L_{opt} as a function of Er^{3+} concentration. The pump power is 20 mW and solid lines correspond to 1470-nm pumping, whereas dashed lines correspond to 980-nm pumping. The fluorescence lifetime τ is shown versus the erbium concentration according to [8].

in Figure 6.5 are determined by the points corresponding to the optimal lengths $L_{opt}(\rho_0)$ given in [8] at $\rho_0 = 10^{25}$ m^{-3} and calculated with the linear law in (6.2).

From Figure 6.5 we find that the simultaneous conditions $\rho_0 \leq 10^{25}$ m^{-3} and the confinement factor down to 0.25 set a lower bound for the optimal length. The accurate value, which in this specific case is around 50 m, is, however, highly design dependent (see [8] for more examples). The specific fiber design used as a basis for the results in Figure 6.5 corresponds to a dispersion-shifted single-mode fiber design. Even though this is not equivalent to the most efficient active fibers discussed later in this chapter, it is used as an example here. The purpose is to point out the problem of incompatibility between the high-performance active fiber and the standard fibers in terms of increased splice loss. However, as addressed in Section 8.4.1, this problem can be partly overcome, but it might still guide the amplifier manufacturer towards a compromise between the highly efficient erbium-doped fibers and a fiber fully compatible with the passive fiber in the system. We will quantify this question further in the following section.

After this discussion of the upper limit of the erbium concentration, a short comment should also be given on the subject of a lower limit on ρ_0. Except for the description of the distributed amplifier in Chapter 11, we assume ρ_0 to be so large (and, as a result, the fibers so short) that the intrinsic fiber losses may be ignored within the fiber amplifier.

In the following analysis, the pump and signal are traveling in the same direction unless it is directly indicated that another pump configuration is applied.

A natural final comment to this section might be that the noise figure for many applications is as important a parameter as the gain. Therefore, it is important to notice that the preceding optimizations may also be seen as a means of improving the population inversion in the fiber amplifier whereby the noise properties are simultaneously improved. However, it is also important to point out that a fiber with gain optimal length is not operating for optimal noise performance, which is a subject we will return to in Chapter 10. Furthermore, it should be noted that amplifiers with noise figures close to the quantum limitation have been demonstrated for gain-optimized fibers (see Chapters 7 and 8).

6.3 IN-LINE AMPLIFIERS AND PREAMPLIFIERS

As described in Section 6.1, the fiber amplifier should either be operated in the small-signal regime or as power booster amplifier. We will initiate the detailed optimization of fiber amplifiers by focusing on 1480-nm-pumped in-line amplifiers or preamplifiers (i.e., small-signal amplifiers). The reason for starting the description with this pump wavelength is basically historical, and we start here in order to underline how important the development of pump sources has been in the development process of fiber amplifiers. The first high-performance LD-pumped amplifiers were demonstrated in the 1480-nm pump band. This was due to the existence of 1480-nm pump sources for Raman fiber amplifiers [9]. Later (in 1990), high-power laser sources at 980 nm were manufactured and offered commercially. A third possibility is 800-nm pumping, but even though this pump band compared to the 1480- or 980-nm band is fundamentally less attractive due to the existence of ESA, it might still find some use because of the low cost of 800-nm pump sources [10]. This section includes the optimization of erbium-doped fibers for all three pump bands, where laser diode pump sources can be supplied. However, it should be stressed that higher energy absorption can also be used (e.g., 665-nm pumping, as described in Chapter 3).

6.3.1 Optimal Fibers for 1480-nm-Pumped Amplifiers

In Section 6.2, we saw that gain optimization independent of the actual pump power is achieved by minimizing the threshold pump power. We will in this section make use of this property and, consequently, Figure 6.6 relates the threshold pump power, with the fiber design represented by the normalized frequency. Four groups of curves are seen corresponding to the four values of NA considered (0.10; 0.15; 0.20; 0.25). The lowest threshold pump power values are for the groups with the largest NA. Each group of curves consists of five curves distinguished by the erbium-

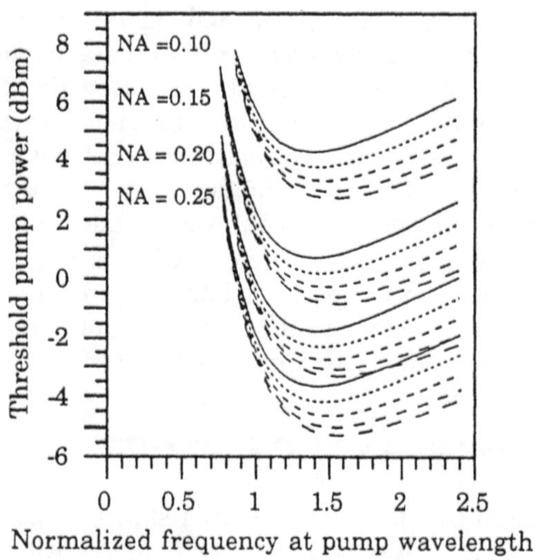

Figure 6.6 1480-nm threshold pump power shown against normalized frequency at 1480 nm for four NA values and five erbium-dopant-radius-to-index core radius ratios [1]. The solid corresponds to the fully erbium-doped core, $a_{dot}/a = 1.0$, and the dashed curves represent the results of reducing the confinement factor in equivalent steps of 0.2. $a_{dot}/a = 0.2$ corresponds to the curve indicated by the longest dash period.

dopant-radius-to-core index radius ratio (also called the confinement factor). Results for the five values of the confinement factor, a_{dot}/a (0.2; 0.4; 0.6; 0.8; 1.0), show that the lower confinement factor, the lower the threshold pump power. The curves have minima around $V = 1.4$ to 1.5, with the minima of $a_{dot}/a = 0.2$ slightly above the minima of $a_{dot}/a = 1.0$. The benefits in changing from $V = 2.4$ to $V = 1.4$ are comparable to changing a_{dot}/a from $a_{dot}/a = 1.0$ to $a_{dot}/a = 0.6$. Note that standard single-mode fibers are operated close to $V = 2.4$.

A fiber in the preform-to-fiber drawing process moves along the curves of Figure 6.6. From this figure, it follows that the highest fiber gain for a given preform is obtained when the preform-to-fiber reduction parameters are chosen, so a cutoff wavelength of 900 nm is the result, in addition to the conclusion drawn in [11], where it is shown experimentally that one should choose $V < 1.55$ ($\lambda_c < 950$ nm).

It is noteworthy that the above optimization has been performed solely with respect to maximum small-signal gain, and the question of compatibility with standard or dispersion shifted fibers has not been included so far. However, before we take up these compatibility questions, it is interesting to establish the correspondence between gain optimal active fibers as described above and fibers optimized with respect to the gain coefficient.

In order to illustrate this, the results in Figure 6.7 are valuable. Here, the pump efficiency is shown versus the cutoff wavelength λ_c for the LP_{11} mode in a step-index Al/Er fiber. Curves for NA = 0.1 to NA = 0.4 in steps of 0.05 are shown for 1480-nm pumping, and as in the previous cases, the signal wavelength is 1530 nm (corresponding to an Al/Er fiber). As seen from the figure, the pump efficiency for a given NA has a maximum with regard to cutoff wavelength, and this optimum cutoff wavelength is independent of the NA. In the present case of 1480-nm pumping, the optimum cutoff wavelength is 900 nm.

It is therefore clear that the optimum design of erbium doped fibers pumped at 1480 nm is the same both with respect to small-signal gain and with respect to maximum gain coefficient. Apart from compatibility concerns, the optimum fiber has a cutoff wavelength close to 900 nm and an NA as high as possible. However, in the choice of a proper EDFA design, the fiber resistivity against bending losses should be considered, as well as its coupling efficiency to the appropriate passive fibers. We will in this connection limit the description to dispersion-shifted fibers, since they currently represent the most likely choice for transmission medium for long-distance communication systems.

The fiber bending resistivity and coupling efficiency can be taken into account by constraining the signal spot size to spot size tolerances of dispersion-shifted fibers [12]. This leads us to our final examples of 1480-nm pumping as presented

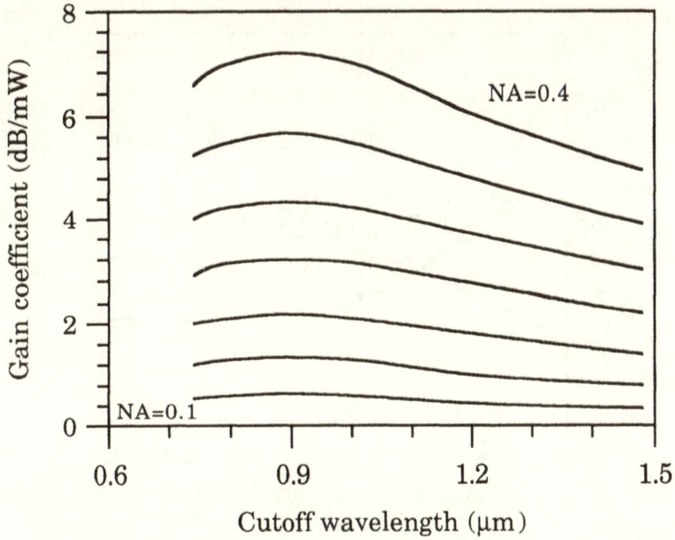

Figure 6.7 Gain coefficient shown against the cutoff wavelength for step-index Al/Er fibers pumped at 1480 nm [3]. Curves are shown for NA = 0.1 to 0.4 in steps of 0.05.

in Figure 6.8, where different designs are compared at a signal spot size value of 3.5 μm. From these curves, it can be seen that the threshold pump power decreases with increasing NA, decreasing normalized frequency V, and decreasing confinement factor a_{dot}/a.

However, since it appears from Figure 6.2 that the fiber gain saturates with small threshold pump powers, there is no advantage in decreasing the threshold pump power below the point of saturation (which depends on the pump power). Considering pump powers exceeding 10 mW, a threshold pump power at −2.0 dBm is found to be appropriate from Figure 6.2. Using these threshold values, the following constraints are found from Figure 6.8 for different choices of the confinement factor:

$$a_{dot}/a = 0.2: \quad NA > 0.18 \quad V < 2.4$$

$$a_{dot}/a = 1.0: \quad NA > 0.20 \quad V < 1.6$$

In this section, we have analyzed the design dependence of gain and gain coefficient for 1480-nm-pumped EDFAs. It was shown that for a fixed NA the fiber gives optimal performance when the cutoff wavelength is 900 nm. Restricting the designs to a spot size of 3.5 μm at the signal wavelength, the gain increases with increasing NA and, equivalently, decreasing normalized frequency. For a spot-size value of

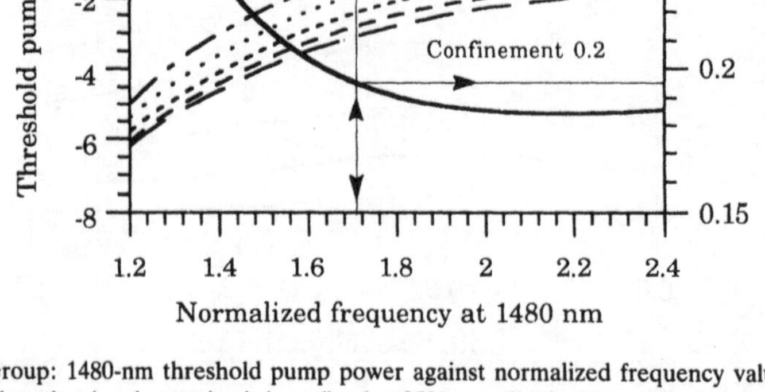

Figure 6.8 Group: 1480-nm threshold pump power against normalized frequency values at 1480 nm, when the signal spot size is kept fixed at 3500 nm. Single: NA against normalized frequency at 1480 nm for fixed signal spot size. (Data from [1].)

3.5 μm, it was shown that an optimum gain implies that the NA exceeds 0.18 when the erbium confinement factor is 0.2, and exceeds 0.2 when the erbium dopant radius and core index radius coincide.

6.3.2 Optimization of 980-nm-Pumped Erbium-Doped Fibers

With the appearance of commercially available 980-nm laser diodes (LDs) and an experimentally demonstrated gain coefficient of 11 dB/mW [13], it became obvious that this pump band was as relevant as the 1480-nm band. The attention was therefore also focused on the 980-nm pump band, and optimal designs as described in the following text were the result.

As was the case for 1480-nm pumping, we consider the signal wavelength of 1530 nm in all calculations for the Al/Er fibers (the wavelength coincides with the emission peak). The fluorescence lifetime has been measured to be 10.6 ms, and the absorption cross section at 980 nm was found from attenuation measurements to be 2.0×10^{-25} m^2 [3].

In Figure 6.9, the gain coefficient is shown versus the LP_{11} cutoff wavelength in step-index Al/Er fibers with erbium doped uniformly in the full core area. Curves for NA = 0.1 to NA = 0.4 are shown in steps of 0.05. The figure shows that for

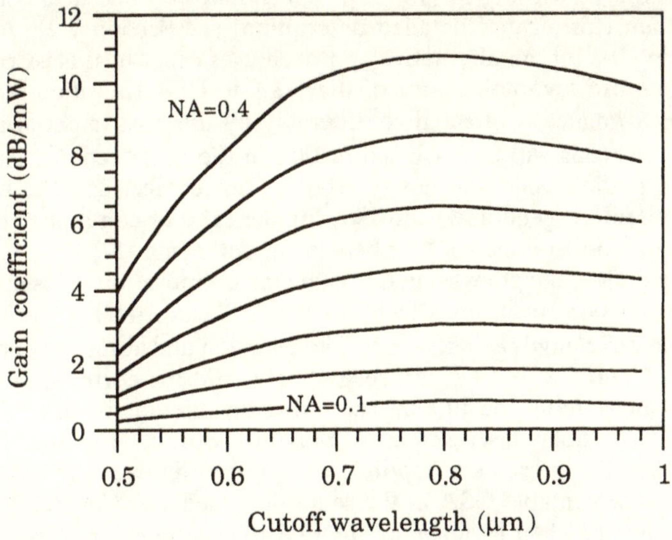

Figure 6.9 Gain coefficient versus cutoff wavelength for step-index Al/Er fibers pumped at 980 nm. Curves are shown for NA = 0.1 to 0.4 in steps of 0.05. (Data from [3].)

980-nm pumping, also, the gain coefficient for a given NA will have a maximum with regard to cutoff wavelength, and the optimum cutoff wavelength is 800 nm independent of the NA.

Besides the optimum gain performance discussed above, design-dependent characteristics such as noise figure or the required pump power to achieve a given gain is of substantial interest. These results, however, become even more illuminating when they are compared with results at other pump wavelengths and when the influence of an index-raising co-dopant is included. Additional description is therefore moved to Section 6.3.4.

6.3.3 800-nm-Pumped Amplifiers—An Optimal Choice?

As illustrated above, EDFAs exhibit high performance and efficiency when pumped at 980 or 1480 nm. However, practical considerations dictate that EDFAs should be pumped by reliable high-power laser diodes, which for cost reasons preferably would be 800-nm pump sources. For Er^{3+}, however, 800-nm pumping leads to inefficient operation due to the low strength of the GSA transition and a considerable ESA band, which overlaps the GSA band (see Chapter 3 for further description of these properties).

The means of achieving high gain performances in spite of these problems has been to avoid the ESA band through pumping in the long-wavelength part of the pump band. According to [14, 15], this should be done at a wavelength of around 820 nm (for a more detailed description, see Section 7.2). As described thoroughly in [10, 16], an alternative approach uses other host glass compositions that provide more favorable ratios of the GSA-to-ESA transition strength. The amplifier performance is affected considerably by the host material surrounding the erbium ions (this will be discussed further in the next section). However, the optimal waveguide design will not depend on this choice, and the optimization results are therefore generally valid. So, for the sake of continuity, we will only consider aluminum-co-doped silica fibers in the following [17].

The signal wavelength was, in the following optimization, chosen to lie at the peak of the emission spectrum, which, for the Al/P silica fiber referred to here, is located at the wavelength $\lambda = 1532$ nm. The emission and absorption cross sections used were 5.5 and 5.2×10^{-25} m^2, respectively, and the corresponding radiative lifetime was measured to be 10.8 ms (note that this lifetime is slightly higher than that for the Al/Er fibers discussed in the previous section). The fiber was pumped at 815 nm, which represents an optimal compromise between reduced ESA and maintaining an acceptable GSA at the same time (see also Chapter 3).

The gain coefficient is shown in Figure 6.10 as a function of the cutoff wavelength and with the NA as parameter. The results are displayed using dashed lines for co-directional pumping and solid lines for bi-directional pumping. In the latter

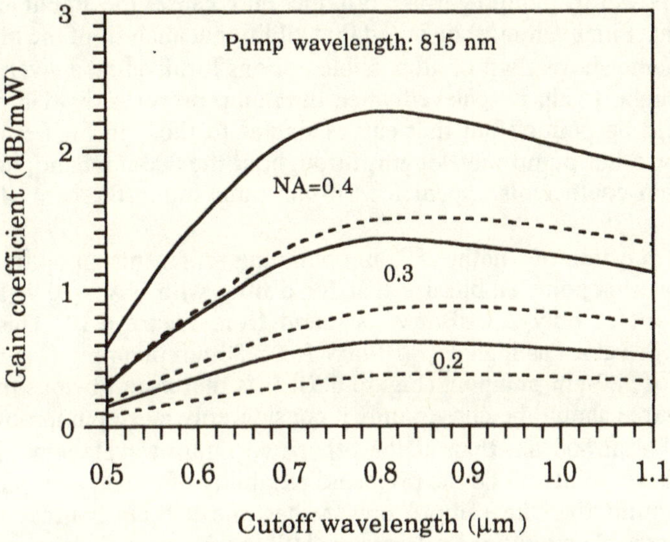

Figure 6.10 Gain coefficient versus cutoff wavelength for step-index Al/P fibers with indicated NAs and with erbium doped uniformly throughout the core. The dashed curves correspond to co-directional and the solid curves to bi-directional pumping. (Data from [17].)

case, equal amounts of co-propagating and counter-propagating pump light are employed.

As can be seen from the figure, the optimal choice of fiber cutoff wavelength is close to 800 nm for the bi-directionally pumped fiber amplifiers and around 850 nm for the co-directional pumping scheme. In the latter, the difference between fibers with cutoff wavelengths at 800 or 850 nm is, however, only a few percent and is therefore not considered to have much significance. Furthermore, it will be advantageous to ensure single-mode operation of the fibers at the pump wavelength, and consequently a cutoff wavelength of λ_c = 800 nm is suggested in all cases. It should be noted that a physical explanation of the difference between bi-directional and co-directional pumping is found in the competition between useful GSA and useless ESA combined with the mode field overlap differences between the two designs.

It is more important to note that for equal NAs the optimal gain coefficient is only 65% as large in the case of co-directional pumping as the value for bi-directional pumping. This difference is due to the ESA, and may be explained by considering the situation in which the pump power is high enough to reach inversion. In this case, a significant fraction of any additional pump power will be attenuated by ESA, and it will therefore be of no use further on in the fiber. Therefore, splitting the power by pumping from both fiber ends causes the amplifier to become more efficient. Finally, it must be noted that additional analysis of the bi-directional pumping scheme shows that, of all possible options for dividing a given total pump power, the highest gain is achieved when the pump power is divided equally.

It should be pointed out that curves similar to those in Figure 6.10 may be calculated for other pump wavelengths throughout the 800-nm band; however, the maximum gain coefficients appear for 815-nm pumping in the case of Al/P silica host glass.

To the question of whether 800-nm pumping represents an optimal choice at all, we can at this point emphasize that for a fiber with NA = 0.4, a maximum gain coefficient of only 2.3 dB/mW is found from Figure 6.10. This has to be compared with values as high as 7 dB/mW for 1480-nm pumping (Figure 6.7) and 11 dB/mW for 980-nm pumping (Figure 6.9). It is therefore obvious that the 800-nm pump source should be chosen only if considerably more pump power may be made available at 800 nm than at the other two pump wavelengths. In the long run, the decisive point will be the price and reliability of the 800-nm pump sources supplying around five times more power than the 980-nm sources required to achieve comparable amplifier performance [18]. In the next chapter, we will elaborate on this point. It should only be noted here that even though 800-nm pumping is less efficient than 980- or 1480-nm pumping, it still gives a very useful amplifier.

6.3.4 Influence of Host Glass Composition

The description in this chapter has so far concentrated on the optimal choice of NA and cutoff wavelength for step index fibers. This optimization has been illustrated solely for aluminum co-doped fibers, which is sufficient because the optimum waveguide designs are independent of the refractive index-modifying co-dopant used in silica glass fibers. The fact that the optimum waveguide parameters are the same for different index-raising co-dopants does not, however, mean that the actual choice of co-doping material is unimportant. The reason for this is that the specific properties of the differently co-doped fibers vary considerably for various combinations of host materials.

Even though we will return to the performance possibilities of different fiber amplifiers in more detail in Chapter 7, the specific performance differences are still relevant in relation to fiber design parameters. In this section, therefore, we will describe how fibers with optimum cutoff wavelengths perform for different choices of index-raising co-dopant material.

The figure of merit for small-signal amplifiers, as described above, is the gain coefficient, which we presented for aluminum co-doped fibers in Sections 6.3.1 and 6.3.2. As discussed in Chapter 3, on the other hand, germanium co-doping will result in narrower emission and absorption spectra than the Al/Er fiber. The spectral emission peak in Ge/Er fibers will also be moved towards higher wavelengths compared to the Al/Er fiber. Furthermore, the spectra for Al/Ge/Er fibers will be combinations of the spectra from the Al/Er fiber and the spectra from the Ge/Er fiber (see figures in Section 3.1.1). In order to demonstrate how these spectral differences influence the maximum gain coefficient values, analysis has been performed for three fibers [3–5]. The first fiber has aluminum as an index-raising co-dopant, the second fiber has both aluminum and germanium as co-dopants, and the third fiber is co-doped with germanium only. The fluorescence lifetimes for the three fibers were measured at 10.6, 11.4, and 12.5 ms, respectively.

The gain coefficients (increasing curves) at the optimum cutoff wavelength and the pump power needed to obtain the gain coefficients are shown in Figure 6.11 as a function of the NA. Curves are presented for Al/Er fibers (solid), Al/Ge/Er fibers (small dashes), and Ge/Er fibers (long dashes). The Al/Ge/Er fiber was fabricated with an aluminum concentration was so low (the Al:Ge ratio below 1:40) that the spectra are significantly affected by the presence of germanium (see also Section 3.1.2). The results for 1480-nm pumping are shown in Figure 6.11 (a), and the results for 980 nm pumping are shown in Figure 6.11 (b). As signal wavelengths are used, the wavelengths at the emission peaks. As can be seen, the highest gain coefficient is obtained when pumping the Ge/Er fiber at 980 nm, where a gain coefficient as high as 23 dB/mW is predicted for an NA of 0.4. This is due to the higher absorption cross section at 980 nm, as well as the higher emission peak at the signal wavelength. The absorption cross section at 980 nm in Ge/Er fibers is more than twice the value in Al/Er fibers. This explains, together with the 60% higher emission peak in Ge/Er fibers, why the maximum gain coefficient is up to 12 dB/mW higher than in Al/Er fibers when pumping at 980 nm.

For Al/Er fibers, the absorption cross section at 980 nm is almost equal to the absorption cross section at 1480 nm. However, when pumping at 1480 nm, the maximum obtainable inversion in the upper laser level is 0.7. For 980-nm pumping, a total inversion is possible. This explains why the gain coefficient for 980-nm pumping is higher than that for 1480-nm pumping.

Experimentally obtained gain coefficients of 11.0 dB/mW (980 nm) and 5.0 dB/mW (1480 nm) has previously been reported for a Ge/Er fiber with an NA of 0.28 [13]. These values agree very well with the values 11.8 dB/mW for 980-nm pumping and 4.3 dB/mW for 1480-nm pumping, which were predicted as shown in Figure 6.11.

For the three fibers investigated, the maximum obtainable inversion in the upper laser level, $n_2^{max} = \sigma_{12}/(\sigma_{12} + \sigma_{21})$, at 1480 nm is almost equal, $n_2^{max} \approx 0.71$. When pumping at 1480 nm (see Figure 6.11), the choice of index-raising co-dopant is not very significant for the gain coefficient. This is because the absorption cross section at 1480 nm increases when the influence of germanium on the Er^{3+} ions decreases. With a fixed n_2^{max}, this increasing absorption cross section compensates for the decreasing emission peak.

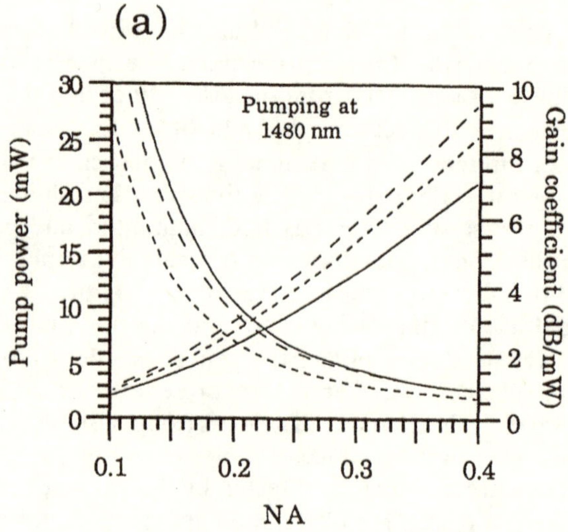

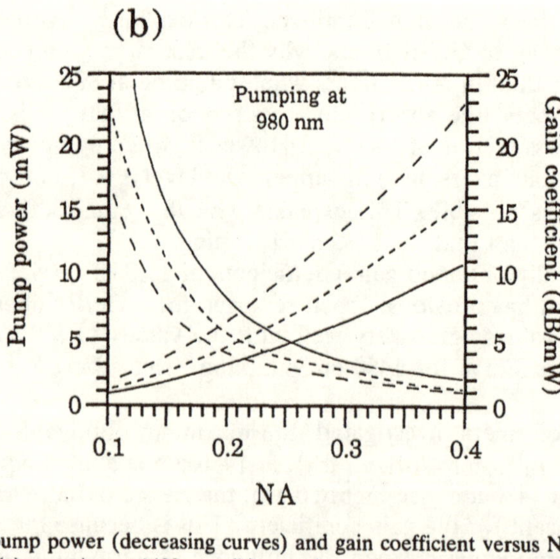

Figure 6.11 Input pump power (decreasing curves) and gain coefficient versus NA in step-index Er^{3+} fibers: (a) pumping at 1480 nm; (b) pumping at 980 nm. The index-raising co-dopants are aluminum (solid), aluminum and germanium (small dashes), and germanium (long dashes). (Figure from [4].)

The curves in Figure 6.11 yield that the NA should be as high as possible, and the reason for this is that the overlap integrals between the erbium profile and the two LP modes increase with the NA. An increasing NA leads to a decreasing spot size. In a practical system, this means that the coupling between the active fiber and the passive fiber will decrease with increasing NA (see Section 6.3.1). However, the above mentioned overlap integrals may be increased even for a fixed NA, and the gain coefficient may therefore still be improved simply by confining the erbium doping radius. Figure 6.12 shows the gain coefficient versus the confinement. The NA is 0.25 and curves are shown for step index Al/Er fibers (solid), Al/Ge/Er fibers (small dashes), and Ge/Er fibers (long dashes). Figure 6.12 shows results for fibers pumped at 980 nm as well as at 1480 nm (marks) for fibers with the three different compositions. As can be seen from the figure, the improved gain coefficient is between 40% and 45% in all cases when decreasing the confinement from 1 to 0.1. In absolute numbers, this means that the improvement can be up to 4.3 dB/mW for the Ge/Er fiber pumped at 980 nm.

To complete the discussion of erbium confinement, it should be noted that in [19] it is concluded that for 1480-nm pumping the optimum cutoff wavelength increases as the erbium becomes more confined in the fiber. Therefore, the fiber

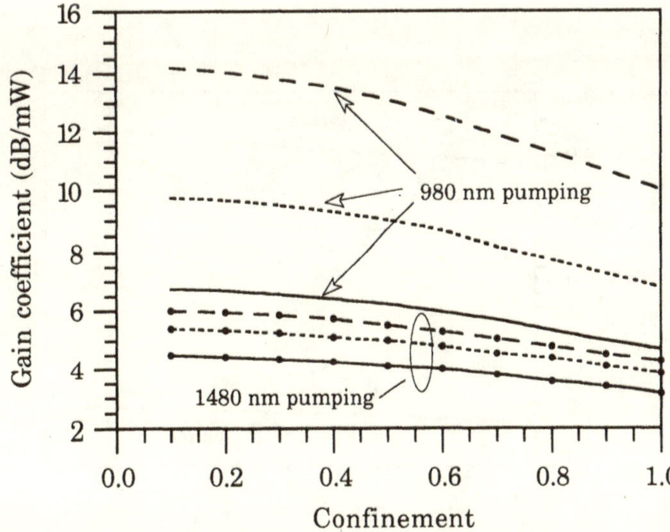

Figure 6.12 Gain coefficient versus erbium confinement in step-index Er^{3+} fibers with an NA of 0.25. Marks: pumping at 1480 nm. No marks: pumping at 980 nm. Solid curves: index raising co-dopant is aluminum. Small dashes: index-raising co-dopants are aluminum and germanium. Long dashes: index raising co-dopant is germanium. (Figure from [4].)

parameters should be chosen in proportion to the Er^{3+}-doped area in order to obtain high signal gain. According to [19], more specifically, this means that for highly confined erbium fibers (with confinement factors below 0.25), the optimum cutoff wavelength should be increased by up to 100 nm.

In the treatment of optimal fibers, we have so far only concentrated on the gain performance. It is, however, essential to also illustrate the noise figures for the pump efficiencies presented in Figure 6.11. These are shown in Figure 6.13 versus the NA. Only fibers co-doped with either aluminum or germanium are considered. As seen from Figure 6.13, the Al/Er fiber and the Ge/Er fiber have almost identical noise figures when pumping at 980 nm. For 1480-nm pumping, the noise figure for the Ge/Er fiber is 1.8 dB higher than that for Al/Er fiber. Furthermore, due to non-ideal population inversion in the 1480-nm-pumped fibers, the noise figures are considerably larger than those for 980 nm pumping.

The results presented above suggest the conclusion that the 980-nm-pumped Ge/Er fiber should be used as an in-line amplifier and preamplifier. Note, however, that this conclusion may be affected if the gain bandwidth is of importance for the amplifier application. In this case, the larger bandwidth of the aluminum-co-doped fibers may make this the preferred choice. Provided there is 1480-nm pumping, Figure 6.13 suggests that aluminum should be chosen as co-dopant.

Before we leave the subject of small-signal gain optimization of erbium-doped fibers, it is of special importance also to discuss the host glass dependence of the

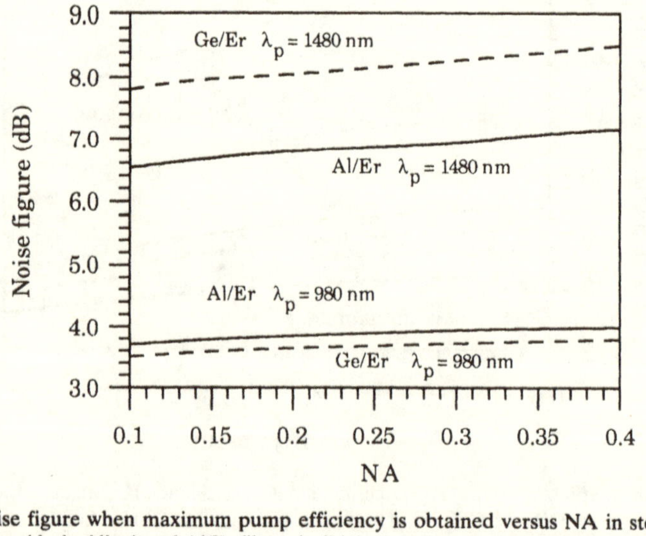

Figure 6.13 Noise figure when maximum pump efficiency is obtained versus NA in step-index Ge/Er fibers (dashed line) and Al/Er fibers (solid line), pumped at 1480 nm and at 980 nm. (Figure from [5].)

gain for 800-nm-pumped amplifiers. The main reason is that not only is the 800-nm GSA band affected by the host glass composition, but the overlapping and relative strong ESA transition will also be spectrally modified by a change of host glass. It must also be noted that both the GSA and the ESA cross sections are strong functions of the pump wavelength, and we will return to this subject in more detail in Chapter 7.

At this point, it is relevant to understand that high gains in the 800-nm-pumped EDFAs are obtained [15, 20] by pumping in the long-wavelength region of the GSA band, where the ESA is less significant. A central wish would therefore be to decrease the ESA at the peak of the GSA because the efficiency of the amplifier would improve, if it were possible to move the pump wavelength to a value with a higher GSA cross section.

As we shall see in the following, a step in the direction of amplifier improvement for 800-nm pumping will be achieved by exchanging the index-raising codopant in silica fibers. However, in the discussion of 800-nm pumping, it is also relevant to include host glasses other than silica, since the GSA and ESA cross-section ratio is shown [21] to improve significantly. More specifically, we will refer in this context to results on Al/P silica, Ge/P silica, and fluorophosphate. As described in [17], the latter may be manufactured with low fluorine content and is one of the most favorable compositions with respect to ESA known today.

As we saw in Section 6.3.3, an optimal erbium-doped fiber for 800-nm pumping will have a cutoff wavelength close to λ_c = 800 nm, and we have consequently chosen designs of this type in the preceding analysis. For such fibers, Figure 6.14(a) shows the gain coefficient as a function of NA. The full core areas of the step-index fibers are uniformly doped with erbium, and results are shown for an Al/P silica fiber pumped at 815 nm, for a Ge/P silica fiber pumped at 810 nm, and for a fluorophosphate fiber pumped at 805 nm. Results are displayed for both co-directional (dashed curves) and bi-directional (solid curves) pumping. The emission and absorption cross sections applied in the calculations were, respectively, 5.5×10^{-25} and 5.2×10^{-25} m^2 at 1532 nm for the Al/P silica, 8.6×10^{-25} and 8.9×10^{-25} m^2 at 1536 nm for the Ge/P silica, and 7.5×10^{-25} and 6.8×10^{-25} m^2 at 1533 nm for the fluorophosphate. The corresponding radiative lifetimes were reported in [16] as 10.8, 11.0, and 8.25 ms, respectively.

As can be seen from Figure 6.14(a), the fluorophosphate fiber is predicted to give the highest gain coefficient, 3.5 dB/mW, for an NA of 0.4. This is because both the GSA cross section σ_{13} and the ratio between GSA and ESA cross sections σ_{13}/σ_{24} are higher for this composition. However, the Ge/P silica fiber has a gain coefficient that is only slightly lower. The reason is that the emission cross section at the signal wavelength is higher for the Ge/P silica and the spontaneous emission rate is 25% lower, resulting in a smaller threshold pump power compared to the Al/P silica fiber. As was seen in Section 6.3.3, the bi-directional pumping scheme improves the gain coefficient by up to 50%.

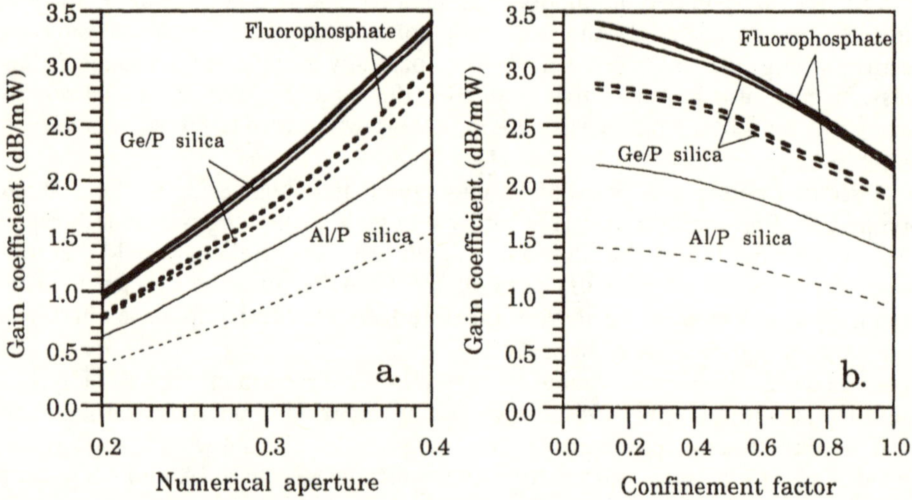

Figure 6.14 (a) Gain coefficient as a function of NA for different host glass compositions. The cutoff wavelength is 800 nm. (b) Gain coefficient shown versus the confinement factor in step-index fibers with an NA of 0.3. Results are presented for bi-directional (solid curves) and co-directional (dashed curves) pumping. (Data from [16, 17].)

We will finally direct our attention towards Figure 6.14(b), which is included here in order to describe how the gain coefficient for different glass host compositions is influenced by the erbium confinement factor. For this purpose, a step-index fiber with NA = 0.3 and a cutoff wavelength of 800 nm is chosen, and the values for signal and pump wavelengths are as described above. Figure 6.14(b) shows that the gain coefficient is improved by more than 50% when the confinement factor is decreased from 1.0 to 0.1. It reaches values close to the maxima found for step-index fibers with an NA of 0.4 and a confinement factor of 1.0. For practical applications, this means that the gain coefficient can be improved without affecting the inherent splice loss between the amplifier fiber and a standard telecommunications fiber. A primary impact of a change in the gain coefficient is on launched pump power and not on the gain. Thus, by decreasing the confinement factor, a reduction of the required pump power for a given gain value will result, and will be in inverse proportion to the increasing gain coefficient. In contrast, when improving the gain coefficient by increasing the NA, the improvement in the ratio of the gain to launched pump power will be countered by the increasing splice loss. As was the case with the 1480-nm pumped amplifiers discussed in Section 6.3.1, it should be noted that when the confinement factor is decreased from 1.0 to 0.1, the optimum cutoff wavelength is reported to increase [16]. In this case, pumping in the 800-nm band, the optimum cutoff wavelength should be changed from 800 to 920 nm for bi-directional pumping and from 830 nm to 950 nm for co-directional pumping.

In the case of 800-nm pumping, three glass host compositions, Al/P silica, Ge/P silica, and fluorophosphate were discussed. It was found that the highest gain coefficient, 3.5 dB/mW, can be obtained for a fluorophosphate fiber with an NA of 0.4.

6.4 POWER BOOSTER AMPLIFIERS

There has been an increased recognition of the importance of power amplifiers, not only as signal boosters in long-haul transmission links but also as power amplifiers for optical distribution systems. As described in [22], these devices are useful for systems with large amounts of splits, as well as for other systems requiring large optical power levels (e.g., WDM, FDM, and AM-VSB systems) [23–25]. The preferred choice of high-power amplifiers is LD-pumped erbium doped fiber. This statement is based on the fact that Er^{3+}doped silica fibers have excellent properties under highly saturated conditions (e.g., large signal output powers, low signal distortion and crosstalk at modulation frequencies of interest, and a broad, flat gain distribution) [23,26,27]. Even though the commercially available diode laser pump sources are manufactured with increasingly large output powers, the applications and reliability still often require two or more pump lasers in the amplifier package. Furthermore, it is strongly desired that the pump-to-signal conversion efficiency in the fiber be as large as possible, and, consequently, the QCE defined in Section 6.2 should be maximized.

6.4.1 980- and 1480-nm-Pumped Amplifiers

The booster amplifier optimization is first considered for the 980- and 1480-nm pump bands, since they represent the two most obvious pump choices. This choice is motivated by the problems of ESA in the 800-nm band and the additional lower power conversion due to larger differences in photon energies between pump and signal. The goal is to maximize the fraction of launched pump photons that are converted into signal photons. In order to accomplish this, we must first understand how the pump photons are lost through competing processes of either generation and amplification of spontaneous emission or residual pump photons exiting the amplifier. The latter is reduced by improving the overlap between the pump mode and the Er^{3+} distribution, which corresponds to minimizing the threshold pump power (see Section 6.2). The spontaneous emission rate, which is proportional to the number of excited erbium ions, can, on the other hand, be reduced by increasing the rate of stimulated emission into the signal mode by maximizing the overlap of the signal mode and the profile of the excited ions. This would lead to a highly signal-saturated amplifier in which the number of ASE photons can be ignored compared to the total number of lost pump photons.

However, it is generally not possible to simultaneously optimize a fiber with respect to both the described processes. The optimal choice, therefore, as shown in [6], will be a compromise between lowering the pump threshold and maximizing the stimulated emission rate. With the aim of finding this compromise, we have shown in Figure 6.15 the QCE as a function of the cutoff wavelength for step-index fibers with erbium uniformly doped throughout the core. The pump power is chosen to be 50 mW and is launched in the same direction as the signal, which has a power of 1 mW at the amplifier input end. For each point on the curve, these co-directionally pumped amplifiers have been optimized by adjusting the fiber length so that maximum QCE was obtained. The considered fibers were Al/Er silica fibers, and results are shown for different NA values and for signal wavelengths at the emission peak (1532 nm) and at 1551 nm.

The optimum cutoff wavelength is observed to be close to 800 nm for 980-nm pumping (see Figure 6.15(a)) and approximately 900 nm for 1480-nm pumping (see Figure 6.15(b)). With pumping of 980 nm, the QCE is observed to be slightly higher for the signal wavelength of 1532 nm compared to 1551 nm. This is because

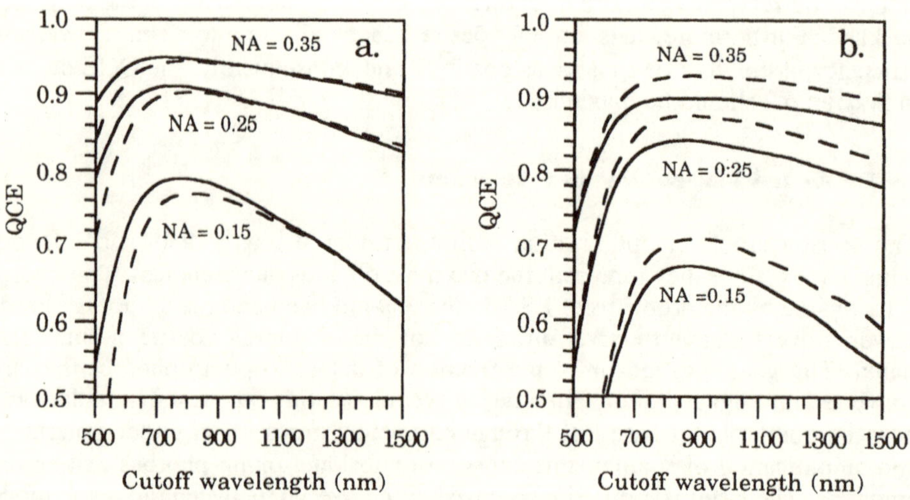

Figure 6.15 QCE shown as a function of the cutoff wavelength for the LP_{11} mode in step-index fibers with erbium doped throughout the core. Results are shown for both co-directional pumping (solid) and bi-directional pumping (dashed). (Figure from [6].) (a) The fibers are pumped at 980 nm. (b) 1480-nm pumping is applied.

a relatively high average inversion can still be obtained, which favors wavelengths with higher cross sections. In contrast to this, 1480-nm-pumped amplifiers will show larger QCE at 1551 nm than at the signal wavelength of 1532 nm. The reason is that 1480-nm pumping limits the obtainable inversion because of the nonnegligible emission cross section. At this lower inversion, the ratio of the emission to the absorption cross section has a significant influence on the amplifier performance, and since the ratio is lower at 1532 nm than at 1551 nm, the latter shows the best QCE.

It is very important to note in Figure 6.15 that the maxima are not very pronounced, and minor deviations from the optimum design parameters will not result in seriously deteriorated power booster performance. A property just as important as the optimum fiber design, however, will be the booster efficiency dependence of the available pump power. To illustrate this, we have included Figure 6.16, which shows the booster efficiency (or QCE) versus pump power for step index EDFAs with a signal input power of 1 mW. Results for 980-nm (Figure 6.16(a)) and for 1480-nm (Figure 6.16(b)) pumping of Ge/Er fibers (dashed curves), as well as Al/Er fibers (solid curves), are presented for NA = 0.1, 0.2, 0.3, and 0.4. Corresponding with previously discussed results, the figure shows that the highest NA results in the highest efficiency. Figure 6.16(a) shows that the Ge/Er fiber is more efficient than the Al/Er fiber when pumping at 980 nm. For 1480-nm pumping (see Figure 6.16(b)), the booster efficiency for the Al/Er fiber is higher than the booster efficiency for the Ge/Er fiber. However, the difference is not significant, and we can therefore conclude that the choice of co-dopant is not important when pumping at 1480 nm.

It should also be noted that later investigations on whether to choose aluminum or germanium as co-dopant in power booster amplifiers do not favor germanium co-doping. This further emphasizes that the choice of co-dopant is more significant for small-signal amplifiers than for booster amplifiers.

Besides the concern for how to choose the best index-raising co-dopant for power booster amplifiers such as that addressed above, a number of additional questions are of interest. This includes the choice of signal wavelength, the influence of erbium confinement, and the noise performance of the power amplifiers. Note that the latter is of interest for applications in television signal distribution systems employing analog modulation schemes [28]. One of the best sources of finding the answer to these questions is [6], and we will therefore briefly describe the major conclusions here.

In order to do so, we have included Figure 6.17, which illustrates both QCE and noise figure F as a function of the pump power. As in the previous situation, curves were shown for different NA values, but here the solid curves are for fibers with erbium doped throughout the core and the dashed curves are for fibers with an erbium confinement of 0.5. Both QCE and noise figure are improved with increasing pump power and can be further enhanced by increasing the NA or

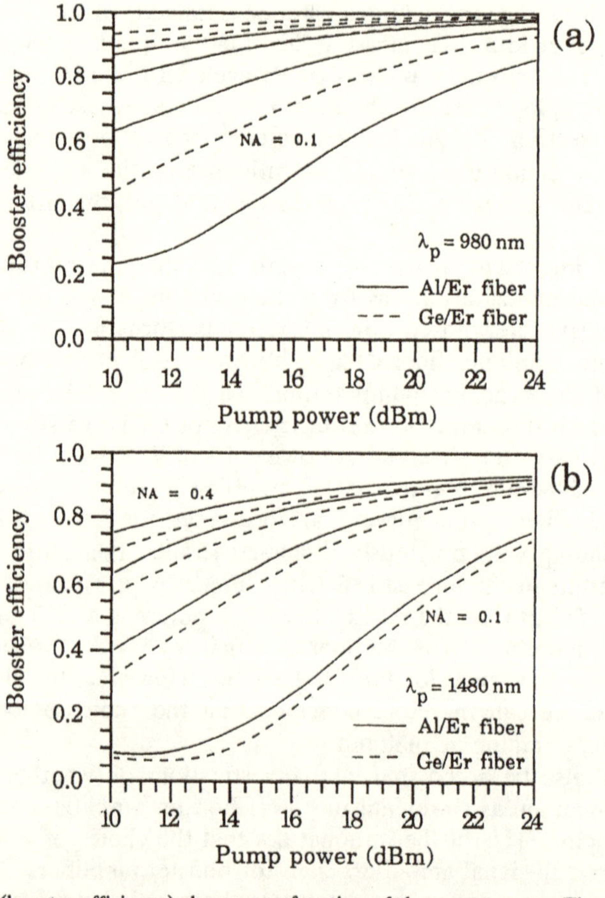

Figure 6.16 QCE (booster efficiency) shown as a function of the pump power. The considered fibers are step index Ge/Er fibers (dashed) and Al/Er fibers (solid). The NA is changed from 0.1 to 0.4 in steps of 0.1. (Figure from [5].) (a) The fibers are pumped at 980 nm. (b) 1480-nm pumping is applied.

confining the erbium within a smaller radius. If realistic pump power levels between 25 and 100 mW are considered, the quantum slope efficiency and thereby the signal output power can be improved by up to 60% by increasing the NA from 0.15 to 0.25. A change in the erbium confinement factor from 1.0 to 0.5 will provide an enhancement of up to 20% for the same pump power levels.

On the subject of noise performance, it must be noted that 1480-nm pumping produces up to 3 dB higher noise figures than 980-nm pumping. This is due to the lower average inversion for 1480-nm resonant pumping. Figure 6.17 also indicates

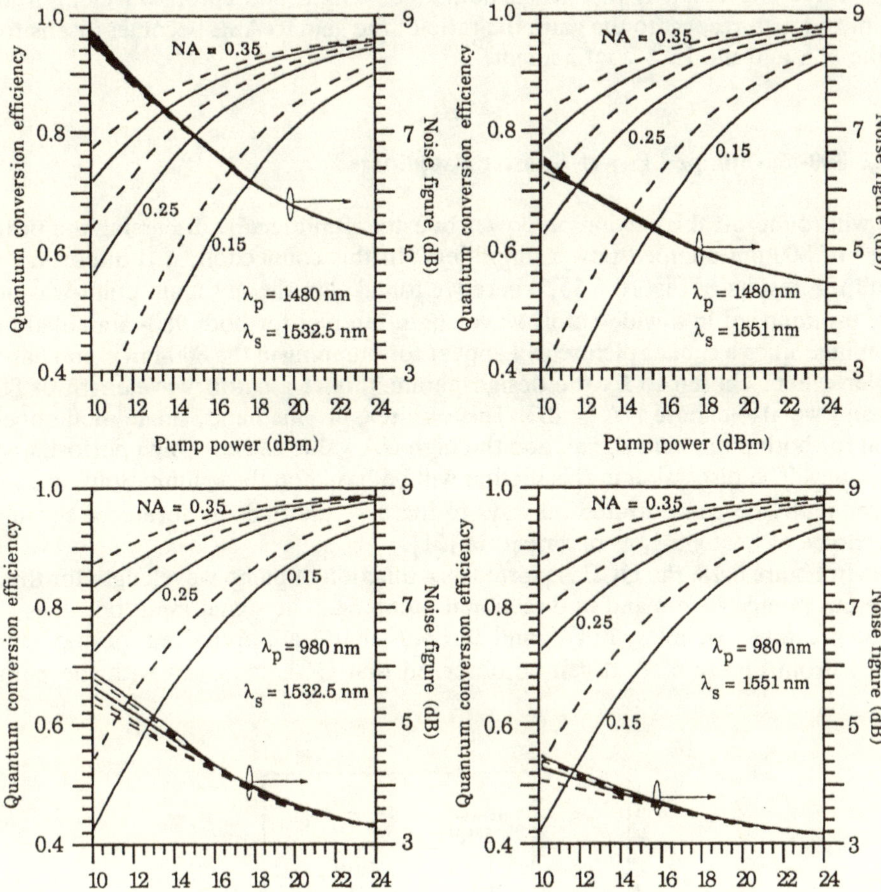

Figure 6.17 QCE and noise figures versus pump power. (Figure from [6].) Solid curves are for erbium doped throughout the core with the optimum cutoff wavelengths determined from Figure 6.15. Dashed curves are for erbium doping radii of half the core radius with 50 nm higher cutoff wavelength.

that the noise figure for a signal wavelength of 1532 nm is up to 2 dB higher than for one at 1551 nm due to the difference in the cross-section ratio. Figure 6.17 further shows that neither NA nor erbium confinement has a significant influence on the noise figure. The reason is that both the gain and the spontaneous emission captured in the fiber are affected by the NA and the erbium confinement in the same way. The result is that when both fiber length and cutoff wavelength are optimized with respect to the gain, the ratio of the gain to ASE becomes insensitive to the NA and the Er^{3+} confinement.

6.4.2 800-nm-Pumped Power Booster Amplifiers

We will round off this section on power booster amplifiers by discussing the properties of 800-nm-pumped power amplifiers. In this connection, it is important to recall the results of Figure 6.15, where we found that the optimum point was not very pronounced in a wide cutoff wavelength interval for both 980- and 1480-nm pumping. Since a similar picture will appear for pumping in the 800-nm pump band, we have here chosen to fix the design parameters at a cutoff wavelength of 800 nm and we also choose NA = 0.3. This ensures, on one hand, single-mode operation for both pump and signal, and the high NA value causes a high performance efficiency. The discussion in this section will be based on these limitations and will subsequently be concentrated on ways to increase the fiber performance through the choice of host glass, as described in [21].

In Figure 6.18, the QCE is plotted as a function of pump wavelength for three different pump powers and bi-directional pumping. The signal input power is, as in the previous examples, 1 mW, and all fiber lengths are optimized for maximum QCE. From Figure 6.18, it can be observed that QCE increases with the pump

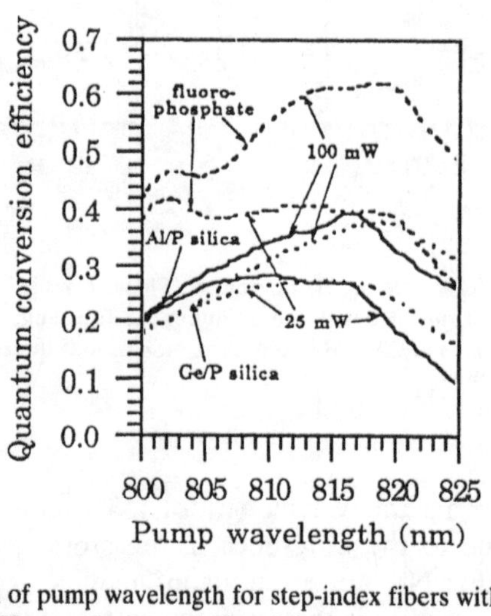

Figure 6.18 QCE as function of pump wavelength for step-index fibers with an NA of 0.3 and a cutoff wavelength of 800 nm. Curves, which originate from [16], are shown for bi-directional pumping with different pump powers.

power. This agrees with the performance in other pump bands, but the similarity ends here due to the existence of ESA in the 800-nm pump band. For high pump power levels, QCE is limited by the ratio $\sigma_{13}/(\sigma_{13} + \sigma_{24})$, and, consequently, the highest values are reached in the long-wavelength tail of the GSA band (see Chapter 3). A maximum QCE of 63% is predicted for the fluorophosphate fiber pumped with 100 mW. This is 60% higher than the maximum value for the Al/P silica fiber.

6.5 INDEX PROFILES DIFFERENT FROM THE STEP-INDEX DESIGN

The optimal fibers presented in the previous sections of this chapter have all been designed with step-index cores with radius a. The Er^{3+} ions have been considered homogeneously distributed within a radius $a_{dot} \leq a$. The principal reason for this is that the step-index active fiber is not only simple, but it also represents the optimal choice of index profile as far as the properties defined in Section 6.2 are concerned.

However, as a result of the limitations of the manufacturing processes, the refractive index profiles will always deviate from the step-index profiles. This primarily means that the fiber manufacturers have to know how much a realistic deviation from the optimal index profile will affect the characteristics of the erbium-doped fiber. Also, it is of interest to discuss how very strict requirements to ultra high NAs can be softened with alternative fiber designs.

In order to get closer to the answer of these questions, we will in Section 6.5.1 discuss the gain properties of fibers with (non-ideal) index profiles. Furthermore, Section 6.5.2 will describe methods for improving fiber properties with a depressed cladding approach.

6.5.1 Alpha Profiles

Since the deviation from an ideal step-index profile affects the gain coefficient disadvantageously, it is of interest to examine how serious this effect is. In order to do so, in Figure 6.19 we have shown the gain coefficient versus the parameter alpha, defined as α in the power-law function $(1 - (r/a_c)^\alpha)$ for the index profile. The erbium concentration profile is assumed to follow the same power-law function. This is considered a very good approximation for the Al/Er fibers (as discussed in Section 3.1.2), but, for the cogency of the description, it is also assumed for the Ge/Er fibers, even though some erbium diffusion might appear in this case. The NA is 0.25, and for each alpha the gain coefficient is calculated as the maximum obtainable. As can be seen from the figure, it will be significant if the index profile deviates more from the step-index profile than described by α-values below 5. As an example, the triangle profile, alpha = 1, has a gain coefficient 3.5 dB/mW lower than the step-index profile.

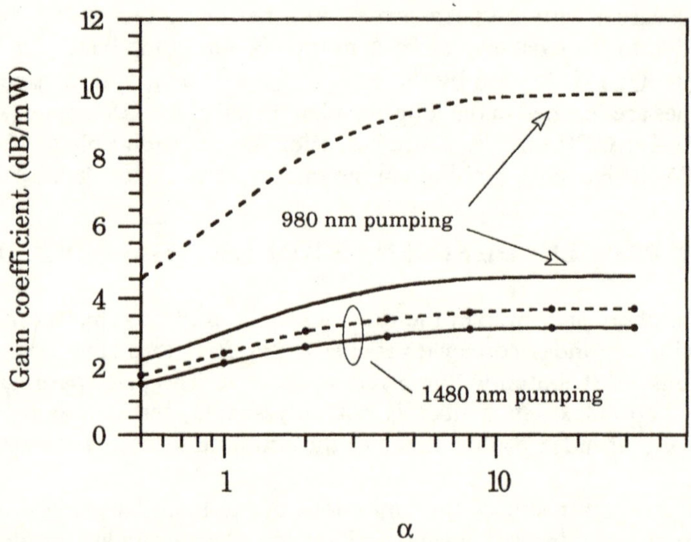

Figure 6.19 Gain coefficient shown versus α (α in the power-law function $1 - (r/a)^{\alpha}$ for the index profile). The NA is 0.25 and both Al/Er fibers (solid lines) and Ge/Er fibers (dashed lines) are considered. The * indicates 1480-nm pumping, whereas the curves with no marks refer to 980-nm pumping. (Figure from [5].)

6.5.2 Fibers With Depressed Cladding Layers

The majority of the description in this chapter has concentrated on the step-index fiber, since it represents the optimum choice with respect to gain efficiency and pump-to-signal conversion. It has also been shown that very-high-performance fibers are obtained by increasing the NA to values at or above 0.3.

This leads to fibers with very small spot-size values which suffer from the problem of severe coupling losses when transmitting light between the active fiber and standard telecommunication fibers. Furthermore, step-index fibers with large NA values are found to suffer from increased scattering losses, which is believed to be due to induced stress in the core-cladding boundary. It is also the primary reason why dispersion-shifted fibers are fabricated with a triangular core index. Increasing the NA above 0.3 is also difficult from a manufacturing point of view, since the above mentioned tension between core and cladding materials increases the possibility of introducing cracks in the glass during the preform collapse and drawing process.

In order to limit some of these problems, it was suggested in [29] that the optimum profile in the core should be flat over the central region and graded in the rest of the core. In this way, a trapeze-shaped core area appears whereby abrupt radial index changes are avoided. Furthermore, it was proposed that a depressed cladding region be made. If the depression of the cladding region is made equal to the core height, a doubling of the gain is reported [29] compared to the fiber without depressed cladding. Note, however, that there is a technological limit to the amount of depression, which can be obtained from fluorine doping, and the most likely way to obtain high-NA fibers will therefore be through an increase of aluminum or germanium doping. It should, however, also be mentioned that depressed claddings with fluorine doping are advantageous for making low-loss splices.

We will not at this point go further into a detailed discussion of the design of such depressed cladding fibers, because the fact is that the step-index core design leads to the most advantageous gain performance. Depending on the actual manufacturing process, it can then be considered to modify the design for the purpose of reducing residual effects such as increased intrinsic loss. This will generally reduce the efficiency of the amplifier, and the example given here represents only a few of the possibilities.

6.6 SUMMARY

The basic requirement for finding a means to optimize erbium-doped fibers is to define the regions for amplifier operation. We have, therefore, in this chapter discussed how the fiber amplifier should be operated either in the small signal regime, where the gain is independent of the signal power and solely given by pump and ASE power, or in the large signal regime. In the latter case, the output power from the amplifier is partly self-adjusting, since gain decreases for increasing signal input power.

The characteristic parameter for optimization was shown to be the fast calculable threshold pump power, or the gain coefficient, which indicates the pump power consumption for a small-signal amplifier designed for a specified gain value. Finally, the conversion of pump photons to amplified signal photons in power booster amplifiers was defined by the QCE.

A detailed analysis was performed on the waveguide parameters for erbium-doped fibers. The optimum cutoff wavelength for step-index profiles was shown to be 800 nm for 980-nm pumping and 900 nm for 1480-nm pumping, independent of the NA. For the less efficient 800-nm-pumped fiber amplifiers, an optimum cutoff wavelength of 800 nm was also found, and the bi-directional pumping scheme was demonstrated to improve the gain coefficient by up to 50% compared to co-directional pumping. The highest gain coefficient was predicted in Ge/Er fibers pumped at 980 nm, where the maximum gain coefficient is 23 dB/mW, predicted for an NA of 0.4. This gain coefficient is significantly higher than the highest value of 11 dB/mW predicted in Al/Er fibers.

The step-index profile with an NA as high as possible was shown to be optimum with regard to the gain coefficient. Furthermore, the use of a confined erbium doping radius has been shown to improve the gain coefficient up to 45%. In this connection, however, it must be noted that the optimum cutoff wavelength should be increased by up to 100 nm if the fiber is designed with a high degree of erbium confinement.

For the optimization of in-line amplifiers or preamplifiers, it was shown that the 980-nm-pumped Ge/Er fiber is the best choice with regard to gain efficiency. The host glass composition, in the case of 1480-nm pumping, was shown to be less significant. In contrast to this, 800-nm pumping will make the host glass choice most critical because of the possibilities of considerably lowering the ESA-to-GSA ratio. The most attractive host glass for 800-nm-pumped fibers is currently considered to be fluorophosphate fibers, in which gain coefficients of 3.5 dB/mW should be obtainable for NAs of 0.4.

For power booster amplifier applications, the results showed that the NA should be 0.20 or higher, and the optimum choices of cutoff wavelengths are the same as those in the case of fibers for small-signal amplifiers. The optimum design, however, is not critical with respect to cutoff wavelength, since the QCE does not show a very pronounced maximum with respect to this parameter. Provided there is enough pump power, the differences between Ge/Er and Al/Er fiber and between 980- and 1480-nm pumping were shown to be insignificant. For 800-nm-pumped amplifiers, the fluorophosphate fiber has 60% higher QCE than Al/P fibers.

Finally, there was a discussion on how seriously the gain properties are affected for fibers, which deviates from the ideal step-index design. It was found that only very small deviations could be seen for fibers with core index profiles described by α-values above 5; but if a triangular core is chosen, the gain coefficient will decrease 3.5 dB/mW compared to the optimum value for step-index fibers.

REFERENCES

[1] Povlsen, J. H., A. Bjarklev, B. Pedersen, H. Vendeltorp-Pommer, and K. Rottwitt, "Optimum Design of Erbium Fibre Amplifiers Pumped With Sources Emitting at 1480 nm," *IEE Electronics Letters*, Vol. 26, No. 17, 1990, pp. 1419–1420.

[2] Bjarklev, A., J. H. Povlsen, and S. Lomholt-Hansen, "Large Signal Modelling of an Erbium Doped Fibre Amplifier," *Proceedings of Society of Photo-Optical Instrumentation Engineers Conference, OE/Fibers '89*, part: Fiber Laser Sources and Amplifiers, Boston, 6–8 Sept. 1989, SPIE Vol. 1171, pp. 118–129.

[3] Pedersen, B., A. Bjarklev, and J. H. Povlsen, "Design of Erbium Doped Fibre Amplifiers for 980 nm or 1480 nm pumping," *IEE Electronics Letters*, Vol. 27, No. 3, 1991, pp. 255–256.

[4] Pedersen, B., A. Bjarklev, O. Lumholt, and J. H. Povlsen, "Detailed Design Analysis of Erbium-Doped Fibre Amplifiers," *IEEE Photonics Technology Letters*, Vol. 3, No. 6, 1991, pp. 548–550.

[5] Pedersen, B., A. Bjarklev, J. H. Povlsen, K. Dybdal, and C. C. Larsen, "The Design of Erbium-Doped Fiber Amplifiers," *IEEE Journal of Lightwave Technology*, Vol. 9, No. 9, 1991, pp. 1105–1112.

[6] Pedersen, B., M. L. Dakss, B. A. Thompson, W. J. Miniscalco, T. Wei, and L. J. Andrews, "Experimental and Theoretical Analysis of Efficient Erbium-Doped Fibre Power Amplifiers," *IEEE Photonics Technology Letters*, Vol. 3, No. 12, 1991, pp. 1085–1087.

[7] Ainslie, B. J., et al., "Optical Analysis of Erbium-Doped Fibres for Efficient Lasers and Amplifiers," *Proceeding of IOOC '89*, paper 20A3-2, 1989.

[8] Desurvire, E., J. L. Zyskind, and C. R. Giles, "Design Optimization for Efficient Erbium-Doped Fibre Amplifiers," *IEEE Journal of Lightwave Technology*, Vol. 8, No. 11, 1990, pp. 1730–1741.

[9] Spirit, D. M., L. C. Blank, S. T. Davey, and D. L. Williams, "System Aspects of Raman Fibre Amplifiers," *IEE Proceedings Part J*, Vol. 137, No. 4, 1990, pp. 221–224.

[10] Zemon, S., B. Pedersen, G. Lambert, W. J. Miniscalco, L. J. Andrews, R. W. Davies, and T. Wei, "Excited-State Absorption Cross Sections in the 800-nm Band for the ErDoped, Al/P-Silica Fibers: Measurements and Amplifier Modeling," *IEEE Photonics Technology Letters*, Vol. 3, No. 7, 1991, pp. 621–624.

[11] Kagi, N., A. Oyobe, T. Morikawa, Y. Sasaki, and K. Nakamura, "Gain Characteristics of Er3+ Doped Fibre With Quasiconfined Structure," *Optical Fiber Communications OFC '90*, San Francisco, 1990, paper FA8, p. 200.

[12] Povlsen, J. H., and S. B. Andreasen, "Analysis on Splice, Microbending, Macrobending, and Rayleigh Losses in GeO_2-Doped Dispersion-Shifted Single-Mode Fibre," *IEEE Journal of Lightwave Technology*, Vol. LT-4, No. 7, 1986, pp. 706–710.

[13] Shimizu, M., M. Yamada, H. Horiguchi, T. Takeshita, and M. Okayasu, "Erbium-Doped Fibre Amplifiers With an Extremely High Gain Coefficient of 11.0 dB/mW," *IEE Electronics Letters*, Vol. 26, No. 20, 1990, pp. 1641–1643.

[14] Suzuki, K., Y. Kimura, and M. Nakazawa, "High Gain Er^{3+}Doped Fibre Amplifier Pumped by 820 nm GaAlAs Laser Diodes," *IEE Electronics Letters*, Vol. 26, No. 13, 1990, pp. 948–949.

[15] Kimura, Y., K. Suzuki, and M. Nakazawa, "High Gain Erbium-Doped Fiber Amplifier Pumped in the 0.8 μm Pump Band," *European Conference on Optical Communication ECOC '90*, Amsterdam, 1990, Proc. Vol. I, pp. 103–106.

[16] Pedersen, B., S. Zemon, W. J. Miniscalco, and M. P. Singh, "Erbium-Doped Fiber Amplifiers Pumped in the 800 nm Band," *Techn. Dig. of Optical Amplifiers and Their Applications OAA '91*, Snowmass Village, CO, 1991, WD4-1, pp. 56–59.

[17] Pedersen, B., S. Zemon, and W. J. Miniscalco, "Erbium Doped Fibres Pumped in the 800 nm Band," *IEE Electronics Letters*, Vol. 27, No. 14, 1991, pp. 1295–1297.

[18] Pedersen, B., S. Zemon, W. J. Miniscalco, and T. Wei, "Power Requirements for Erbium-Doped Fiber Amplifiers Pumped at 800, 980, and 1480 nm," *Optical Fiber Communications OFC '92*, San Jose, 2–7 February 1992, paper TuG3, p. 35.

[19] Ohashi, M., and M. Tsubokawa, "Optimum Parameter Design of Er3+-Doped Fiber for Optical Amplifiers," *IEEE Photonics Technology Letters*, Vol. 3, No. 2, 1991, pp. 121–123.

[20] Horiguchi, M., M. Shimuzu, M. Yamada, K. Yoshino, and H. Hanafusa, "Highly Efficient Optical Fibre Amplifier Pumped by a 0.8 μm Band Laser Diode," *IEE Electronics Letters*, Vol. 26, No. 21, 1990, pp. 1758–1759.

[21] Pedersen, B., S. A. Zemon, and W. J. Miniscalco, "Analysis of Erbium-Doped Fiber Amplifiers Pumped at 800 nm," *Proceedings of Society of Photo-Optical Instrumentation Engineers Conference, OE/Fibers '91*, part: Fiber Laser Sources and Amplifiers III, Boston, September 1991, SPIE Vol. 1581, pp. 293–302.

[22] Dakss, M. L., B. Pedersen, G. R. Joyce, and W. J. Miniscalco, "Signal Output Power and Noise in Erbium-Doped Fibre Power Amplifiers," *Proceedings of Society of Photo-Optical Instrumen-*

tation Engineers Conference, OE/Fibers '91, part: Fiber Laser Sources and Amplifiers III, Boston, September 1991, SPIE Vol. 1581, pp. 236–250.

[23] Massicott, J. F., R. Wyatt, B. J. Ainslie, and S. P. Craig-Ryan, "Efficient, High Power, High Gain Er3+ Doped Silica Fiber Amplifier," *Proceedings of Society of Photo-Optical Instrumentation Engineers Conference, OE/Fibers '90*, part: Fiber Laser Sources and Amplifiers II, Boston, September 1990, SPIE Vol. 1373, pp. 93–102.

[24] Gabla, P. M., V. Lemaire, H. Krimmel, J. Otterbach, J. Auge, and A. Dursin, "35 AM-VSB TV Channels Distribution With High Signal Quality Using a 1480 nm Diode-Pumped Erbium-Doped Fiber Postamplifier," *IEEE Photonics Technology Letters*, Vol. 3, No. 1, 1991, pp. 56–58.

[25] Lin, C., W. Way, and M. A. Saifi, "Optical-Fiber Amplifiers Make Broadband Fiber Networks Practical," *Laser Focus World*, 1991, pp. 161–171.

[26] Desurvire, E., C. R. Giles, and J. Simpson, "Gain Saturation Effects in High-Speed, Multichannel Erbium-Doped Fiber Amplifiers at $\lambda = 1.53\ \mu m$," *IEEE Journal of Lightwave Technology*, Vol. LT-7, No. 7, 1989, pp. 2095–2104.

[27] Laming, R. I., J. E. Townsend, D. N. Payne, F. Meli, G. Grasso, and E. J. Tarbox, "High-Power Erbium-Doped-Fiber Amplifiers Operating in the Saturation Regime," *IEEE Photonics Technology Letters*, Vol. 3, 1991, pp. 253–255.

[28] Way, W. I., M. M. Choy, A. Yi-Yan, M. Andrejco, M. Saifi, and C. Lin, "Multichannel AM-VSB Television Signal Transmission Using an Erbium-Doped Optical Fiber Power Amplifier," *IEEE Photonics Technology Letters*, Vol. 1, No. 10, 1989, pp. 343–345.

[29] Holmes, M. J., D. M. Spirit, and F. P. Payne, "Optimization of Refractive Index Profile for High Gain Fibre Amplifiers," *IEE Electronics Letters*, Vol. 26, No. 22, 1990, pp. 1873–1874.

Chapter 7
Pump Wavelength Choice for 1550-nm Amplifiers

The first results from optical fiber amplifiers based on an Er^{3+}-doped fiber [1,2] made use of physically large pump lasers (e.g., gas-laser-pumped dye lasers), which were only practical for laboratory demonstrations. However, the rapid development that took place in the late 1980s, especially within the area of semiconductor pump lasers, gave a tremendous push forward to the development of fiber amplifiers for optical communication systems. We have, therefore, reached a situation in which a number of possibilities exist in the choice of practical pump sources, and different amplifier requirements may determine the preferred pump wavelength.

In this chapter, we will describe typical properties and limitations of amplifiers pumped at different wavelengths. We will first describe characteristics of the available semiconductor pump sources, and provide a short description of fiber splicing problems and coupler requirements. Second, we have included a short discussion of the possibility of pumping at visible light wavelengths. Then we focus the description on the three most promising candidates for pumping the EDFA (i.e., the 800-nm, 980-, and 1480-nm pump bands). Finally, the chapter includes a comparison between the different pump bands.

7.1 COMPONENT TECHNOLOGY FOR THE ERBIUM-DOPED FIBER AMPLIFIER

For a complete description of the EDFA, it is necessary not only to look at the rare-earth-doped fiber, but also at pump lasers, fiber splices, and passive components such as couplers. Although it is not the purpose of this book to provide a thorough description of these additional components, we will include a short discussion of them. Sufficient background will be provided for the following discussion of advantages and disadvantages of the different pump bands, and, in addition, a number of references will be listed in which more detailed information may be found.

7.1.1 Semiconductor Pump Sources for Fiber Amplifiers

A newly emerging use of semiconductor lasers in optical communications is as pump sources for EDFAs. Here, diode lasers emitting around 800, 980, and 1480 nm come into question. In an EDFA application, the demand placed on the pump diode laser is for large amounts of optical power (tens to hundreds of milliwatts) [3]. Often, the pump laser must operate in environments where laser driving current and heat sinking are limited. Therefore, efficiency in the conversion from electrical to optical power and high-power capabilities of the laser diode are important considerations.

Also, the question of the price of the semiconductor pump source may come into question, and this issue is an argument that has been used in support of operating the GaAlAs laser diodes in the wavelength range around 800 to 830 nm. It should, however, be considered that compared to the total cost of a long-haul optical communication system, the price of the pump sources is negligible. Therefore, the price consideration first becomes relevant in cases where the overall system cost is limited. Examples are applications in which the amplifier is used to supply a single or a few users, or in cases where the EDFA is a part of optical measurement equipment. On the other hand, if the 800-nm pumped amplifiers provide sufficient gain and noise properties for the given application, there will be no reason to choose another (more expensive) pump source. The 800-nm pump sources are therefore still relevant, and as an example we refer here to the buried twin-ridge substrate structure described in [4]. This laser diode operates around 820 nm with a single transverse mode and has a spectrum width of approximately 2 nm. According to [4], it is possible to obtain gain above 20 dB using such commercially available and inexpensive GaAlAs laser diodes.

For fiber amplifier pump purposes, *quantum well* (QW) lasers are particularly attractive because of the lower threshold current and increased slope efficiency compared to bulk lasers [3]. This has the combined result that the laser electrical-to-optical conversion efficiency is higher while the laser's power dissipation is smaller. Furthermore, the QW lasers have the attractive feature of low internal losses, which, according to [3], are typically three times smaller than in bulk lasers. The importance of this property is that it allows a longer QW laser chip, in which both the injected current and the dissipated heat may be spread over more surface area. The result is a lower series resistance that increases the conversion efficiency even more, and a lower thermal impedance that reduces laser temperature and allows higher optical output powers.

Extremely efficient pumping has been demonstrated at both 980 nm [5] and 1480 nm [6], with gain exceeding 20 dB produced by less than 5 mW of pump power. As we shall see in this and other chapters, 10 to 100 mW are typically needed to ensure adequate output power. This requirement is in no way a deterrent, since semiconductor diode lasers suitable for pumping EDFAs can emit hundreds of milliwatts at 980 or 1480 nm (e.g., [7–11]). For 980-nm lasers, strained *indium gallium arsenide* (InGaAs) quantum wells are grown on *gallium arsenide* (GaAs). For 1480-nm pumps, the material used, either InGaAs quantum wells or InGaAs phosphide grown on indium phosphide, is very similar to the material used for standard communication lasers [7]. The shorter wavelength 980-nm diode lasers require less injection current and less cooling, thereby reducing the electrical power requirements.

Currently, 1480-nm pump sources are more widely used than 980 nm pump lasers. This is mainly because they are readily available, and there is thus greater confidence in their reliability. This picture may change if 980-nm pump lasers, over a long period of operation, show behavior as good as the 1480-nm sources. Note, however, that for applications such as remote pumped or distributed fiber amplifiers (see Chapter 11), 1480-nm amplifiers will under all circumstances be needed for optimal system performance. The reason is that they represent examples of systems where the pump transmission losses must be minimized. Let us return to the question of reliability by stressing that this is especially important for pump lasers, which have to deliver high levels of power over long periods of time. Power amplifier applications will require the highest pump power, and pump reliability should be a key issue in their development.

As we shall see in the following part of this chapter, there are numerous physical arguments for choosing one or the other pump wavelengths for a given application. However, the unanswered question of pump laser reliability makes it difficult to predict whether the preferred pump wavelength is going to be 980 or 1480 nm. We will therefore not elaborate further on this point in the present description, but only note that power emitted from 980-nm laser diodes of 350 mW or more has been demonstrated [12]. Reliability testing of such components indicates that 980-nm strained QW lasers show considerable promise as a practical low-noise pumping source for EDFAs [13]. For 1480-nm lasers, the fiber excitation allowance for wavelength detuning is fairly large (up to as much as ±10 nm, depending on the specific application). Output powers of 250 mW have been reported using *multi-quantum well* (MQW) epitaxial layer structures [14,15].

Before we leave the subject of pump sources for EDFAs, it should additionally be noted that reports on solid-state laser pump sources emitting at 1064 nm have been presented lately [16]. These pumps need, however, semiconductor pumping themselves, and an ytterbium-co-doped EDFA must be used (for further details, see Section 3.3.1).

7.1.2 Splicing Considerations for Fiber Amplifiers

The discussion of the optimal waveguide parameters for erbium-doped fibers in Chapter 6 showed a need for core diameters and spot sizes much smaller than those in conventional fibers. A typical example is that whereas present EDFAs have 2.5-μm cores and spot sizes on the order of 4 μm, the conventional fiber has a spot size between 7 and 8 μm (the mentioned values are diameters). Even for such large differences in spot size, splice losses of 0.1 to 1.0 dB between the erbium fiber and the conventional fiber can be obtained. This may be done by (additionally) heating the fibers to produce a fusion splice with a low-loss transition region [7]. If sufficiently low splice losses cannot be obtained by this procedure, more advanced methods may be used. We will return to this subject in Section 8.4.1.

7.1.3 WDM Couplers for Fiber Amplifiers

The pump and signal are combined by a WDM, which is typically either a fused fiber coupler or a miniature bulk optic device. The latter device uses lenses to couple light in and out of the fiber and includes appropriate mirrors and filters [7]. In these WDMs, the pump and signal typically experience a loss of about 0.5 dB. The fiber WDM couplers are, however, much more important for practical EDFA construction. Their important characteristics include insertion loss of both signal and pump and the variation of these losses with temperature, polarization, and wavelength. These properties have been studied for 1480-nm pumping, and results are compared in [17] for commercially available packaged fused tapered couplers and fiber-tailed interference filter multiplexers. Both types of components offer advantages for long-haul amplifier applications, and the conclusions should be drawn forward. Lord et al. points out that fused couplers can have lower insertion loss (a typical value is 0.1 dB compared to 0.4 dB for the interference filter) and may be cheaper to produce. The interference filter couplers, on the other hand, exhibit a flat transmission response for both signal and pump, whereas the fused coupler has an intrinsic raised cosine transmission response. The interference filter coupler will therefore provide a transmission loss independent of variations in pump and signal wavelengths, and, in addition, they are inherently polarization- and temperature-insensitive. According to [17], a standard fused coupler exhibits polarization dependence of $\approx$ 0.5 dB, which, however, may be reduced to $\approx$ 0.2 dB by the introduction of a twist after the fusion process [18]. Finally, the intrinsic raised cosine wavelength response of the fused coupler causes the possibility of light coupling back into the pump laser, and this feedback can, in practice, lead to instabilities or even lasing. Even though the conclusions of [17], therefore, point towards the interference filter coupler as the preferred choice, practical development over the last few years favors the fused coupler in the majority of applications.

This is also a result of the simplicity and the potential lower cost of fused couplers. Commercially available EDFA modules today include fused couplers, so we may conclude that the problems listed above can be solved for most applications. The low-loss fusion spliced coupler is therefore likely to remain a key component in fiber amplifiers for the forseeable future.

Note finally that other passive components such as filters and isolators normally are also needed to obtain optimal amplifier performance. We will, however, not discuss the significance of these components further at this point, since they are thoroughly discussed in Chapter 8.

7.2 PUMPING THE ACTIVE FIBER AT VISIBLE LIGHT WAVELENGTHS

One of the possibilities of pumping the EDFA at visible light wavelengths, and still maintain a compact pump setup, is by using the second harmonic of a LD-pumped, intracavity frequency-doubled Nd:YAG laser [19,20]. The characteristics of a high-gain, high-output power performance of an EDFA pumped at 532 nm by such a pump source were presented in [20]. In the investigation, the signal wavelength was chosen to be at the gain peak of 1536.4 nm. The gain is shown in Figure 7.1(a) as a function of the pump power coupled into the erbium-doped fiber for two different fiber lengths. We note that the gain is saturating at a value of 41 to 42 dB with a coupled pump power of 40 mW or more, but a gain of 40 dB may be obtained for pump powers around 25 mW. The gain coefficient (see Figure 6.3) is about 1.6 dB/mW. In [20], the signal output saturation properties were also examined, and an output saturation power of 10 dBm was obtained for a coupled pump power of 60 mW. It should finally be mentioned that a noise figure characterization of the specific amplifier has not been presented. However, since the Er^{3+} ions work as a three-level laser system at this specific wavelength, it should be possible to obtain a high population inversion within the erbium-doped fiber. Consequently, noise figures of 3 to 4 dB may be expected for 532-nm pumping.

We will now turn to another possibility of pumping in the visible wavelength interval; namely, in the 640- to 685-nm wavelength interval. Investigations of this pump band were presented in [21], and even though the pump source was a non-compact Ar^+-ion-pumped DCM dye laser, some of the results are reviewed here for comparison with other pump bands. Furthermore, the 650-nm pump band is relevant for laboratory use. The dependence of the fiber gain on pump power and wavelength is shown in Figure 7.1(b). The threshold pump power is slightly above 10 mW, and a gain of 25 dB is available for an approximately 40-mW pump absorbed in the erbium-doped fiber. Thereafter, the gain saturates, indicating the high degree of inversion in the three-level Er^{3+} system. The maximum gain observed was 28 dB. It should also be mentioned that a 3-dB compression from the linear gain was

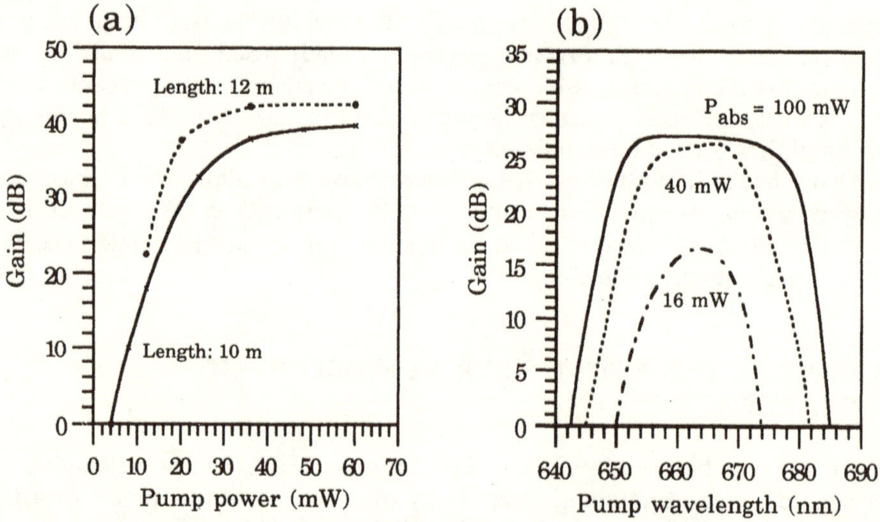

Figure 7.1 (a) Gain versus coupled pump power (at 532 nm) for an input signal at the gain peak wavelength of 1536.4 nm for two different erbium-doped fiber lengths. The signal input power is -32 dBm, and the erbium-doped fiber has a core radius a = 2.15 m and NA = 0.21. (Data from [20].) (b) Amplifier gain (around 1536 nm) versus pump wavelength with the pump power as parameter. The fiber is specified by NA = 0.2, cutoff wavelength of 1400 nm, and erbium dopant concentration $\approx 10^{25}$ m^{-3}. (Data from [21].)

observed for an output power of +7 dBm. With respect to the noise properties, a low-noise figure can also be expected for modest pump powers in this wavelength range.

7.3 PUMPING IN THE 800-nm WAVELENGTH BAND

Some of the first semiconductor laser-pumped fiber amplifiers were realized by the use of 800-nm pump sources [22]. However, the first results showed only modest gain values (e.g., 6 dB obtained for 15-mW pump at 807 nm), and maximum gain values around 15 dB were estimated [22]. The major cause of this gain limitation (as described in Section 3.1.3) is the existence of ESA in the 800-nm pump band. As the progress of 1480- and 980-nm pump sources became evident, interest in the 800-nm pump band declined [23]. Some interest for the 800-nm pump band, however, remained because inexpensive, high-output GaAlAs laser diodes could be used as pump sources. In the beginning of 1990, the confidence in 800 nm-pumped amplifiers was renewed, since high gain amplification was reported. For this amplifier, a Ti:sapphire pump laser operating around 820 nm was used [24]. Shortly after this, several reports on semiconductor laser pumping followed [4, 25, 26], and gain values as high as 39 dB and gain coefficients between 0.86 and 1.3 dB/mW were described.

All of the high gain results from 800-nm-pumped EDFAs were obtained by choosing a pump wavelength around 820 nm (i.e., in the high-wavelength end of the absorption band). The reason is that the ratio between GSA and ESA becomes more favorable as the pump wavelength is changed from 807 to 820 nm. Based on these results, the 800-nm-pumped amplifiers became a realistic alternative to 980- and 1480-nm-pumped EDFAs, and investigations on noise properties [27] and accurate amplifier optimizations have followed [28–31]. In this section, we will review the major results of 800-nm pumping of EDFAs, with special focus on the pump wavelength optimization. Furthermore, the pump configuration becomes of special significance due to the ESA problem, and so we will discuss the most favorable amplifier design from this point of view.

7.3.1 Optimum Choice of Pump Wavelength in the 800-nm Band

To describe the pump wavelength dependence of the amplifier properties when pumped in the 800-nm band, Figure 7.2 has been included. This shows the small-signal gain as a function of pump wavelength for four different pump powers launched co-directionally with a 1551-nm signal [32]. The Ge/Al/P/Er-doped fiber

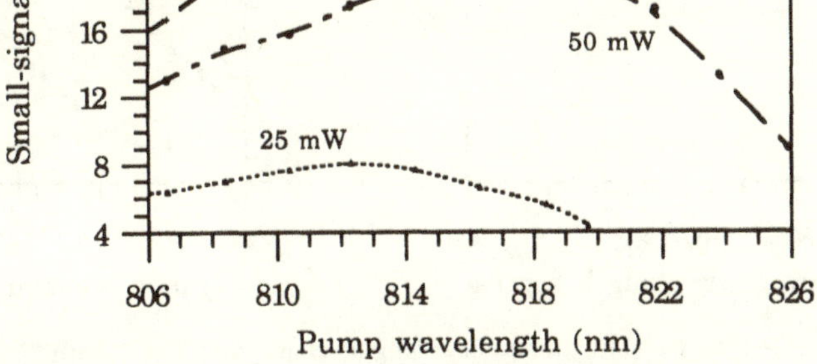

Figure 7.2 Measured small-signal gain versus pump wavelength for different pump powers. Each point is for the fiber length that gave the highest gain. (Data from [32].)

used in these experiments had a cutoff wavelength of 940 nm and an NA of 0.18. For 800-nm pumping, excitation in the fundamental pump mode was predominant because the launching setup for the pump used a fiber that was single-moded at 800 nm. Fiber lengths between 7 m and 40 m (in steps of 3 m) were used, and each point represents the fiber length giving the maximum gain for the respective pump power. It should here be noted that the pump wavelength providing the highest gain increases with the pump power. The reason for this is that, in order to make use of the increased pump power, a longer fiber length is needed, and consequently the significance of the ESA becomes stronger. It is therefore necessary to move towards longer wavelengths where the ESA-to-GSA ratio becomes smaller, even though the absolute GSA at the same time decreases.

For a more detailed discussion of the EDFA properties, calculated small-signal gain is displayed in Figure 7.3(a) as a function of pump wavelength for co-directional (dashed curves) and bi-directional (solid curves) pumping [31]. The total launched pump power is 50 mW, and the specific step index fiber has an

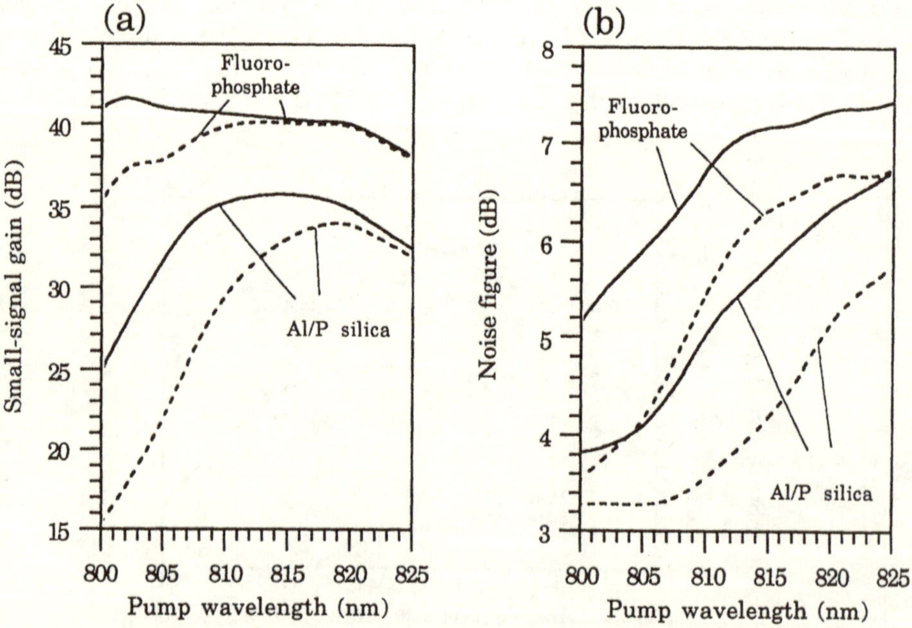

Figure 7.3 (a) Small-signal gain at optimum fiber length versus pump wavelength for a step-index fiber with NA = 0.3, cutoff wavelength = 800 nm, and erbium doped uniformly throughout the core. The total pump power is 50 mW. (b) Noise figure shown versus pump wavelength corresponding to cases treated in (a). Results are shown for co-directional pumping (dashed lines) and bi-directional pumping (solid lines). (Figures from [31].)

NA = 0.3 and a cutoff wavelength of 800 nm. The core is considered uniformly doped with erbium, and two different glass host compositions have been considered as indicated in the figure. All fiber lengths have been optimized with respect to maximum gain. As a consequence of the ESA, both types of fiber show higher gain for bi-directional pumping than for co-directional. For the latter scheme, the pump power is high enough in the initial section of the fiber to generate high population inversion, which leads to a large and unbleachable attenuation coefficient due to ESA at the pump wavelength. This inversion-dependent ESA coefficient is lowered by dividing the pump power equally and pumping from both ends, whereby a longer fiber length is permitted. Furthermore, Figure 7.3(a) shows that the bi-directional pumping scheme leads to a flatter gain characteristic over the pump band, which indicates a quite effective compensation of the larger ESA at shorter pump wavelengths. A comparison between the two glass hosts also shows a flatter spectrum for the fluorophosphate, which is due to a less important ESA for this glass host. The maximum small-signal gain for the fluorophosphate is 41 dB, which, as described in [31], is 6 dB higher than in Al/P silica glass. However, it should be noted that a practical fluorophosphate fiber still has to be fabricated to verify this prediction.

The noise figure is shown in Figure 7.3(b) as a function of the pump wavelength, and each value at a particular pump wavelength corresponds to the gain shown in Figure 7.3(a). We note that the calculated noise figure values increase by up to 3 dB as the pump wavelength is increased from 800 to 825 nm. This is remarkable, since the gain spectrum is relatively flat over much of this region (see (5.33)), but may be explained by the decreasing GSA cross section. It may also be observed that the fluorophosphate has a noise figure up to 2 dB higher than that of Al/P silica. This is a consequence of the lifetime of the $^4I_{13/2}$ level being 25% shorter in the fluorophosphate host [31]. This is a significant difference, which may very well lower the interest in fluorophosphate host glasses in small-signal applications. For booster amplifier applications, however, the fluorophosphate host seems much more promising, and a 60% efficiency improvement is predicted [31] (see also Section 6.4.2).

The theoretical results shown in Figure 7.3 are in good agreement with experimental values published in [27]. It was demonstrated here that when the pumping wavelength was in the range of 818 to 824 nm and the signal wavelength was longer than 1530 nm, noise figures of 3.9 to 4.7 dB were obtainable. These results are, as we shall see in the following, similar to those obtained with a 1480-nm pumping scheme [33,34].

7.3.2 Pump Configurations for 800-nm-Pumped Amplifiers

As already demonstrated in the previous description, the main obstacle to efficient pumping in the 800-nm band is the ESA. In Section 7.2.1, it was shown that pumping in the long-wavelength tail of the 800-nm band can reduce the significance of the ESA. How the amplifier properties could be improved by changing from co-directional to bi-directional pumping was also discussed. To examine possible improvements, the three different pump configurations in Figure 7.4(a) were investigated in [35], and we will refer to the results here. Since this analysis was made before accurate ESA and GSA cross-section measurements were available, results are only shown for 807-nm pumping. However, the qualitative information is also valid at higher wavelengths in the 800-nm pump band. The considered fiber has an NA = 0.18 and a core radius of 3.2 μm, within which the erbium doping material is uniformly distributed. All results are calculated for gain maximal fiber lengths, where it is noted that in configuration B the first and second erbium-doped fiber are of equal length. The total launched pump power is 100 mW in all cases, and the analysis is performed as described in Section 5.1.

In Figure 7.4(b), the gain and the noise figure are shown as a function of the signal input power. We find that, at small-signal input levels, configuration B gives a slightly smaller gain than configuration A. This may be explained by the influence of ASE on the population inversion. We first note that the gain in the first amplifier in configuration B is 18.5 dB, with a signal input power of -50 dBm. When the

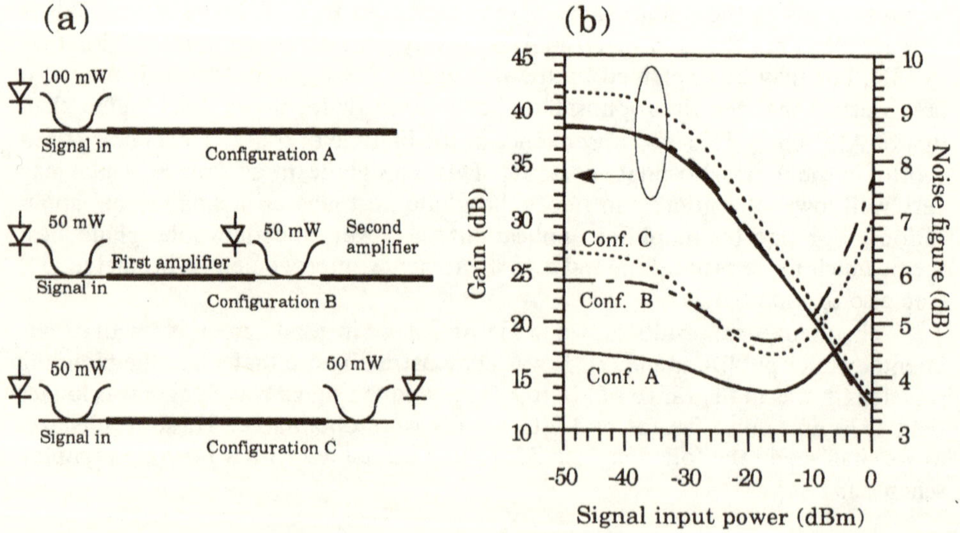

Figure 7.4 (a) Three configurations for 807-nm pumping with the pump powers indicated. (b) Gain and noise figure versus signal input power for the three configurations shown in (a). Erbium concentration is 2×10^{24} m^{-3}; it is assumed that $\sigma_{24} = \sigma_{13} = 6.8 \times 10^{-26}$ m^2 at the pump wavelength. (Data from [35].)

second amplifier in configuration B is completely removed, the gain rises to 28 dB. The reason is that the backward propagating ASE from the second amplifier reduces the population inversion, and thereby the gain of the first amplifier. The maximum gain in configuration B is 38 dB, meaning that the gain in the second amplifier is larger than in the first. This can be understood from the fact that the forward ASE power from the first amplifier (−11.5 dBm) is smaller than the backward ASE power from the second amplifier (-7 dBm). The maximum gain in configuration C is 41.6 dB, which is 3.3 dB more than in configuration A. This illustrates again that a more efficient pump power distribution can be obtained with bi-directional pumping. In all cases, the noise figures show a minimum with regard to signal input power at approximately −17 dBm. The explanation is that, at small-signal input levels, the large ASE decreases the population inversion, but as the signal input power is increased, it will suppress the ASE. The noise figure therefore becomes better until the point where the signal power itself starts to deteriorate the population inversion, leading to increasing noise figures for large signal operation. We find that the noise figure of the co-directionally pumped amplifier is 1 to 2 dB smaller than for the other two configurations. This is because the pump power distribution in configuration A ensures the highest population inversion, where the signal power is smallest (see (5.34)). For small-signal operation, configuration B shows noise figures, which are 0.5 dB smaller than for the bi-directional pumping.

7.4 980-nm PUMPING

The previous section discussed how the existence of ESA in the 800-nm pump band lowered the amplifier efficiency. The only way to overcome this problem was to make use of other pump bands, and a large effort was therefore invested in the investigation of the 980-nm pump band in the late 1980s. Among the first reports, which identified the 980-nm pump band as very advantageous and free of ESA, was the 24-dB gain result of [36]. This was obtained with only 10.5 mW of pump power and was therefore well within the reach of semiconductor pump diodes. The dye laser pump was soon replaced by a 978-nm strained GaInAs/AlGaAs QW pump laser [37]. A 24-dB gain was then experienced by a 1535.3-nm signal, and the required pump power was only 6.2 mW, corresponding to a gain coefficient of 3.9 dB/mW. This gain coefficient was not only the record as of the publication date, but it is also a realistic value for practical high-performance amplifiers fabricated today. The pumping at 980 nm had therefore shown to be both practical and very attractive, and these results were confirmed by the large number of publications that followed [5, 38–45]. In the following part of this section, consequently, we will describe the most significant properties of EDFAs pumped at around 980 nm.

7.4.1 High-Performance Gain/Gain-Coefficient Results for 980-nm Pumping

We will start the more detailed description of the 980 nm-pumped amplifier by concentrating on the small-signal properties. One way to evaluate the amplifier performance for this type of applications is, as described in Chapter 6, to focus on the gain coefficient. In Figure 7.5, we showed this as a function of the signal wavelength for three erbium-doped fibers. The difference between them is the composition of index-raising co-dopants, since one is purely aluminum co-doped, another is purely germanium-co-doped, and the third is fabricated with both aluminum and germanium (in a ratio that ensures spectral properties, which are influenced by both aluminum and germanium co-doping). The NA of the considered fibers is 0.25. From Figure 7.5, it can be seen that the gain coefficient becomes more sensitive to the signal wavelength when the influence of germanium on the Er^{3+} ions becomes more dominant [46]. Figure 7.5 also indicates that very high gain coefficients, such as the 11.0-dB/mW record value published in [5], require purely germanium co-doped fibers, and under practical circumstances it may only be obtained for signal wavelengths around 1535 nm. Furthermore, Figure 7.5 confirms that aluminum should be used as an index-raising co-dopant when broadband gain characteristics are most desirable.

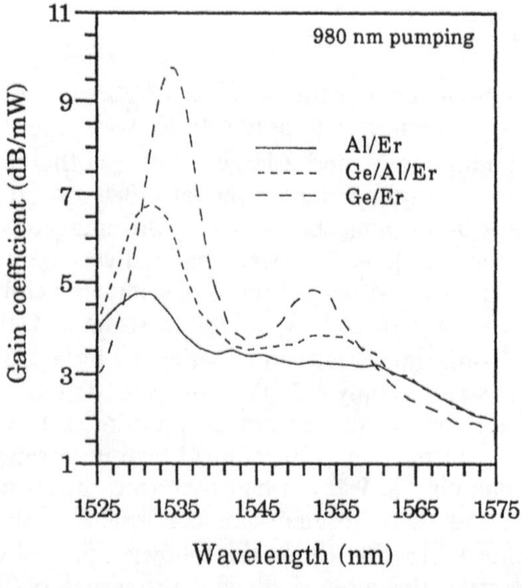

Figure 7.5 The gain coefficient versus signal wavelength in a step-index erbium-doped fiber with an NA = 0.25. The results are shown for fibers purely co-doped with either aluminum or germanium as index-raising material and for an Al/Ge fiber. (Figure from [46].)

If we instead want to look directly at the gain of the 980-nm-pumped EDFA, some of the early [37] (but typical) results are reviewed in Figure 7.6. These results are measured on a germanium-co-doped silica fiber with NA = 0.17, an LP_{11} cutoff wavelength of 955 nm, and the erbium localized in the central region of the core. Figure 7.6 shows results of diode-pumped gain as well as results obtained by dye laser pumping [37] as a function of signal output power. We first note that the semiconductor pumping and dye laser pumping with comparable amounts of coupled pump power give similar gain values. Second, gain values in excess of 40 dB are found for only 20 mW of pump power at 980 nm. Note in this connection, however, that such results may only be obtained provided that reflections do not limit the gain to lower values. These effects are described in detail in [47], where the highest reported gain value to date, 51.4 dB, was presented.

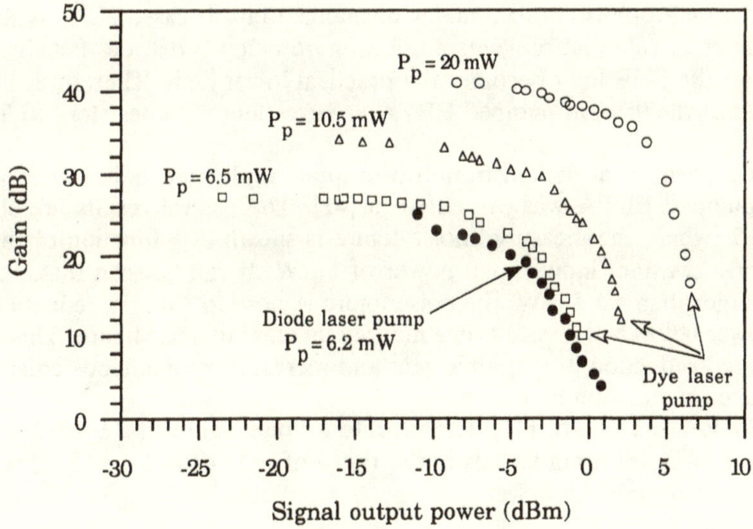

Figure 7.6 Gain versus signal output power for different pump powers in the 980-nm band. (Data from [37].)

7.4.2 Low-Noise Properties for 980-nm Pumping

For some amplifier applications, the noise properties are as important as the gain, and the 980-nm pump band has highly favorable properties in this respect. To understand the full extent of this statement, it is advantageous to consider the minimum value for the noise figure. An expression for this may be found from the rate equation by considering that infinite pump powers are available. If this is done as described in [48], a minimum noise figure limit can be expressed as

$$F(G) \geq 1 + \frac{1+\gamma}{1-\gamma}\left(1 - \frac{1}{G}\right) \qquad \gamma = \frac{\sigma_e(\lambda_p)\cdot\sigma_a(\lambda_s)}{\sigma_a(\lambda_p)\cdot\sigma_e(\lambda_s)} \tag{7.1}$$

where σ_e and σ_a are the emission and absorption cross section values, respectively, determined at either the signal wavelength λ_s or the pump wavelength λ_p. In the 980-nm pump band, where $\gamma = 0$, (7.1) shows a 3-dB limit for large gain values. Note, however, that if the gain value $G \approx 1$, we find that $F(G) \geq 1$. This shows that the noise figure may be smaller than the 3-dB limit, which was derived for large gain values. Since this sometimes may seem a little confusing, it is relevant to try to clarify the point further. We will therefore look at the noise figure expression given in (5.34) and note that the integral term vanishes if the fiber length is reduced to zero. In this case, a noise figure $F = 1$ is found, which is not surprising, since the signal-to-noise ratio remains unchanged in this case. There is therefore no fundamental rule that prevents a noise figure below 3 dB, but for all practical gain values the 3-dB limit becomes the practical lower limit. Therefore, the noise properties of the 980-nm-pumped EDFA are excellent if values close to 3 dB are obtained.

An experimental demonstration of a quantum-limited noise operation of a 980-nm-pumped EDFA was presented in [41]. The central results are shown in Figure 7.7, where the measured noise figure is shown as a function of the pump power for a constant input signal power of 1 μW. It can be seen that, for pump powers larger than ≈5.5 mW, the noise figure is close to 3 dB. A reduction of the pump power below 5 mW results in a marked increase in noise figure. This is owing to a marked reduction in amplifier gain and increased spontaneous emission due to a reduced population inversion.

Additional results of [41] demonstrate a noise figure of 2.9 ± 0.4 dB in a high-gain EDFA for input signals in the range of −35 dBm to −15 dBm and for pump powers of only 5.8 mW.

7.4.3 High Output Power Performances of the 980-nm-Pumped Erbium Fibers

We will now turn to the power amplifier applications of 980-nm-pumped EDFAs, and, more specifically, examine the QCE and noise properties as described in [49].

Figure 7.8(a) illustrates the dependence of the QCE and noise figure on the signal power input for a step-index fiber with the Er^{3+} ions uniformly distributed in the core area. The NA = 0.15 and 100 mW of pump power is launched co-propagating with the signal. Two signal wavelengths are considered: one at the emission peak location (i.e., 1532 nm (solid curves)) and one at 1555 nm (dashed

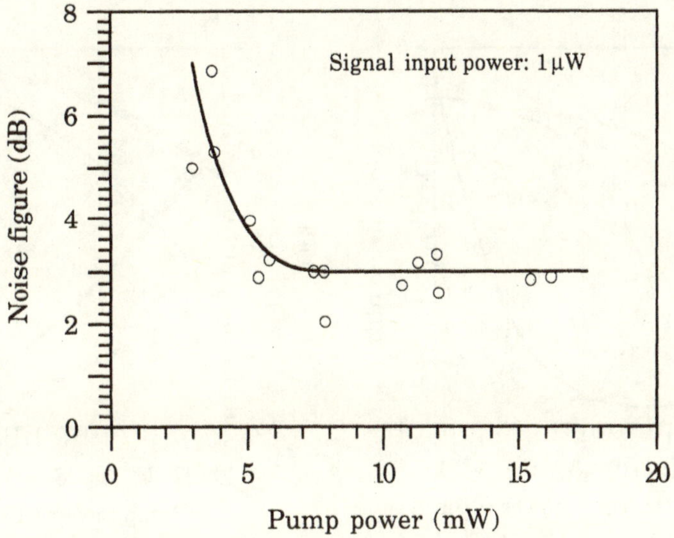

Figure 7.7 Measured amplifier noise figure as a function of 980-nm pump power for a fixed input signal of 1 μW at 1535 nm. (Figure from [41].)

curves). It should also be noted that the cutoff wavelength and the fiber length have been optimized for each point on the curve. We find an improvement both in noise figure and QCE for increasing signal powers until about −5 dBm. If the signal input power is increased above this value, the QCE levels off and a noise figure degradation is seen. This is due to a population inversion decrease in the signal input end of the EDFA. We further find that the QCE is very low (around 20%) for small signal 1555-nm signals because the signal amplification has to compete with a large ASE power around 1532 nm. Note also that the QCE at all points is largest for the 1532-nm signal, but the noise figure for the 1532-nm signal experiences the most significant degradation for signal power increases above −5 dBm.

It is noteworthy that the high QCE values shown in Figure 7.8(a) can actually be achieved in practice, and a good example is presented in [44]. Here, a signal wavelength of 1533.3 nm was chosen for a 978-nm-pumped amplifier. For only 100 mW of pump power, a 54-mW signal output power was measured (the input signal power was 0.5 mW). According to (6.1), this corresponds to a QCE of 84%, which is very close to the value illustrated in Figure 7.8(a).

Figure 7.8(b) shows the calculated QCE as a function of pump power for a signal input power of 1 mW. Also shown are the corresponding noise figures, and both parameters are analyzed for the signal wavelengths of 1532 and 1555 nm. The fibers are somewhat different in the two cases, since the cutoff wavelength is fixed

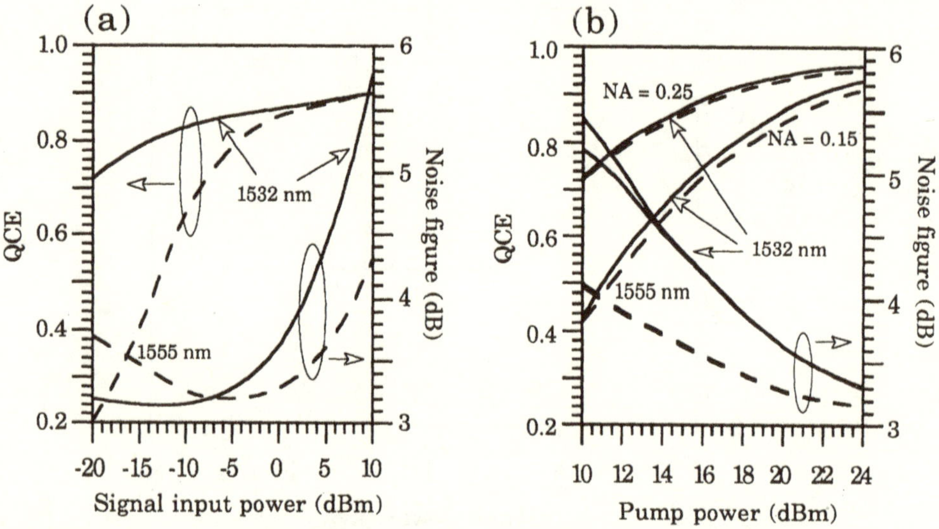

Figure 7.8 (a) Quantum conversion efficiency and noise figure shown versus signal input power for 100-mW pump power at 980 nm. The fiber has an NA = 0.15 and the full core is erbium-doped. The pump and signal are co-propagating. (b) QCE and noise figure shown versus pump power at 980 nm. Curves are shown for signal wavelengths of 1532 nm (solid curves) and 1555 nm (dashed curves). (Data from [49].)

at $\lambda_c = 750$ nm for the signal wavelength of $\lambda_s = 1532$ nm, and at $\lambda_c = 800$ nm for the signal wavelength of 1555 nm. The fiber lengths are optimized with respect to maximum QCE.

It can be seen in Figure 7.8(b) that in the region from 25 to 100 mW (i.e., 14 to 20 dBm), which corresponds to the range of pump powers that are typically used for packaged amplifiers at present [49], the QCE can be improved by up to 52% by increasing the NA from 0.15 to 0.25. For applications where the noise figure is important, it is relevant to note that reductions of more than 1 dB can be obtained by going from a signal wavelength of 1532 nm to a signal wavelength of 1555 nm. It is, however, also noteworthy that the noise figure is insensitive to the choice of NA except at small pump powers. The reason is that increasing the NA has two counter balancing effects on the noise figure [49]. First, the stimulated emission rate is increased, leading to a larger gain and a lower noise figure. Second, the fraction of spontaneous emission captured by the fiber is increased, tending towards a higher noise figure. In the case where the fiber length and cutoff wavelength are optimized with respect to QCE, these two contributions apparently cancel each other out.

7.4.4 Choice of Pump Wavelength in the 980-nm Pump Band

In the two previous sections, we have discussed the small-signal and power amplifier properties of 980-nm-pumped EDFAs, and very good performance has been documented. However, all these investigations do not include the fact that the 980-nm absorption band of erbium is relatively narrow (typical values for the peak seen in Figure 3.2 are FWHM ≈ 16 nm for Al/Er and ≈ 11 nm for Ge/Er silica [50]). This narrow linewidth gave rise for a long time to serious concerns and problems for users and manufacturers of semiconductor pump sources for EDFAs. The demand was, namely, only for devices that emitted light within a few nanometers from the absorption peak. These strict requirements led to a low yield and increased cost of 980-nm pump sources. However, detailed experimental and theoretical investigations of EDFAs pumped between 960 and 1000 nm have lately shown that previously believed specification limits may be loosened significantly [40, 50–52]. We will therefore review some of the most significant results of this work.

We will start by looking at Figure 7.9, which shows the signal output power as a function of pump wavelength for a Ge/Al/P/Er-doped silica fiber with an NA = 0.18 and a cutoff wavelength of 940 nm. The signal input power is −1.4 dBm, and the results are obtained for three different fiber lengths, each pumped with 40 and 80 mW. The theoretical curves are found from a model as described in Section 5.1, and details of the experimental setup can be found in [50, 53]. Also, the GSA cross-section spectrum is indicated in Figure 7.9. It is found that the signal output power and QCE increase with pump power, reaching 16.3 dBm and 0.83, respectively, for the 16 m long fiber pumped with 80 mW at 979 nm. Figure 7.9 also illustrates that the magnitude of the gain penalty for pump wavelengths away from 979 nm is significantly reduced at high pump powers (e.g., at 80 mW the penalty is <0.5 dB for a pump wavelength range of ±9 nm about the peak). The corresponding range for 40 mW is only ±5 nm. As observed in [11, 40, 50, 52], the pump wavelength detuning results in a decreased absorption coefficient, which, however, is partly compensated for by the increased fiber length.

With a model application, the above results were extended to a thorough examination of the effects of fiber length and NA on a power amplifier pumped away from the absorption peak [50]. One of the results of this work is shown in Figure 7.10, which may be used to quantify the tradeoffs between pump wavelength, amplifier gain, and noise figure. The specific design has an NA = 0.2, and a pump power of 50 mW is used to amplify a 1551-nm signal with an input power of 1 mW. The solid curves are contours for progressively larger gain penalties ranging from 0 to 1.0 dB in steps of 0.2 dB. The dashed curves are for the different noise figure penalties, also shown in steps of 0.2 dB. Both penalties are given relative to the optimum operating point found for 979-nm pumping (here, the gain is 14.5 dB and the noise figure is 3.7 dB).

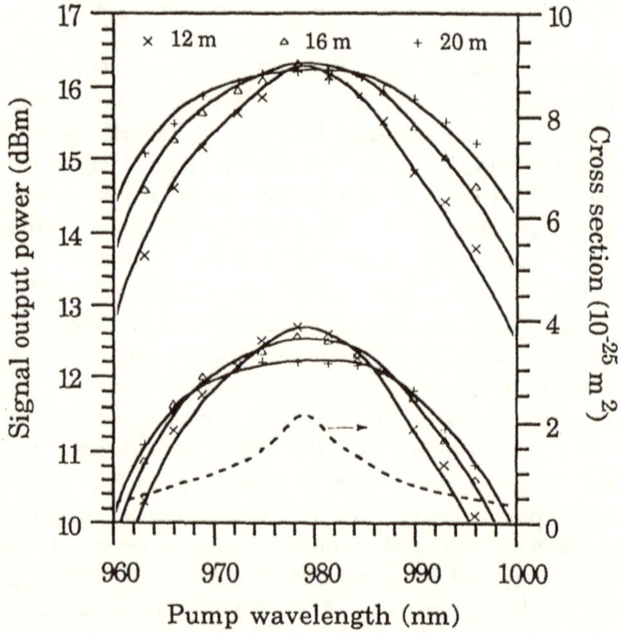

Figure 7.9 Signal output power versus pump wavelength for three different fiber lengths pumped by 40 and 80 mW (the higher pump power resulting in the higher output). The 1551 nm signal input is −1.4 dBm. The experimental data are indicated by marks and the curves are numerical simulations. (Figure from [50].)

Figure 7.10 makes it possible to find the fiber length that maximizes the pump wavelength tolerance, and it can be used to establish gain and noise figure at any pump wavelength and fiber length. An example, for which the pump wavelength is 985.5 nm, is indicated by an open circle in Figure 7.10. We find here that the highest gain may be obtained for a fiber length that is 50% longer than the reference fiber. Compared to the reference values, the gain is lowered by 0.2 dB and the noise figure increased by 0.5 dB. From Figure 7.10, it may be concluded that the fiber length, which gives the maximum gain, has a minimum with respect to pump wavelength at 979 nm. This is also the wavelength, which provides a minimal noise figure. Furthermore, we see that if only gain is considered, it is better to use a too-long fiber length than a too-short one because this allows a wider pump wavelength variation.

The gain penalty contour curves, furthermore, are strongly sensitive to the NA, whereas the noise figure penalty is less pronounced. If we therefore focus on the gain penalty, Figure 7.11 can be used to quantify the dependence of the NA.

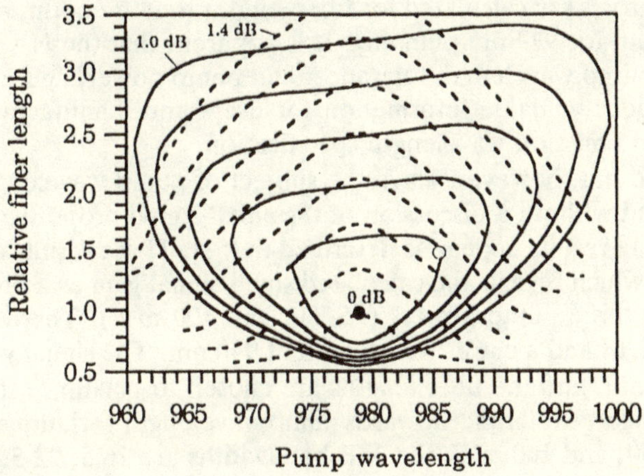

Figure 7.10 Gain and noise figure versus relative fiber length and pump wavelength. The fiber has an NA = 0.2 and all other conditions are the same as those in Figure 7.9. The solid curves represent different gain penalties ranging from 0 to 1 dB in steps of 0.2 dB. The dashed curves are noise figure penalties ranging from 0 to 1.4 dB, also in steps of 0.2 dB. (Figure from [50].)

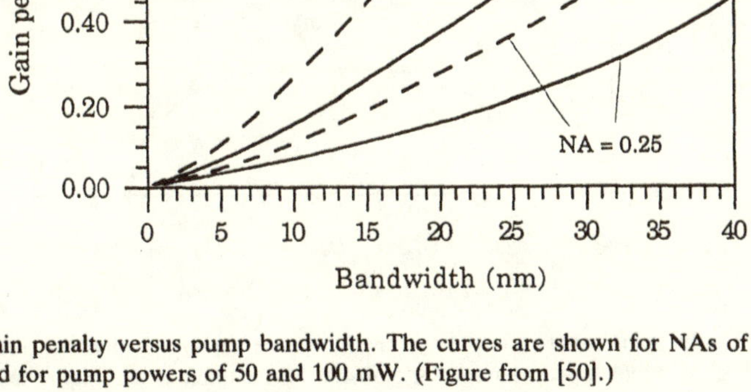

Figure 7.11 Gain penalty versus pump bandwidth. The curves are shown for NAs of 0.15 and 0.25, and for pump powers of 50 and 100 mW. (Figure from [50].)

Note that the curves are calculated for fiber lengths that are optimized with respect to maximum gain for 979-nm pumping. It is apparent that the NA is more critical for increasing pump wavelength tolerance than pump power, and Figure 7.11 consequently provides valuable information for users and manufacturers of 980-nm LDs in terms of emission wavelength specification.

We should not, however, leave the subject of pump wavelength variation in the 980-nm band without a discussion of the small-signal properties because some highly interesting results appear as described in [51]. These results are reproduced in Figure 7.12, which reveals the calculated small-signal gain as a function of pump wavelength for three pump powers (25, 50, and 100 mW). The considered fiber has an NA = 0.18 and a cutoff wavelength of 940 nm. The signal wavelengths are 1532 and 1551 nm, and the fiber lengths are chosen to maximize the gain. Figure 7.12 illustrates a large tolerance towards pump wavelength variations, and for pump powers of 25, 50, and 100 mW, the 3dB bandwidths are 16.5, 22.5, and 27 nm for a signal at 1532 nm. The respective linewidths are 33, 47.5, and 56 nm for the 1551-nm signal. The wider bandwidths for the 1551-nm case are due to a more favorable emission-to-absorption cross-section ratio, which permits gain with lower excited-state populations. It is equally interesting to find that for a signal at the peak of the stimulated emission cross-section spectrum (1532 nm for the present fiber), the optimum pump wavelength will always be at the peak of the pump absorption (979 nm). For a signal at the high-wavelength shoulder of the emission spectrum

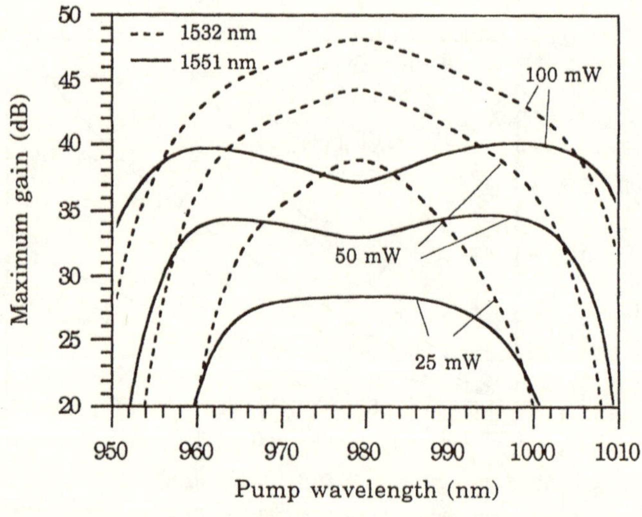

Figure 7.12 Maximum gain versus pump wavelength for the signal wavelengths and pump powers indicated. Fiber lengths have been adjusted to maximize gain. (Figure from [51].)

(1551 nm), however, higher gain may be obtained by detuning the pump wavelength as much as ≈20 nm to each side of the 979-nm peak.

The explanation for this phenomenon is given in [51] as follows. A co-directionally pumped amplifier is considered. In the first part of the erbium-doped fiber, both pump and backward-propagating ASE is strong, which leads to a significant recycling of the erbium ions through the stimulated emission and absorption. As a consequence, there is a significant power transfer from the pump to the backward propagating ASE. In the middle section of the amplifier, both forward- and backward-propagating ASE powers are low and the inversion is high, since the only deexcitation process is spontaneous emission. The gain, therefore, increases most rapidly in this middle section of the amplifier, and the pump attenuation is lowest. In the third part, the forward propagating ASE has reached a level where stimulated emission has again become important, but the low pump intensity is not sufficient to maintain inversion and the gain per unit length falls off. Pumping at 1000 nm provides higher gain than pumping at the absorption peak because less pump power is lost to backward-propagating ASE in the first section of the fiber. This is a result of the lower absorption cross section at 1000 nm, which reduces the rate at which the erbium ions are re-excited and then de-excited by stimulated emission into the backward-propagating ASE. Although the gain is the same for both pump wavelengths, calculations in [51] show that roughly twice as much pump power remains at 1000 nm than at 979 nm. The effect is that the pump power in the detuned case is transferred to the second region of the fiber, where it is more efficient at producing gain. This is also consistent with the higher noise figure of the 1000-nm-pumped amplifier, since the lower pump rate, which reduces loss in backward propagating ASE, also means a lower population inversion in the critical first section of the amplifier. More specifically, a noise figure penalty of 0.6 dB per 10 nm of detuning is estimated in [51].

7.5 1480-nm PUMPING

Pumping EDFAs in the $^4I_{15/2}$ to $^4I_{13/2}$ absorption band (around 1480 nm) was found attractive at an early stage of the amplifier development. This was because ESA is absent and since the development of laser diode pumps could take advantage of the already advanced technology for InGaAsP 1.5 μm laser diodes [54]. High gain performance of fiber amplifiers was therefore already demonstrated in the beginning of 1988 [55]. Since then, numerous results have been published [56,70,71] using laser diodes with emission spectra centered between 1470 and 1490 nm. Among some of the most remarkable results are gain values of 46.5 dB [57] and gain coefficients of 6.3 dB/mW [58] obtained by 1480-nm pumping. In the discussion that follows, we will go into further detail on the subject of pumping in the 1480-nm band.

7.5.1 Gain and Noise Figure Dependence of Pump Wavelength Between 1460 and 1510 nm

As may be expected from a simple inspection of the emission and absorption spectra of Er^{3+} ions in silica (see Figure 3.4), the pump band centered around 1480 nm is relatively broad. This appears in Figure 7.13(a), which shows gain as well as noise figure measurements as a function of the pump center wavelength [59]. The fiber amplifier was, in this situation, pumped in a co-propagating configuration with a tunable color center laser having a narrow spectral width of 0.1 nm. The fiber consisted of a 15-m-long germano-silicate erbium-doped fiber characterized by an NA of 0.17 and a cutoff wavelength at 1050 nm. The gain values plotted in Figure 7.13(a) are measured for a signal wavelength at 1536 nm, and a maximum gain of 29.6 dB was obtained with 54-mW pump power. We also notice that the gain variation is within ≈2 dB for pump wavelengths between 1460 and 1490 nm. At the lower pump power of 21 mW, the gain variation with pump wavelength is more pronounced, with the peak gain occurring near 1480 nm.

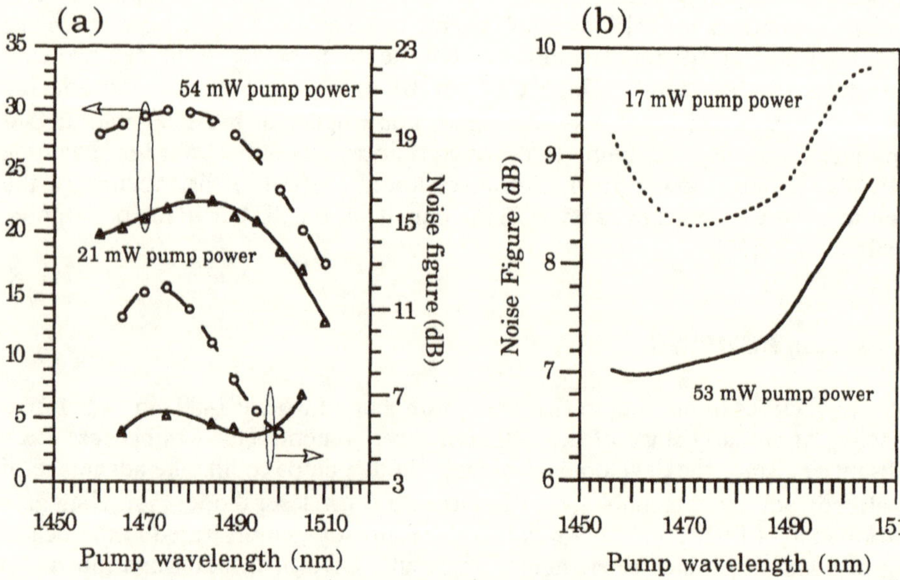

Figure 7.13 (a) Fiber amplifier gain and noise figure shown versus pump wavelength. Curves are shown for two different pump powers of 21 and 54 mW. The signal input power was −34 dBm. (Data from [59].) (b) Calculated pump wavelength dependence of noise figure. (Results from [62].)

The noise of the fiber amplifier was measured at the signal wavelength using a heterodyne receiver [59]. The amplifier noise figure was obtained by comparing the *carrier-to-noise ratio* (CNR) measured with the amplifier with that obtained without the amplifier. The principal assumption here is that the local-oscillator spontaneous beat noise is dominant, after which noise figures may be derived as described in [59, 60]. The pump wavelength-dependent results, which have an estimated accuracy of ±0.5 dB, are plotted in Figure 7.13(a). With 54-mW pump power, a noise figure of 12 dB was obtained at 1475 nm. The minimum noise figure of 5.2 dB was obtained at a pump wavelength of 1500 nm, where the gain is ≈7 dB below the maximum value. With a lower pump power of 21 mW, a minimum noise figure of 5.2 dB was also obtained at a pump wavelength of 1500 nm, and there is much less variation with the pump wavelength. A general comment on the noise figures of Figure 7.13(a) is that the pump dependence is somewhat surprising and anomalous. However, the results have been included here anyway to point out that many properties have to be taken into account in the evaluation of fiber amplifier data, and we will here try to give a possible explanation of the results. To understand the higher noise figure measured for the higher pump power, it must be remembered that the fiber length is fixed in all experiments. For the co-propagating pump configuration, Figure 5.6 illustrates that no serious increase is seen in the noise figure, even if the fiber length is extended above the gain maximal length. This is because the population inversion in the fiber input end is maintained relatively high. This means that the noise figure shown in Figure 7.13(a) for the 21-mW pump power may very well be measured on a fiber that is too long with respect to gain. As the pump power is increased, the forward and backward ASE is increased, resulting in a larger noise figure in the majority of the considered pump wavelength range.

The noise figure results shown in Figure 7.13(a) were measured on a germanium-co-doped silica fiber. However, as pointed out in [61], some erbium glass compositions (e.g., Al_2O_3-SiO_2) are more suitable for pumping near 1480 nm. The reason is that the ratio $\sigma_e(\lambda_p)/\sigma_a(\lambda_p)$ in such host glasses can be made smaller (see Figure 3.4), whereby maximum inversion may be achieved. Also, the noise figure dependence of pump wavelength in the 1480-nm band can be found to be affected by the host glass composition.

A more typical variation of the noise figure with pump wavelength is shown in Figure 7.13(b). The results are obtained for a Al/Er-doped fiber with a core radius of 4.5 μm and an NA of 0.175. The input signal power is −40 dBm, and curves are shown for two pump power levels of 17 mW and 53 mW, respectively. A fixed fiber length has been chosen. From Figure 7.13(b) it can be found that the noise figure varies only very little (±0.2 dB) within a typical bandwidth of 20 nm. It can also be seen that this noise optimal pump wavelength band depends on the pump power level. Finally, it should be noted that minor differences between theoretical and experimental results can be seen unless the spectral bandwidth of the pump source is considered.

In Figure 7.14, we have shown the gain coefficient as a function of the signal wavelength for three erbium-doped fibers with a different composition of index-raising co-dopants. Each point is calculated for gain-optimized fiber length. As in the discussion of Figure 7.5, one of the fibers is purely aluminum-co-doped, another is purely germanium co-doped, and the third is fabricated with both aluminum and germanium. The NA of the considered fibers is 0.25. For the considered fiber design, a maximum gain coefficient of 4.6 dB/mW is found, which is significantly smaller than for 980 nm pumping [46]. We also notice that, in agreement with the suggestions of [61], the aluminum-co-doped fiber appears to have the most favorable properties, here reflected through a broader and flatter variation of the gain coefficient with the pump wavelength.

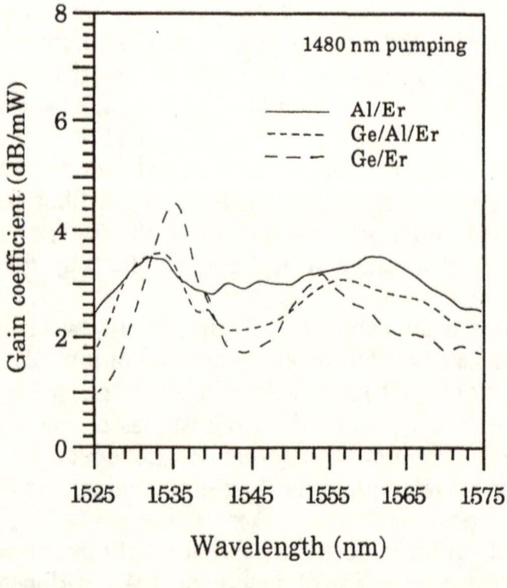

Figure 7.14 Gain coefficient versus signal wavelength in a step-index erbium-doped fiber with an NA = 0.25. The results are shown for fibers either purely co-doped with aluminum or germanium as index-raising material and for an Al/Ge fiber. (Figure from [46].)

7.5.2 1480-nm-Pumped Power Booster Amplifiers

To describe the properties of the 1480-nm-pumped EDFA in booster applications, comparisons similar to those presented in Section 7.3.3 for 980-nm pumping are included. To facilitate the comparisons, the only difference has been the exchange of pump wavelength, and for the curves in Figure 7.15(b), the cutoff wavelength was chosen to be 900 nm in all cases.

For the signal input power dependence as described in Figure 7.15(a), we find that, for signal inputs larger than −10 dBm, the QCE for 1555 nm is larger than that for 1532 nm, while the opposite is true for any input signal levels below −10 dBm. As in the case of 980-nm pumping, the QCE levels off for input signal powers above a certain level, which for 1480-nm pumping is above −5 dBm. It should also be noted that the noise figure is lowered 0.8 dB or more by choosing the 1555-nm signal wavelength compared to that at 1532 nm. This means that, for 1480-nm pumping, the 1555-nm signal is preferable with respect to QCE as well as noise figure for signal powers above −10 dBm.

The advantages of choosing the higher signal wavelength are also evident from Figure 7.15(b), where the dependence of pump power is displayed. Noise figure reductions of more than 2 dB are found here. As with QCE, the lower inversion achieved by pumping at 1480 nm favors the longer signal wavelength for the noise figure because of the larger emission-to-absorption ratio. With respect to the large noise figure reductions, it should also be mentioned that the values can be up to 3 dB worse for 1480-nm pumping than it is for 980-nm pumping.

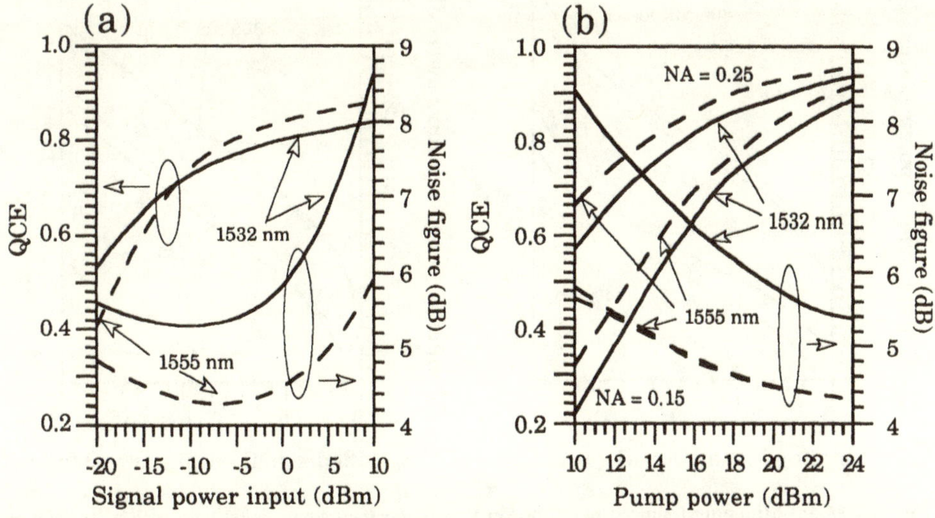

Figure 7.15 (a) QCE and noise figure shown versus signal input power for 100-mW pump power at 1480 nm. The fiber has an NA = 0.15 and the full core is erbium-doped. The pump and signal are co-propagating. (b) QCE and noise figure shown versus pump power at 1480 nm. Curves are shown for signal wavelengths of 1532 nm and 1555 nm. (Data from [49].)

7.6 COMPARISON OF DIFFERENT PUMP BANDS

In the previous sections, a number of characteristics have been described for pumping EDFAs in different pump bands, but even though other pump sources may find some use in future applications, the most promising are the semiconductor pumps. It is therefore highly relevant to compare the properties of EDFAs excited in either the 800-, the 980-, or the 1480-nm pump band. It is useful to initiate this discussion with reference to the recent work of [30, 32].

For small-signal applications, it is interesting to consider the results shown in Figure 7.16(a), where the minimum required pump power for a specified small-signal gain is found. The considered erbium-doped Al/P silica fiber has an NA of 0.25 and the erbium is uniformly distributed throughout the core. The fiber length and the cutoff wavelength (see Section 6.3) are optimized to give the minimum required pump power for each point in the figure. Also, the pump wavelengths have been optimized in the 800-nm band, as described in Section 7.2.1. Curves are shown for co- and bi-directional pumping in the 800-nm band and for co-directional pumping at 980 nm and at 1480 nm [32]. When pumping bi-directionally in the 800-nm pump band, approximately 1 dB less pump power is required as compared

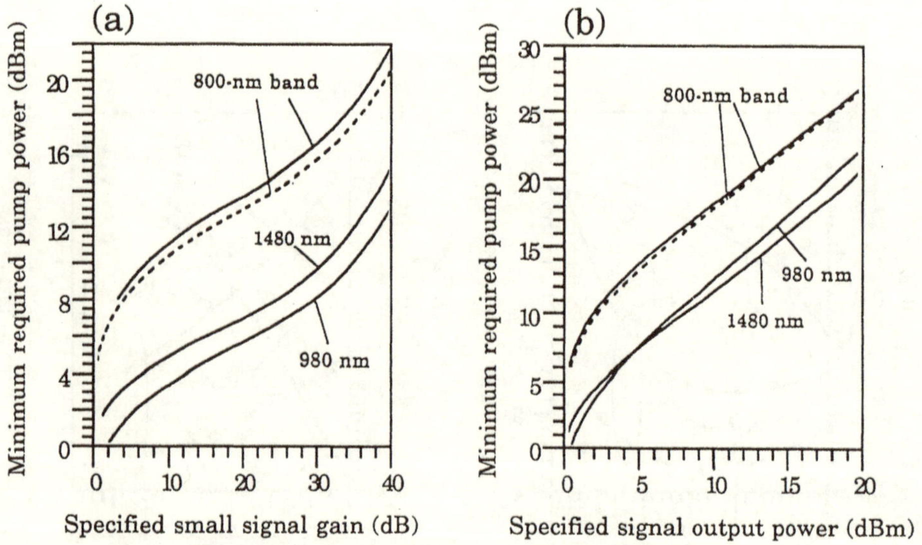

Figure 7.16 (a) Minimum required pump power versus specified small-signal gain with a 1-mW input signal. (b) Minimum required pump power versus specified output signal power with a 1-mW input signal. In both figures, the dashed curves are for bi-directional pumping and the solid curves are for co-directional pumping. (Figures from [32].)

to co-directional pumping. The most efficient pump band is at 980 nm, since it requires ≈ 2 dB less pump power compared to 1480 nm pumping and 7 to 8 dB less pump power than that needed in the 800-nm pump band to provide the same small-signal gain. Whereas the lower efficiency of the 800-nm pump band is due to pump ESA, the 1480-nm pumping suffers from the inability of obtaining population inversions above ≈ 0.7. Note finally that, in the discussion of Figure 7.16(a), the curves have been experimentally verified for pumping in the 800- and 980 nm pump bands [30].

Another central application of the EDFA is as a power booster, where the objective is to maximize the signal output power. To compare the three pump bands with respect to this goal, Figure 7.16(b) has been included. The required pump power is shown here as a function of the specified signal output power for EDFAs with a signal input power of 1 mW at 1532 nm. The same conditions as those described for Figure 7.16(a) are valid. Figure 7.16(b) shows that the 800-nm pump band is also in this case 6 to 7 dB less efficient than the other two. However, for a specified output power above 5 dBm, the pump power requirements are smallest for 1480-nm pumping. The difference with the 980-nm pumping is as much as 1.7 dB, which is because less photon-to-photon power is lost in the conversion from 1480 to 1532 nm than in the case of 980-nm pumping. Because the quantum efficiency of 980-nm LDs is approximately twice that of 1480-nm LDs [14], however, the 980-nm LDs will require less drive current to provide the same output power from the booster amplifier [32].

For many applications of EDFAs, the noise figure is as important as the gain, and it may often be advantageous if the fiber length is cut back from the maximum gain length [48] to improve the noise figure. The specified gain may then only be retained if the pump power is increased [32]. In Figure 7.17, the required pump power is plotted as a function of the specified noise figure for the same pumping case as that treated for Figure 7.16(a). Three curves are shown at each pump band for different specified small-signal gains of 15, 25, and 35 dB.

We first note that noise figures below 5 dB are not practical for 1480-nm pumping, since resonant pumping leads to a minimum F of 4.8 dB for a 1532-nm signal. This specific minimum value may vary somewhat for different cross-section spectra and choices of pump wavelength as indicated by (7.1). It is further observed in [32] that pump powers slightly above the minimum required can significantly improve the noise figure. It was also found that when the specified noise figure is below 4.0 dB, the best pump wavelength in the 800 nm band is 806 nm (±3 nm) and the best pump configuration is the co-directional one. However, the pump power required to obtain similar amplifier performance is 7 to 8 dB higher when pumping in the 800-nm band as compared to pumping at 980 nm.

We have now compared the three currently most relevant pump bands with respect to both gain, signal output power, and noise figure, and have specifically focused on the pump power consumption of the amplifier. A slightly different

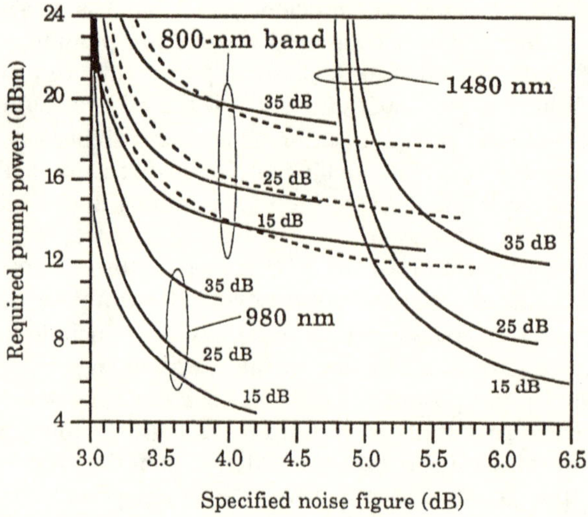

Figure 7.17 Required pump power versus specified noise figure for specified small-signal gains of 15, 25, and 35 dB. Dashed curves are for bi-directional pumping and solid curves are for co-directional. (Figure from [32].)

approach is taken for the results presented in Figure 7.18, where each point indicates a significant experimental result published in the literature. The illustration shows gain versus pump power, and, in addition, lines representing constant gain coefficients are included. Besides the results for 800-, 980-, and 1480-nm pumping, a few values obtained by 532- and 665-nm pumping are also included. It should be noted that the different points are obtained by pumping at different wavelengths within each pump band, and the signal wavelengths can be varied from point to point. However, this detailed information is not considered necessary here, since the objective is to give an overall view of the best experimental results presented in the literature.

Figure 7.18 clearly indicates that more pump power is needed for 800-nm-pumped EDFAs than for 980- and 1480-nm pumping. It can also be seen that one of the highest gain coefficients obtained in the 800 nm is 1.3 dB/mW [25]. It should also, however, be noted that the larger energy difference between pump and signal photons diminishes the obtainable gain coefficient at a given pump wavelength.

Besides one result on a 532-nm-pumped EDFA, all gain coefficients above 2 dB/mW are published for either 980- or 1480-nm pumping with values of 11.0 dB/mW as the present record [5] achieved by 980-nm pumping. It should also be noted that most of the plotted gain values in Figure 7.18 are between 20 dB and 35 dB, which represent absolutely satisfactory gain values for most practical small-

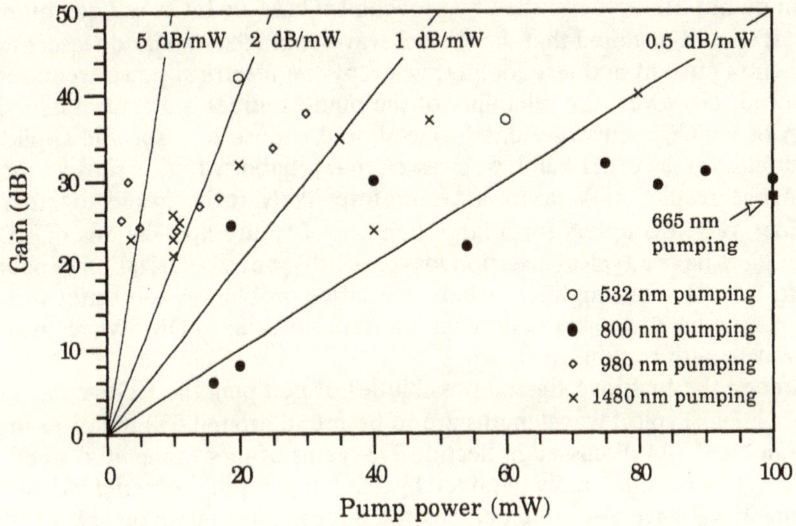

Figure 7.18 Experimentally obtained gain values shown versus pump power for pumping at 532 and 665 nm, and in the 800-, 980-, and 1480-nm pump band. (Data published in [2, 5, 20–22, 24–28, 38, 40, 43, 57, 58, 63–69, 70, 72].)

signal amplifiers. The more modest gain coefficients of 1 to 2 dB/mW will, in the same way, also be sufficient for most applications. The results collected in Figure 7.18, therefore, do not supply any additional or decisive arguments for choosing either 1480- or 980-nm pump sources. On the contrary, the many high-performance results indicate that all the listed pump bands are likely to be found useful in future EDFA applications.

7.7 SUMMARY

This chapter was initiated by a short review of the present status of semiconductor pump sources. It was discussed how gains in excess of 20 dB can be obtained by inexpensive GaAlAs laser diodes that emit light around 820 nm. The principal point was, however, that QW lasers are particularly attractive for fiber amplifier pumping because of low threshold currents and increased slope efficiencies compared to bulk lasers. The use of such sources has led to demonstrations of highly efficient amplifier pumping, obtaining gain values of more than 20 dB with less than 5 mW of pump powers at 980 nm, as well as at 1480 nm. The 980-nm laser diode pumps typically use strained InGaAs quantum wells grown on GaAs, whereas

1480-nm pumps are accomplished by growing InGaAs or InGaAsP quantum wells on InP. It should be noted that the shorter wavelength 980-nm diode lasers require less injection current and less cooling, whereby the electrical power requirements are reduced. However, the reliability of the pump sources is a central issue in the decision of which pump wavelength one should choose in a specific application, but it should, on the other hand, be stressed that reliability testing shows promising performance results. QW lasers are therefore likely to be found together with fused-fiber WDM couplers for a large number of future applications of EDFAs. Such couplers have a typical insertion loss of 0.1 dB, but a polarization dependence of 0.2 to 0.5 dB. For amplifiers where the latter problem is too serious, it may, however, be advantageous to apply an interference filter WDM, which also has a flatter wavelength response.

Among the first investigated possibilities of pumping the EDFA was the use of visible light. Typical wavelengths are to be found around 532 nm or in the 640- to 685-nm band. As discussed in Section 7.2, gains of more than 40 dB and noise figures of 3 to 4 dB are readily obtained by 532-nm pumping. Around 665-nm gains exceeding 25 dB have also been demonstrated for modest pump powers of 40 mW. However, for both these possible pump bands, the fundamental problem is the lack of semiconductor laser diodes. This will probably limit the use of these pump bands to laboratory setups, where they may, on the other hand, prove very useful.

It was shown that an optimal pump wavelength exists in the long-wavelength tail of the 800-nm pump band. The specific location, which for realistic pump power levels of 50 to 100 mW is between 815 and 825 nm, is dependent on both the pump power level as well as the pump configuration. The optimal pump wavelength is determined in the high-wavelength tail as the most favorable tradeoff between reduced GSA and increased GSA-to-ESA ratio. It may furthermore be concluded that the smallest noise figures are obtained for co-directional pumping, whereas the highest gain values are achieved for bi-directional pumping. The reason is that the high population inversion in the signal input end of the co-directionally pumped fiber on one side leads to a low noise figure, but at the same time results in increased ESA.

The 980-nm pump band may in almost all respects be considered an extremely efficient and advantageous choice. The reasons for this are that we may not only obtain noise figures around the quantum limit of 3 dB, but gain values in the excess of 40 dB may be obtained for only 20 mW of pump power. The efficiency of the 980-nm pump band is also expressed through the obtained gain coefficient record of 11.0 dB/mW and the highest demonstrated gain values of more than 50 dB. Also, as the power booster amplifier, the 980-nm-pumped EDFA has remarkably good properties. QCEs of more than 85% may be expected for practical EDFAs with corresponding noise figures below 3.5 dB. It is noteworthy that significant QCE improvements can be obtained by increasing the NA of the fiber, whereas the noise figure is almost insensitive to variations of the NA except for small pump powers (below 15 dBm). Note finally that in applications where the noise figure is relevant, it is advantageous to choose a signal wavelength around 1550 nm and not around the emission cross section peak.

The tolerances on pump wavelengths and fiber lengths for power EDFAs pumped in the 980-nm band showed that even though fiber length and pump power play an important role, a higher fiber NA was found to be the dominant consideration. It may be concluded, therefore, that the higher the NA, the less sensitive the gain is to length and the wider the range of pump wavelength is that can be tolerated for a given fiber length. Although the noise figure is not directly affected by NA, for a given pump wavelength, increasing the NA permits the desired gain to be obtained with a shorter fiber. This will, in turn, improve the noise figure.

The most widely used pump wavelength today is at 1480 nm. This has several reasons, among which the most significant are as follows: the 1480-nm band suffers from no ESA and the pump laser diode technology was rather mature at the time of the first LD-pumped EDFAs. In addition, very good performance results have been demonstrated, and the pump reliability testing is further developed than for 980-nm pumping. Finally, 1480-nm pumping is necessary for applications, where a low pump power attenuation is significant (e.g., for distributed or remote pumped amplifiers).

Among the most significant results for 1480-nm pumping are gain values above 45 dB and a gain coefficient of 6.3 dB/mW. However, for more practical applications, gain values of 20 to 30 dB may be obtained with 1480-nm pump powers of 10 to 30 mW. Since the 1480-nm-pumped amplifier fundamentally works as a two-level laser system, total population inversion cannot be obtained. This means that noise figures much below 5 dB are unreachable in this pump band. However, for many applications, this limitation may not be a problem.

Comparisons of the 800-, 980-, and the 1480-nm pump band were performed with respect to gain and noise figure. It was found that when pumping in the 800-nm band, pump power of approximately 7 to 8 dB higher is required to obtain the same small-signal amplifier performance as when pumping at 980 nm. To obtain equal small-signal gain values, the 1480-nm pumping sets a requirement of 2 dB of additional pump power compared to 980-nm pumping. Minimum noise figures of $\approx$ 5 dB are obtainable with 1480-nm pumping, whereas 980-nm pumping may allow noise figures slightly above 3 dB for practical amplifiers. For power amplifiers, the same specified signal output power requires pump power 6 to 7 dB higher for the 800 nm band. With respect to optical power consumption, 1480-nm pumping seems most favorable for booster amplifiers with signal output powers above 5 dBm. However, recent comparisons of the electrical power consumption of 980- and 1480-nm semiconductor pump lasers lean in favor of 980-nm pumping.

However, not included in the present comparisons are questions such as cost, and, more important, the pump source reliability. These issues, though, are far from clear at this point of EDFA development, and only future demands for pump sources and their actual performance in operating communication systems or measurement setups may provide these answers. It is important to add that the properties of EDFAs today have already been shown to be so highly advantageous that amplifiers also, which are not working under absolute optimal conditions, may still add considerable system improvements. It is therefore far too early to rule out future applications of 800-nm pumping.

REFERENCES

[1] Poole, S. B., D. N. Payne, R. J. Mears, M. E. Fermann, and R. I. Laming, "Fabrication and Characterization of Low-Loss Optical Fibers Containing Rare-Earth Ions," *IEEE Journal of Lightwave Technology*, Vol. 4, No. 7, 1986, pp. 870–876.

[2] Mears, R. J., L. Reekie, I. M. Jauncey, and D. N. Payne, "High-Gain Rare-Earth-Doped Fiber Amplifier at 1.54 μm," *Optical Fiber Communications OFC '87*, U.S., 1987, Proc. Vol. 1, paper WI2.

[3] Geusic, J. E., R. L. Hartman, U. Koren, W. T. Tsang, and D. P. Wilt, "Quantum Well Lasers in Telecommunications," *AT&T Technical Journal*, January/February 1992, pp. 75–83.

[4] Suzuki, K., Y. Kimura, and M. Nakazawa, "High Gain Er^{3+}Doped Fibre Amplifier Pumped by 820 nm GaAlAs Laser Diodes," *IEE Electronics Letters*, Vol. 26, No. 13, 1990, pp. 948–949.

[5] Shimizu, M., M. Yamada, M. Horiguchi, T. Takeshita, and M. Okayasu, "Erbium-Doped Fibre Amplifiers With an Extremely High Gain Coefficient of 11.0 dB/mW," *IEE Electronics Letters*, Vol. 26, No. 20, 1990, pp. 1641–1643.

[6] Zyskind, J. L., D. J. DiGiovanni, J. W. Sulhoff, P. C. Becker, and C. H. Brito Cruz, "High Performance Erbium-Doped Fiber Amplifier Pumped at 1.48 μm and 0.97 μm," *Techn. Dig. of Optical Amplifiers and Their Applications OAA '90*, Monterey, 6–8 Aug. 1990, postdeadline paper PDP6.

[7] Zyskind, J. L., C. R. Giles, J. R. Simpson, and D. J. DiGiovanni, "Erbium-Doped Fiber Amplifiers and the Next Generation of Lightwave Systems," *AT&T Technical Journal*, January/February 1992, pp. 53–62.

[8] Wu, M. C., Y. K. Chen, M. Hong, J. P. Mannaerts, M. A. Chin, and A. M. Sergent, "A Periodic Index Separate Confinement Heterostructure Quantum-Well Laser," *Applied Physics Letters*, Vol. 59, No. 9, 1991, pp. 1046–1048.

[9] Tanbun-Ek, T., R. A. Logan, N. A. Olsson, H. Temkin, A. M. Sergent, and K. W. Weckt, "High Power Output 1.48–1.51 μm Continuous Graded Index Separate Confinement Strained Quantum Well Lasers," *Applied Physics Letters*, Vol. 57, No. 3, 1990, pp. 224–226.

[10] Koren, U., et al., "High Power Laser-Amplifier Photonic Integrated Circuit for 1.48 μm Wavelength Operation," *Optical Fiber Communications OFC '91*, U.S., 1990, Proc. Vol. 2, postdeadline paper PD10.

[11] Wada, M., K. Yoshino, M. Yamada, and J. Temmyo, "0.98-μm Strained Quantum Well Lasers for Coupling High Optical Power Into Single-Mode Fiber," *IEEE Photonics Technology Letters*, Vol. 3, No. 11, 1991, pp. 953–955.

[12] Major, J. S., W. E. Plano, D. F. Welch, and D. Scifres, "Single-Mode InGaAs-GaAs Laser Diodes Operating at 980 nm," *IEE Electronics Letters*, Vol. 27, No. 6, 1991, pp. 539–541.

[13] Wada, M., J. Temmyo, and M. Shimizu, "High-Power Semiconductor Lasers for Er-Doped Fiber Amplifiers," *Optical Fiber Communications OFC '92*, San Diego, 1992, invited paper WF2, p. 115.

[14] Mito, I., and K. Endo, "1.48 μm and 0.98 μm High-Power Laser Diodes for Erbium-Doped Fiber Amplifiers," *OSA Techn. Dig. 13, Optical Amplifiers and Applications OAA '91*, Snowmass, CO, 24–26 July 1991, pp. 22–25.

[15] Kamei, H., M. Yoshimura, H. Kobayashi, N. Tatoh, and H. Hayashi, "High-Power Operation of 1.48 μm GaInAsP/GaInAsP Strained-Layer Multiple Quantum Well Lasers," *OSA Tech. Dig. 13, Optical Amplifiers and Applications OAA '91*, Snowmass, CO, 24–26 July 1991, pp. 30–33.

[16] Grubb, S. G., P. A. Leilabady, K. L. Sweeney, W. H. Humer, R. S. Cannon, and S. W. Vendetta, "High Output Power Er^{3+}/Yb^{3+} Co-doped Optical Amplifiers Pumped by Diode-Pumped Nd^{3+} Lasers," *Techn. Dig. of Optical Amplifiers and Their Applications OAA '92*, Santa Fe, 24–26 June 1992, paper FD1, pp. 202–205.

[17] Lord, A., I. J. Wilkinson, A. Ellis, D. Cleland, R. A. Garnham, and W. A. Stallard, "Comparison of WDM Coupler Technologies for Use in Erbium Doped Fibre Amplifier Systems," *IEE Electronics Letters*, Vol. 26, No. 13, 1990, pp. 900–901.

[18] Wilkinson, I. J., and C. J. Rowe, "Close-Spaced Fused Fibre WDMs With Very Low Polarisation Sensitivity," *IEE Electronics Letters*, Vol. 26, No. 6, 1990, pp. 382–384.

[19] Farries, M. C., R. I. Laming, P. R. Morkel, T. A. Birks, D. N. Payne, and E. J. Tarbox, "Efficient High-Gain Erbium-Doped Fiber Amplifier Pumped With Frequency-Doubled Nd:YAG Laser," *Optical Fiber Communications OFC '89*, Houston, 1989, Proc. Vol. 1, paper TUG5.

[20] Choy, M. M., C. Y. Chen, M. Andrejco, M. Saifi, and C. Lin, "A High-Gain, High-Output Saturation Power Erbium-Doped Fiber Amplifier Pumped at 532 nm," *IEEE Photonics Technology Letters*, Vol. 2, No. 1, 1990, pp. 38–40.

[21] Mears, R. J., L. Reekie, I. M. Jauncey, and D. N. Payne, "Low-Noise Erbium-Doped Fibre Amplifier Operating at 1.54 μm," *IEE Electronics Letters*, Vol. 27, No. 8, 1991, pp. 628–629.

[22] Whitley, T. J., "Laser Diode Pumped Operation of Er^{3+}Doped Fibre Amplifier," *IEE Electronics Letters*, Vol. 24, No. 25, 1988, pp. 1537–1539.

[23] Laming, R. I., S. B. Poole, and E. J. Tarbox, "Pump Excited-State Absorption in Erbium-Doped Fibers," *Optics Letters*, Vol. 13, 1988, pp. 1084–1086.

[24] Nakazawa, M., Y. Kimura, and K. Suzuki, "High Gain Erbium Fibre Amplifier Pumped by 800 nm Band," *IEE Electronics Letters*, Vol. 26, No. 8, 1990, pp. 548–550.

[25] Horiguchi, M., M. Shimizu, M. Yamada, K. Yoshino, and H. Hanafusa, "Highly Efficient Optical Fibre Amplifier Pumped by a 0.8 μm Band Laser Diode," *IEE Electronics Letters*, Vol. 26, No. 21, 1990, pp. 1758–1759.

[26] Nakazawa, M., Y. Kimura, E. Yoshida, and K. Suzuki, "Efficient Erbium-Doped Fibre Amplifier Pumped at 820 nm," *IEE Electronics Letters*, Vol. 26, No. 23, 1990, pp. 1936–1938.

[27] Kimura, Y., K. Suzuki, and M. Nakazawa, "Noise Figure Characteristics of Er^{3+}-Doped Fibre Amplifier Pumped in 0.8 μm Band," *IEE Electronics Letters*, Vol. 27, No. 2, 1991, pp. 146–148.

[28] Delavaque, E., T. Georges, and J.-F. Bayon, "Pump Wavelength Optimisation of Erbium-Doped Fibre Amplifier in 800 nm Band," *IEE Electronics Letters*, Vol. 27, No. 16, 1991, pp. 1421–1422.

[29] Bastien, S. P., and H. R. D. Sunak, "Pump Wavelength Independent Gain with Bidirectionally Pumped Alumino-Silicate Erbium-Doped Fiber Amplifiers in the 800 nm Pump Band," *IEEE Photonics Technology Letters*, Vol. 3, No. 9, 1991, pp. 825–828.

[30] Pedersen, B., W. J. Miniscalco, and S. Zemon, "Evaluation of the 800-nm Pump Band for Erbium-Doped Fiber Amplifiers," *IEEE Journal of Lightwave Technology*, Vol. 10, No. 8, 1992, pp. 1041–1049.

[31] Pedersen, B., S. Zemon, and W. J. Miniscalco, "Erbium-Doped Fibres Pumped in 800 nm Band," *IEE Electronics Letters*, Vol. 27, No. 14, 1991, pp. 1295–1297.

[32] Pedersen, B., B. A. Thompson, S. Zemon, W. J. Miniscalco, and T. Wei, "Power Requirements for Erbium-Doped Fiber Amplifiers Pumped in the 800, 980, and 1480 nm Bands," *IEEE Photonics Technology Letters*, Vol. 4, No. 1, 1992, pp. 46–49.

[33] Giles, C. R., E. Desurvire, J. L. Zyskind, and J. R. Simpson, "Noise Performance of Erbium-Doped Fiber Amplifier Pumped at 1.49 μm, and Application to Signal Preamplification at 1.8 Gbit/s," *IEEE Photonics Technology Letters*, Vol. 1, No. 11, 1989, pp. 367–369.

[34] Yamada, M., M. Shimuzu, M. Okayasu, T. Takeshita, M. Horoguchi, Y. Tachikawa, and E. Sugita, "Noise Characteristics of Er^{3+}-Doped Fiber Amplifiers Pumped by 0.98 and 1.48 μm Laser Diodes," *IEEE Photonics Technology Letters*, Vol. 2, No. 3, 1990, pp. 205–207.

[35] Pedersen, B., A. Bjarklev, H. Vendeltorp-Pommer, and J. H. Povlsen, "Erbium Doped Fibre Amplifier: Efficient Pumping at 807 nm," *Optics Communications*, Vol. 81, No. 1–2, 1991, pp. 23–26.

[36] Laming, R. I., M. C. Farries, P. R. Morkel, L. Reekie, D. N. Payne, P. L. Scrivener, F. Fontana, and A. Righetti, "Efficient Pump Wavelengths of Erbium-Doped Fibre Optical Amplifier," *IEE Electronics Letters*, Vol. 25, No. 1, 1989, pp. 12–14.

[37] Vodhanel, R. S., R. I. Laming, V. Shah, L. Curtis, D. P. Bour, W. L. Barnes, J. D. Minelly, E. J. Tarbox, and F. J. Favire, "Highly Efficient 978 nm Diode-Pumped Erbium-Doped Fibre Amplifier With 24 dB Gain" *IEE Electronics Letters*, Vol. 25, No. 20, 1989, pp. 1386–1388.

[38] Yamada, M., M. Shimuzu, T. Takeshita, M. Okayasu, M. Horiguchi, S. Uehara, and E. Sugita, "Er^{3+}-Doped Fiber Amplifier Pumped by 0.98 μm Laser Diodes," *IEEE Photonics Technology Letters*, Vol. 1, No. 12, 1989, pp. 422–424.

[39] Miniscalco, W. J., B. A. Thompson, E. Eichen, and T. Wei, "Very High Gain Er^{3+} Fiber Amplifier Pumped at 980 nm," *Optical Fiber Communications OFC '90*, San Diego, 1990, Proc. Vol. 1, paper FA2, p. 192.

[40] Becker, P. C., A. Lidgard, J. R. Simpson, and N. A. Olsson, "Erbium-Doped Fiber Amplifier Pumped in the 950–1000 nm Region," *IEEE Photonics Technology Letters*, Vol. 2, No. 1, 1990, pp. 35–37.

[41] Laming, R. I., and D. N. Payne, "Noise Characteristics of Erbium-Doped Fiber Amplifier Pumped at 980 nm," *IEEE Photonics Technology Letters*, Vol. 2, No. 6, 1990, pp. 418–421.

[42] Way, W. I., D. Chen, M. A. Saifi, M. J. Andrejco, A. YiYan, A. Von Lehman, and C. Lin, "High Gain Limiting Erbium-Doped Fibre Amplifier With Over 30 dB Dynamic Range," *IEE Electronics Letters*, Vol. 27, No. 3, 1991, pp. 211–213.

[43] Shimizu, M., M. Horiguchi, M. Yamada, I. Nishi, J. Noda, T. Takeshita, M. Okayasu, S. Uehara, and E. Sugita, "Compact and Highly Efficient Fiber Amplifier Modules Pumped by a 0.98 μm Laser Diode," *IEEE Journal of Lightwave Technology*, Vol. 9, No. 2, 1991, pp. 291–296.

[44] Laming, R. I., J. E. Townsend, D. N. Payne, F. Meli, G. Grasso, and E. J. Tarbox, "High-Power Erbium-Doped-Fiber Amplifiers Operating in the Saturated Regime," *IEEE Photonics Technology Letters*, Vol. 3, No. 3, 1991, pp. 253–255.

[45] Smart, R. G., J. L. Zyskind, J. W. Sulhoff, and D. J. DiGiovanni, "An Investigation of the Noise Figure and Conversion Efficiency of 0.98 μm Pumped Erbium-Doped Fiber Amplifiers Under Saturated Conditions," *IEEE Photonics Technology Letters*, Vol. 4, No. 11, 1992, pp. 1261–1264.

[46] Pedersen, B., A. Bjarklev, J. H. Povlsen, K. Dybdal, and C. C. Larsen, "The Design of Erbium-Doped Fiber Amplifiers," *IEEE Journal of Lightwave Technology*, Vol. 9, No. 9, 1991, pp. 1105–1112.

[47] Hansen, S. L., K. Dybdal, and C. C. Larsen, "Upper Gain Limit in Er-Doped Fiber Amplifiers Due to Internal Rayleigh Backscattering," *Optical Fiber Communications OFC '92*, San Jose, 1992, Proc. Vol. 1, paper TuL4, p. 68.

[48] Vendeltorp-Pommer, H., B. Pedersen, A. Bjarklev, and J. H. Povlsen, "Noise and Gain Performance for an Er^{3+} Doped Fibre Amplifier Pumped at 980 nm or 1480 nm," *Proceedings of Society of Photo-Optical Instrumentation Engineers Conference, OE/Fibers '90*, part: Fiber Laser Sources and Amplifiers II, San Jose, September 1990, pp. 254–265.

[49] Dakss, M. L., B. Pedersen, G. R. Joyce, and W. J. Miniscalco, "Signal Output Power and Noise in Erbium-Doped Fiber Power Amplifiers," *Proceedings of Society of Photo-Optical Instrumen-*

tation Engineers Conference, OE/Fibers '91, part: Fiber Laser Sources and Amplifiers III, Boston, Sept. 1991, Vol. 1581, pp. 236–250.

[50] Pedersen, B., J. Chirravuri, and W. J. Miniscalco, "Gain and Noise Penalty for Detuned 980-nm Pumping of Erbium-Doped Fiber Power Amplifiers," *IEEE Photonics Technology Letters*, Vol. 4, No. 4, 1992, pp. 351–353.

[51] Pedersen, B., J. Chirravuri, and W. J. Miniscalco, "Gain and Noise Properties of Small-Signal Erbium-Doped Fiber Amplifiers Pumped in the 980-nm Band," *IEEE Photonics Technology Letters*, Vol. 4, No. 6, 1992, pp. 556–558.

[52] Percival, R. M., S. Cole, D. M. Cooper, S. P. Craig-Ryan, A. D. Ellis, C. J. Rowe, and W. A. Stallard, "Erbium-Doped Fiber Amplifier With Constant Gain for Pump Wavelengths Between 966 and 1004 nm," *IEE Electronics Letters*, Vol. 27, No. 14, 1991, pp. 1266–1268.

[53] Pedersen, B., M. L. Dakss, B. A. Thompson, W. J. Miniscalco, T. Wei, and L. J. Andrews, "Experimental and Theoretical Analysis of Efficient Erbium-Doped Fiber Power Amplifiers," *IEEE Photonics Technology Letters*, Vol. 3, No. 12, 1991, pp. 1085–1087.

[54] Snitzer, E., H. Po, F. Hakimi, R. Tumminelli, and B. C. McCollum, "Erbium Fiber Laser Amplifier at 1.55 μm with Pump at 1.49 μm and Yb Sensitized Er Oscillator," *Optical Fiber Communications OFC '88*, U.S., 1988, postdeadline paper PD2.

[55] Hagimoto, K., et al., "A 212 km Non-repeatered Transmission Experiment at 1.8 Gb/s Using LD Pumped Er^{3+}Doped Fiber Amplifiers in an IM/Direct Detection Repeater System," *Optical Fiber Communications OFC '89*, U.S., 1989, postdeadline paper PD15.

[56] Inoue, K., H. Toba, N. Shibata, K. Iwatsuki, A. Takada, and M. Shimizu, "Mutual Signal Gain Saturation in Er^{3+}-Doped Fibre Amplifier Around 1.54 μm Wavelength," *IEE Electronics Letters*, Vol. 25, No. 9, 1989, pp. 594–595.

[57] Kimura, Y., K. Suzuki, and M. Nakazawa, "46. 5 dB Gain in Er^{3+}-Doped Fibre Amplifier by 1.48 μm GalnAsP Laser Diodes," *IEE Electronics Letters*, Vol. 25, No. 24, 1989, pp. 1656–1657.

[58] Kashiwada, T., M. Shigematsu, T. Kougo, H. Kanamori, and M. Nishimura, "Characteristics of 1.48 μm Pumped Erbium-Doped Fiber Amplifier With High Efficiency," *Techn. Dig. of Optical Amplifiers and Their Applications OAA '91,* Snowmass, CO, 24–26 July 1991, ThE2-1, pp. 166–169.

[59] Vodhanel, R. S., A. Yi-Yan, B. Enning, and R. I. Laming, "Gain and Noise Figure Variations of an Erbium Doped Fiber Amplifier for Pump Wavelengths Between 1460 nm and 1510 nm," *Proceedings of the Optical Fiber Communication Conference OFC '90*, San Francisco, 22–26 Sept. 1990, paper WL4.

[60] Olsson, N. A., "Lightwave Systems With Optical Amplifiers," *IEEE Journal of Lightwave Technology*, Vol. 7, No. 7, 1989, pp. 1071–1082.

[61] Desurvire, E., "Analysis of Noise Figure Spectral Distribution in Erbium Doped Fiber Amplifiers Pumped Near 980 and 1480 nm," *Applied Optics*, Vol. 29, 1990, pp. 3118–3125.

[62] Lumholt, O., "Erbium Doped Active Fibre Pumped Around 1480 nm," Master's Thesis E419 (in Danish), Electromagnetics Institute, Technical University of Denmark, 1990.

[63] Fevrier, H., J. Auge, V. Parlier, A. Dursin, J. F. Marcerou, and B. Jacquier, "Erbium-Doped Fibre Amplifier With Outstanding Gain Characteristics," *European Conference on Optical Communication ECOC '89*, Göteborg, Sweden, 1989, Proc. Vol. 1, paper TuA5-2, pp. 66–69.

[64] Desurvire, E., "Analysis of Erbium-Doped Fiber Amplifiers Pumped in the ${}^4I_{15/2}$-${}^4I_{13/2}$ Band," *IEEE Photonics Tehnology Letters,* Vol. 1, No. 10, 1989, pp. 293–296.

[65] Zyskind, J. L., C. R. Giles, and J. R. Simpson, "Optimal Pump Wavelength in the ${}^4I_{15/2}$-${}^4I_{13/2}$ Absorption Band for Efficient Er^{3+}-Doped Fiber Amplifiers," *IEEE Photonics Technology Letters,* Vol. 1, No. 12, 1989, pp. 428–430.

[66] Zyskind, J. L., C. R. Giles, E. Desurvire, and J. R. Simpson, "Optimal Pump Wavelength for Efficient Er^{3+} Doped Fiber Amplifiers," *Optical Fiber Communications OFC '90*, San Diego, 1990, Proc. Vol. 1, invited paper FA4, p. 196.

[67] Massicott, J. F., R. Wyatt, B. J. Ainslie, and S. P. Craig-Ryan, "Efficient, High Power, High Gain, Er^{3+} Doped Silica Fibre Amplifier," *IEE Electronics Letters*, Vol. 26, No. 14, 1990, pp. 1038–1039.

[68] Mikkelsen, B., and T. Durhuus, "High-Gain at 1553 nm in Er-Doped Fibre Amplifier Pumped by Semiconductor Lasers," *European Conference on Optical Communication ECOC '90*, Amsterdam, 1990, Proc. Vol. I, pp. 617–620.

[69] Ronarc'h, D., M. Guibert, H. Ibrahim, M. Monerie, H. Poignant, and A. Tromeur, "30 dB Optical Net Gain at 1.543 μm in Er^{3+} Doped Fluoride Fibre Pumped Around 1.48 μm," *IEE Electronics Letters*, Vol. 27, No. 11, 1991, pp. 908–909.

[70] Nakazawa, M., and Y. Kimura, "Amplification and Transmission at 1.53–1.55 μm With Er^{3+} Doped Fibres," *Optical Fiber Communications OFC '90*, San Diego, 1990, Proc. Vol. 1, invited paper FA5, p. 197.

[71] Suyama, M., S. Watanabe, I. Yokota, and H. Kuwahara, "Bidirectional Transmission Scheme Using Intensity Modulation of 1.48 μm Pump Laser Diode for Erbium-Doped Fibre Amplifier," *IEE Electronics Letters*, Vol. 27, No. 1, 1991, pp. 89–91.

[72] Gabla, P. M., E. Leclerc, and C. Coeurjolly, "Practical Implementation of a Highly Sensitive Receiver Using an Erbium-Doped Fiber Preamplifier," *European Conference on Optical Communication ECOC '91*, Paris, 1991, Proc. Vol. I, paper WeC93, pp. 589–592.

Chapter 8
Improved 1550-nm Amplifiers in Advanced Configurations

The description of fiber amplifiers for the 1550-nm wavelength band has so far concentrated on the basic properties of the erbium ions located in silica host material. The preceding chapters have furthermore described optimization of the index profile and erbium distribution, and different pump wavelength choices have been evaluated. The properties of fiber amplifiers in what can be regarded as more straightforward applications and configurations have been presented. As a central example, the simultaneous processes of signal amplification and ASE buildup have, therefore, until this point been treated as properties that could only be affected by the fiber design. This was naturally done in a combination with appropriate choices of pump wavelength and pump power.

However, by introducing additional components within the fiber amplifier and/or more advanced fiber designs, it becomes possible to adjust and improve the amplifier properties considerably. The subject of this chapter is to describe and discuss a number of the most promising approaches for amplifier improvement through more advanced amplifier configurations.

First, we will see how performance improvements can become possible by using a filter or an isolator as ASE-suppressing components within erbium-doped fibers. The two components for ASE suppression are compared, and simultaneous high-gain and near-quantum-limited noise figures will be demonstrated. The noise figure improves with this strategy for input signal powers below −5 dBm. A noise figure improvement of 2.0 dB with a simultaneous gain increase of 4.1 dB is measured relative to a gain-optimized fiber. The optimum isolator location is evaluated for different pump and signal wavelengths in both an Al/Er-doped and Ge/Er-doped fiber for pump and signal power variations and different pump configurations. In addition to the already mentioned advantages of inclusion of ASE-suppressing components, the possibility of obtaining a flat gain spectra will also be evaluated, and it should be mentioned that filters are also used for bandwidth control.

Another of the promising candidates with respect to improving the pump efficiency is an amplifier configuration in which the signal passes the erbium fiber twice by applying a mirror and an optical ring isolator at the fiber ends. In Section 8.2, configurations of this type will be analyzed with respect to gain and low-noise performance. Furthermore, we will also discuss amplifiers in which only the pump is reflected.

The third possibility of designing a low-noise component, while maintaining high amplifier gain, can be achieved by increasing the population inversion (see Chapter 5) in the fiber input end with a longitudinally varied fiber design. More specifically, this means using tapered active fibers as described in Section 8.3. The properties of tapered fibers in power booster applications are also discussed.

Finally, Section 8.4 will include a description of how to overcome coupling difficulties due to spot-size mismatch between optimal active fibers and standard single-mode fibers. The possible advantages of simultaneous pumping at two different pump wavelengths are also presented.

8.1 FILTERS AND ISOLATORS

The rapid development within the field of optimizing the gain efficiencies of EDFAs has resulted in a tremendous increase of gain efficiency to 11 dB/mW [1], and a gain as high as 51 dB has recently been reported [2]. However, even for such highly optimized fibers, most of the pump photons are lost to amplify spontaneously radiated photons. This results in gain saturation and a noise figure that departs considerably from the quantum limit. The latter is due to the strong dependence between noise figure and population inversion in the input fiber end that is heavily diminished by the amount of backward-traveling ASE. A low-noise figure can be obtained either by shortening the fiber well below its optimum length [3] or by increasing the input signal power to the level where neither the ASE nor the signal saturation will be determinative upon the population inversion [4]. However, both possibilities result in a considerable decrease in gain efficiency.

Essential demands for preamplifier or in-line-amplifier operation of the EDFA are to achieve simultaneous high gain (normally on the order of magnitude of 30 dB) and a noise figure close to the quantum limit. Obviously, a way to comply with both requirements is to insert either a filter or an isolator as an ASE-attenuating component within the active fiber length [5]. Reports on amplifier measurements have shown setups where an isolator is inserted between two EDFAs in succession, but for quite different intentions. In [6], two different pump wavelengths are used with the isolator to prevent gain saturation by the 980-nm pump amplifying the 1480-nm light (see Section 8.4). Another example is shown in [7], where gain equalization is performed by using a filter between fibers of different co-doping. An isolator is inserted in order to suppress reflections from the filter.

The aim of this section is to analyze the amplifier performance improvements obtained by including filters or isolators as population inversion-improving components. The optimum locations of these ASE-suppressing components are evaluated and the improvements are analyzed for both 980- and 1470-nm pumping. Note that there is no principal difference between 1480- and 1470-nm pumping; the latter is merely chosen for easy comparison with measurements. The analysis is performed for different signal wavelengths, and an experimental verification of the method is presented.

8.1.1 Background for Placing Wavelength- and Direction-Selective Components in the Amplifier

In order to describe the impact from an isolator inserted within an EDFA, we will first direct our attention to the fundamental characteristics of a traditional forwardpumped EDFA, as illustrated with dashed curves in Figure 8.1. This shows the variation within the EDFA of the pump power, the ASE power in both directions, and the population inversion (see the description in Chapter 5). The signal and pump wavelengths of this example are 1533 and 980 nm, respectively, and the EDFA is Er/La-doped [8]. The fiber length is adjusted to a maximum gain that is 45.3 dB. A characteristic, almost symmetric shape of the population inversion and

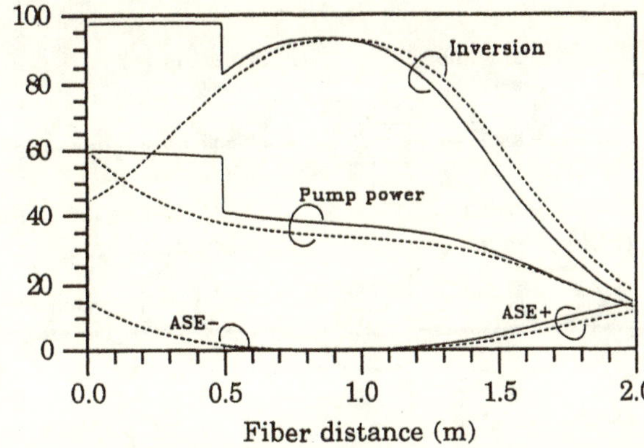

Figure 8.1 Co-propagating pump power, ASE, and population inversion within an erbium-doped fiber. The dashed curves correspond to the situation without an isolator and the solid curves represent the amplifier with an isolator (placed 0.5 m from the signal input end). (Figure from [9].)

thereby the small-signal gain coefficient appears, showing a maximum near the middle of the amplifier. At the fiber ends, the population inversion is depressed by the very large ASE power levels. It is especially remarkable that the pump light leaving the fiber end is far above the pump threshold. This wasted pump power would therefore have a potential for driving an extra small-signal amplifier. However, when the length of the erbium-doped fiber is increased above its optimum gain length, the gain increase related to the extra length of fiber is weaker than the gain reduction in the input end of the fiber caused by generation of a stronger backwards traveling ASE. This effect can obviously be weakened by the insertion of an isolator within the EDFA, thereby making it possible to increase the gain above the optimum gain value for a standard amplifier configuration.

By inserting an isolator 0.5 m from the signal input end, and with a pump bypass coupling as described below in Figure 8.2, the solid curves appear. We notice that the backward-traveling ASE (ASE-) is almost totally removed by the isolator. Consequently, the population inversion increases dramatically in this fiber section, yielding a much higher gain and a noise figure which, as we will see, is close to the theoretical lower 3-dB limit. Several decibels of pump power are seen to be saved due to the lower ASE power within the first half-meter of the fiber, but the isolator insertion loss in this example reduces the pump to almost the same level as previously.

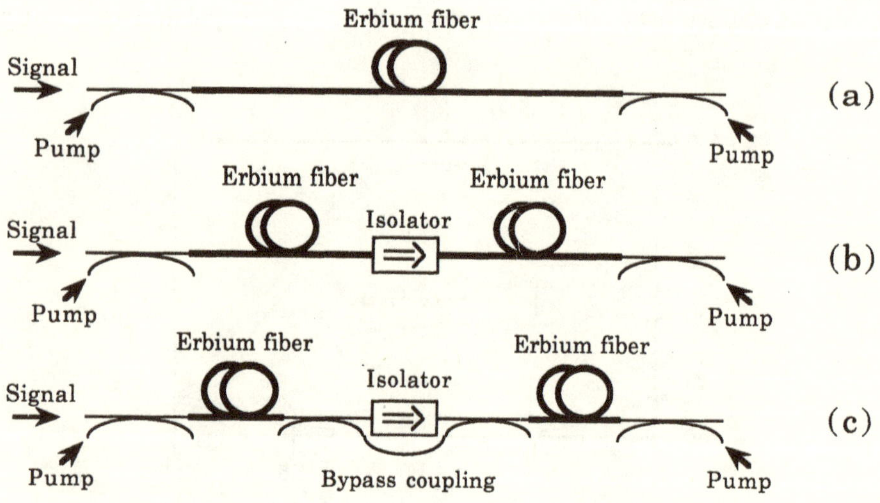

Figure 8.2 (a) A bi-directionally pumped erbium-doped fiber with gain-optimized length; (b) An isolator for 1480-nm pumping inserted within the active fiber with fiber length L1 to the signal input end and followed by fiber length L2; (c) A bypass coupling for 980-nm pumping.

As indicated by the example above, the insertion of an ASE suppressing component such as an isolator will be slightly complicated for 980-nm pumping. When pumping at 1480 nm, the isolator can be inserted directly within the fiber, and the co-propagating pump light will be transmitted through it, only suffering a loss comparable with the signal attenuation. However, high insertion losses of typically 35 dB for the co-propagating and 25 dB for the counter-propagating pump at 980 nm require that the pump has to be guided past the isolator by two WDM couplers. This is shown in Figure 8.2(c).

Before we start a more detailed discussion of the use of filters and isolators as ASE-attenuating components, it is instructive to consider that ASE reduction under certain circumstances occurs by itself. This is the case when the signal wavelength is near the long-wavelength emission peak at 1550 nm. For this signal wavelength and for a 980-nm pump source, it can be shown that the bi-directional pumping scheme enhances the gain with several decibels compared to either the co- or the counter-directional pumping schemes. Similar calculations for signal wavelengths around the 1530-nm emission peak show, on the other hand, that the gain is much less dependent on the degree of backwards pumping. Namely, one will find that even though the bi-directional pumping scheme is still best, the enhancement becomes less than 1 dB. The fact that the large gain enhancements only appear in the case of 1550-nm signals can be understood from Figure 8.3, which shows the internal variation of gain, ASE, and pump power for a Ge/Er-doped fiber at optimum gain length. The amplifier is bi-directionally pumped with 980-nm sources, and it has signal wavelength at 1550 nm. The length of the fiber is 8 m, and it can be seen that most pump power is absorbed within the first 2 m from the ends, leaving 4 m in the middle, where only ASE and signal are effectively present. In the center part of the EDFA, the ASE acts as a pump source on the signal. This is possible only because the signal wavelength at 1550 nm is on the long-wavelength side of the ASE peak at 1530 nm, and one should note here the substantial difference between these two signal wavelengths. In the 1530-nm case, the signal photons have higher energy than the spontaneously radiated photons, whereby the signal may lose energy by amplification of ASE. In the 1550-nm case, the reverse happens; the signal photons have lower energy than the spontaneously radiated photons. Consequently, the coamplified spontaneous emission will lose energy by pumping the signal (counter-propagating ASE is still unwanted due to saturation in the beginning of the fiber).

The bi-directional pumping scheme may be viewed as three series-connected amplifiers, with the middle amplifier being pumped by forward- and backward-propagating ASE generated in the surrounding amplifiers. However, only a very minor part of the gain improvement (up to 10 dB in the specific case) is directly added by the extra gain from the middle amplifier, which, as seen in Figure 8.3, is very modest. The major part of the gain improvement is indirectly achieved through the strong, 20-dB, ASE attenuation in the middle part of the EDFA. The

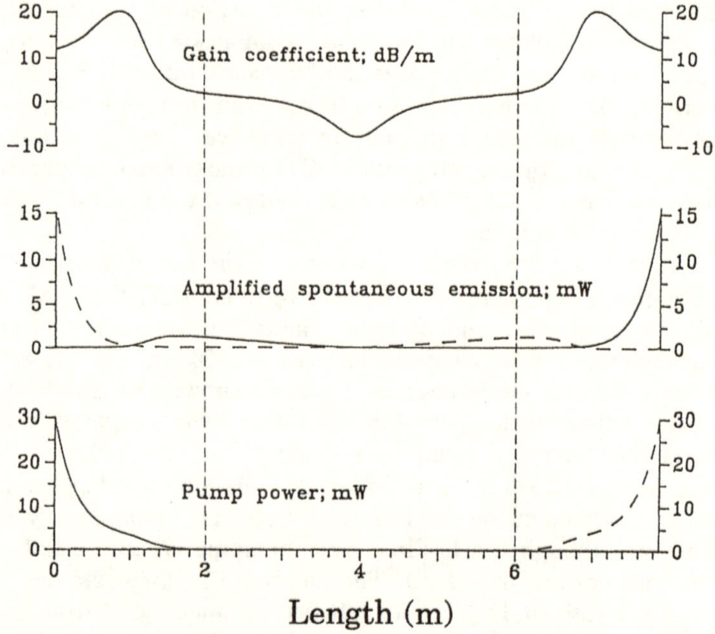

Figure 8.3 Variation of gain coefficient, ASE, and pump power inside a bi-directionally pumped Ge/Er-doped fiber amplifier at optimum gain length. The signal wavelength is 1552 nm, signal power is −60 dBm, pump wavelength is 980 nm, and pump power is 60 mW. Solid curves: forward propagation; dashed curves: backward. (Figure from [10].)

effect of this is that fewer pump photons amplify the spontaneous emission, thereby increasing the pump efficiency.

Even though ASE attenuation is strongly present in the above situation, it will still be advantageous to implement an isolator. The reason is that the ASE attenuation in Figure 8.3 occurs in the middle of the fiber, while the optimum placement of the isolator will be shown to be closer to the signal input end of the EDFA. Additional gain increase will therefore still be possible in 1550-nm amplifying fibers pumped at 980 nm. Another advantage is that the noise accumulates more slowly when the inversion becomes closer to unity, as in the case of isolator inclusion. As we shall see, noise figures very close to the 3-dB limit will be obtainable, which is approximately 1.5 dB better than in optimized situations without an isolator. It should also be noted that the isolator divides the EDFA into two EDFA pieces, each having significant gain. Therefore, another advantage is that the EDFA with an implemented isolator becomes more resistant against lasing.

After surveying the main differences between the ASE buildup for different source wavelengths, we will now focus on the improvements obtainable with the inclusion of a filter.

8.1.2 Gain and Noise Figure Improvements by Inserting Filters

The two obvious candidates for wavelength-suppressing components within the erbium-doped fiber are filters and isolators. As already indicated in Section 8.1.1, the most efficient EDFA improvement can be achieved by including isolators. However, as will be show in this section, filters alone or in combination with isolators are predicted to lead to considerable property enhancements compared to a reference amplifier without ASE-suppressing components.

In order to illustrate this, we will investigate results on an Er/Al/Ge-doped silica fiber. For this example, the emission cross section and absorption cross section have coinciding peaks localized at 1534 nm, while a shoulder at the emission cross section causes a gain peak to appear around 1554 nm, when the inversion decreases. This means that for a fixed pump power and varying fiber lengths, short fibers have gain maximum at 1534 nm, while the longer fiber lengths show a peak around 1554 nm.

For the example, we have chosen a high-NA fiber optimized to high gain operation (see Chapter 6). For each of the four combinations of signal wavelengths (1534 and 1554 nm) and pump wavelengths (980 and 1480 nm) considered here, the gain and noise figures are calculated for pump powers between 1 and 50 mW.

In Figure 8.4, the gain and noise figure for a 1534-nm signal are shown for a reference fiber amplifier, which is an amplifier in standard configuration (without ASE-suppressing components) and with the length optimized with respect to gain. This is illustrated together with gain calculated for the situation in which an internal filter with a bandwidth of 1 nm around the signal wavelength is inserted. The fiber length is optimized for maximum gain. It is assumed that both the pump and the signal are attenuated 0.5 dB in the filter.

For both the 980- and 1480-nm pump wavelength, it can be seen that the filter gives an improvement in gain of 3 to 5 dB compared to the reference situation. Looking at the noise figures, the improvements are most conspicuous for a 980-nm pump source, where improvements of approximately 0.8 dB are obtained. For the 1480-nm pump source in Figure 8.4, the noise figure is improved by approximately 0.2 dB. Note that the noise figures in all these cases increase with increasing pump power, because the amplifiers are designed for maximum gain performance. This means that in the high-gain regime, which is represented by the large pump powers, a large backward-propagating ASE also appears, resulting in low population inversion in the signal input end of the fiber.

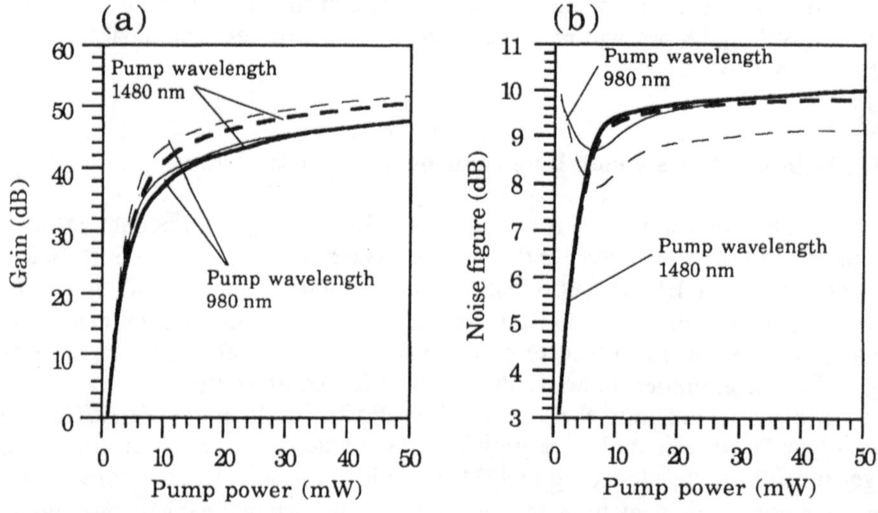

Figure 8.4 Gain (a) and noise figure (b) versus pump power for gain optimized fiber lengths. For the case in which a filter is included, it is placed at the gain optimal position. The signal wavelength is 1534 nm. The Ge/Al erbium-doped fiber has a core radius of 1.4 μm, a core-to-cladding index difference of $\Delta n = 0.03$, and an erbium concentration of 2.0×10^{-24} that is confined to 80% of the core radius. Solid lines: without filter; dashed lines: with filter. (Data from [5].)

For the case of filter inclusion, the conditions for improvements in the second gain peak around 1554 nm are shown in Figure 8.5. Comparing the reference curves in Figure 8.5, it can be seen that the 1480-nm pump source gives 7 dB more gain than the 980-nm pump source. The difference appears because in the latter case the gain saturates in the beginning of the fiber due to a strong backwards ASE power around 1534 nm.

This ASE peak does not grow up as strongly with 1480-nm pump sources, since the stimulated emission at the pump wavelength pushes the ASE spectrum towards longer wavelengths. For this reason, the gain improvement by inclusion of an ASE-suppressing filter also becomes most significant in the case of 980-nm pumping, where a gain increase of around 12 dB might be achieved. For the 1480-nm pumped amplifiers, the filter leads to an improvement of approximately 8 dB.

With regard to the noise figure, we notice that for 980-nm pumping an improvement of more than 4 dB is achievable, resulting in noise figures around 5 dB. For 1480-nm pumping, a noise figure reduction of 1 dB is seen.

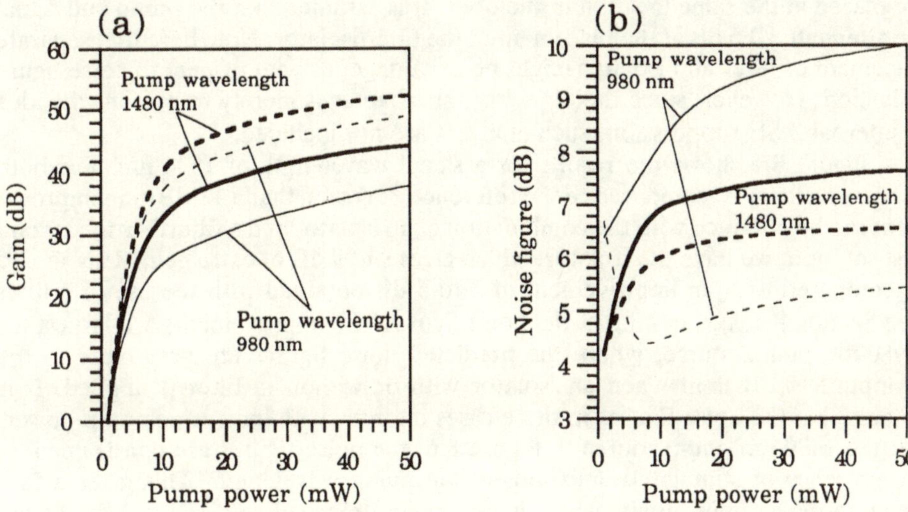

Figure 8.5 Gain (a) and noise figure (b) versus pump power for gain optimized fiber lengths. For the case in which a filter is included, it is placed at the optimum position. The signal wavelength is 1554 nm. The erbium-doped fiber is the same as that considered in Figure 8.4. Solid lines: without filter; dashed lines: with filter. (Data from [5].)

8.1.3 Gain and Noise Figure Improvements by Isolator Insertion

As discussed in Section 8.1.1, an important difference exists in how much performance improvement can be obtained by using either a filter or an isolator within the fiber as ASE suppressing component. The major difference is that the isolator completely blocks the backward-propagating ASE, and, as we shall see in this section, will therefore be a more efficient component with respect to noise figure improvement than the filter. The section is organized so that we will first consider the principal differences that appear when pumping at the two most relevant pump wavelength bands for EDFA operation, namely, the 980- and the 1480-nm band. The improvements will be exemplified by high-gain operation results, which are comparable to those discussed in Section 8.1.2.

In order to be able to make direct comparisons with the case of amplifier improvements by filter inclusion, the same conditions as those described in the previous section have been chosen. The fiber length has still been optimized with respect to maximum gain and an internal 30-dB isolator has been placed at a position in the fiber that is optimal with respect to this goal. Also, the combination of filter and isolator has been considered. However, in this case, both the filter and isolator are placed at the same location in the fiber. It is assumed that the pump and signal are attenuated 0.5 dB at the placement of the filter/isolator. Note here that separate placement of filter and isolator might be advantageous with respect to noise figure reduction. However, since the object of this chapter is merely to present the idea of internal ASE suppression, such analyses are not included.

Figure 8.6 shows the results for a signal wavelength of 1534 nm. For both pump wavelengths, compared to the reference, it is seen that a 12-dB gain improvement can be obtained with the combination of an isolator and a filter. As the second best solution, we have the isolator, which gives 8 to 9 dB of extra gain. This should be compared with an improvement of 3 to 5 dB obtained with the use of a filter (see Section 8.1.2). Looking at the noise figures, the improvements are largest for a 980-nm pump source, where the predicted noise figures are very close to the asymptotic 3-dB limit when an isolator with or without a filter is applied. It is notable that the noise figures in these cases improve with increasing pump power. For the 1480-nm pump source in Figure 8.6, the noise figures are constrained by the presence of stimulated emissions at the pump wavelength. This gives a fundamental noise figure limit, which is very nearly reached with the isolator or the combination of isolator and filter.

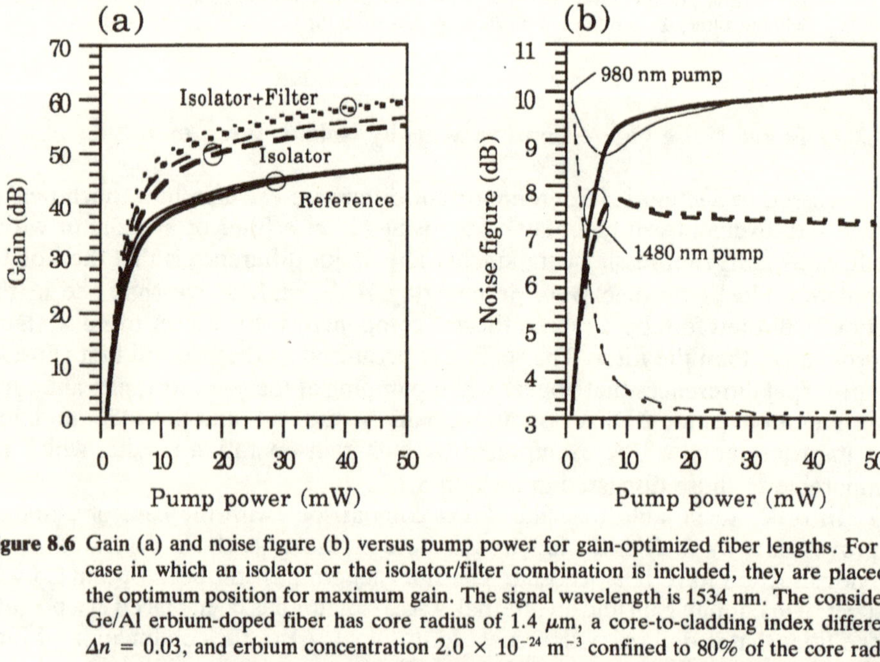

Figure 8.6 Gain (a) and noise figure (b) versus pump power for gain-optimized fiber lengths. For the case in which an isolator or the isolator/filter combination is included, they are placed at the optimum position for maximum gain. The signal wavelength is 1534 nm. The considered Ge/Al erbium-doped fiber has core radius of 1.4 μm, a core-to-cladding index difference Δn = 0.03, and erbium concentration 2.0 × 10^{-24} m^{-3} confined to 80% of the core radius. Solid curves: reference; dashed curves: isolator; dotted curves: isolator + filter; thick curves: 1480-nm pump. (Data from [5].)

The conditions for improvements in the second gain peak around 1554 nm are shown in Figure 8.7. Insertion of an isolator shows a drastic gain improvement with 18 dB for the 980-nm source and 13 dB for the 1480-nm pump. In both pump situations, the isolator brings the noise figures close to their theoretical limits. When a combination of a filter and an isolator is inserted, the 980-nm pump case of Figure 8.7 does not improve as much as it did with just the isolator. The reason for this is that the presence of the filter will prevent the ASE peak around 1534 nm to pump the signal at the output side of the filter.

After the principal description of potential gain and noise figure improvements for very-high-gain amplifiers realized by the use of Ge/Er-doped fibers, a number of questions on amplifier prototype that are more design-related are natural. Among these are questions of where to place the isolator, in which gain range the scheme is applicable, and at which signal and pump power levels the idea is valid. The satisfactory answer to these questions will depend on the chosen components. However, to indicate some overall tendencies, we will for the completeness of this description refer optimization results for an Al/Er-doped fiber amplifier.

The specific data of the fiber considered in this example are given in Chapter 3 for an Er/La-doped fiber with an erbium concentration as high as 23×10^{24} m^{-3}.

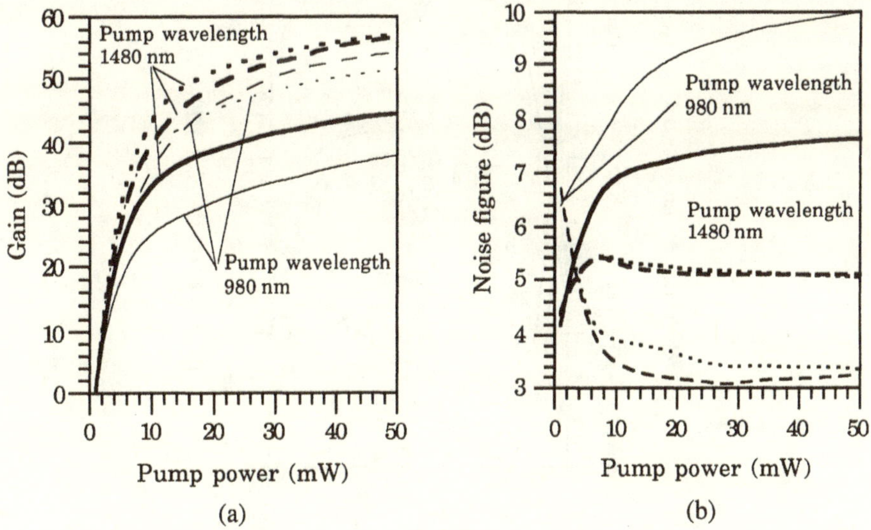

Figure 8.7 Gain (a) and noise figure (b) versus pump power for gain-optimized fiber lengths. For the case in which an isolator or the isolator/filter combination is included, the ASE-suppressing components are placed at the gain-optimized position. The signal wavelength is 1554 nm. The erbium-doped fiber is the same as that considered in Figure 8.6. Solid curves: reference; dashed curves: isolator; dotted curves: isolator + filter; thick curves: 1480-nm pump. (Data from [5].)

As shown previously (Chapter 3), the emission and absorption spectra show a large similarity with those of a fiber without La-doping. Both 1470- and 980-nm pumping are considered, and in the latter case two WDM-couplers are used, as discussed in Section 8.1.1, to guide the pump past the isolator. Two times the insertion loss of the WDM-coupler is included here, and it may typically be measured to 1.5 dB in each direction. The signal and the ASE will be guided through the isolator with a loss of 2 to 3 dB in co-propagating direction and around 40 dB in the counter-propagating direction, including the extra losses from the WDM couplers and splice losses. In this example, a signal wavelength at the fluorescence maximum at 1530 nm and a signal at 1550 nm are considered.

The question of where to place the isolator is answered through the results in Figure 8.8, where gain and noise figure improvements are shown as a function of the relative isolator location. The reference gain and noise figure are found for a gain-optimized, isolator-free "standard" amplifier. Even compared to this amplifier, we find in the specific example that gain improvements of between 4 and 5 dB can be achieved.

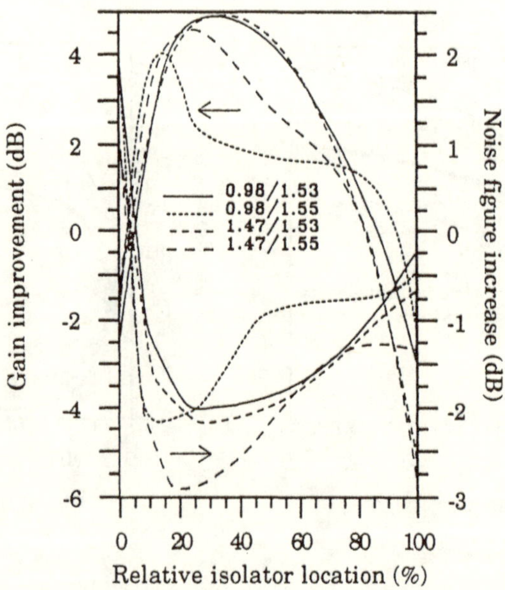

Figure 8.8 Gain and noise figure improvements versus the percentage of relative isolator location for the four indicated combinations of pump and signal wavelengths. Results are shown for an Er/La-doped EDFA, and a bi-directional pumping scheme is considered. The NA of the considered fiber is 0.31 and the cutoff wavelength is 980 nm. (Figure from [10].)

Concerning the noise figure, it can be seen in Figure 8.8 that a combination of 1470-nm pump and 1550-nm signal will benefit most from the insertion of an isolator in an aluminum co-doped EDFA. A decrease of 2.8 dB is achieved. However, the favorite configuration with respect to the lowest attainable noise figure for a high-gain amplifier is the 980-nm pumping of the 1530-nm signal. It yields a noise figure smaller than 3.2 dB, which is very close to the theoretical 3.0-dB lower limit. The largest gain improvements of 4.9 dB are obtained for 980- and 1470-nm pumping of the 1530-nm signal, and, in addition, the overall highest gain of 50.5 dB is calculated for the former.

It was recently reported [11] (and also referred to in Chapter 7) that pumping at the absorption peak does not give the highest gain in the 980-nm band when signal wavelengths around 1550 nm are considered. The gain hole is caused by ASE saturation of the amplifier, so when the emission cross section exceeds a certain level, the hole can be eliminated by inserting the ASE-attenuating isolator in the fiber. Calculations with the isolator at the optimum location also show full elimination of the gain hole, and both the maximum gain and minimum noise figure will in this case be obtained for pumping at the fluorescence peak at 980 nm.

For the pump configurations mentioned above, the optimum isolator location within the erbium fiber was analyzed for EDFAs operated in the small-signal regime and using a fixed pump power of 60 mW. However, when either the signal power or the pump power is varied, the ASE buildup and thereby the optimum isolator location will change. The value of the ASE suppression will therefore also change, and it becomes highly relevant to decide which applications (small- or large-signal amplifiers) the idea of ASE-suppression are valid for.

We will now clarify the operational range of the isolator inclusion with respect to the signal power level. Calculations of the gain and noise figure are therefore shown in Figure 8.9 for increasing input signal power in the erbium fiber. Signal wavelengths at 1530 and 1550 nm are considered, together with 60 mW of launched pump power at 1470 nm. The dashed reference lines illustrate the performance of a gain-optimized fiber without an isolator, while the solid lines show the results for a gain-optimized fiber with an isolator inserted at its optimum location.

For both signal wavelengths, an input signal power below -5 dBm allows for a net improvement in noise figure by the insertion of the isolator, while a noise figure increase arises for higher signal input power levels. Improvements of 2.1 and 2.8 dB are respectively achieved at 1530 and 1550 nm in the small-signal regime, where the large amount of ASE causes deterioration of the population inversion of the reference fiber. For increasing power levels, the reference noise figure will improve because the signal saturation rather than the ASE will determine the population inversion. However, in the very-large-signal regime the signal saturation alone will be determinative. Therefore, no improvement is found from using the isolator, which in this case only adds an additional loss within the EDFA. For both

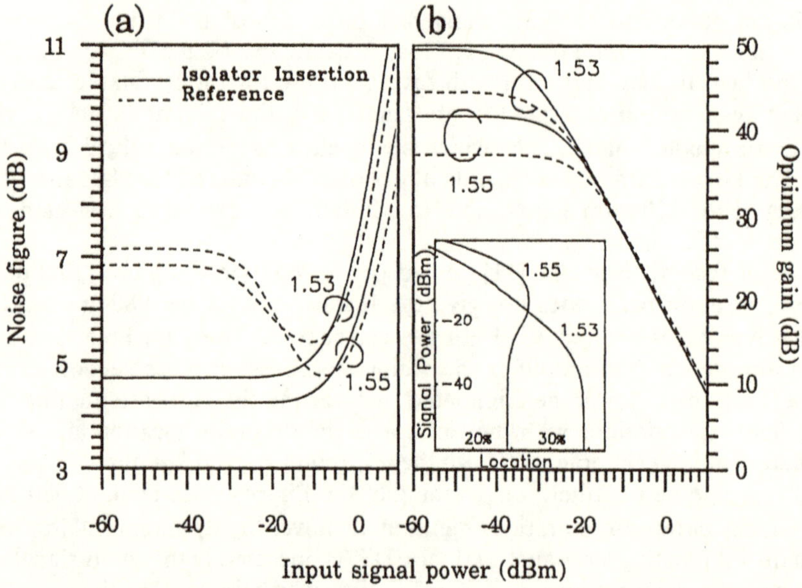

Figure 8.9 Optimum gain (a) and corresponding noise figure (b) versus the input signal power to the erbium-doped fiber. Signals at both 1530 and 1550 nm are considered. Dotted lines show the results of the reference fiber, while the solid lines illustrate the results when a pure isolator is considered. Inserted: optimum isolator location versus input signal power; λ_p = 1470 nm and P_p = 60 mW. (Figure from [10].)

the reference and the amplifier with the included isolator, the noise figure increases rapidly with signal input power in the large-signal regime.

Gain improvements up to 4.9 dB for 1530-nm and 4.6 dB for 1550-nm signals are achieved for signal powers below the 3-dB saturation point: −28 dBm and −18 dBm, respectively. However, it is seen that the input signal has to be approximately 10 dB lower than the 3-dB saturation point to allow for gains within a few tenths of a decibel from the value in the extremely-small-signal region. In the large-signal regime, the isolator insertion loss has a weightier influence upon the amplifier performance than the moderate profit of ASE reduction, resulting in a minor gain decrease.

The optimum isolator location within the fiber is illustrated versus the input signal power in the lower left of Figure 8.9(b). It shows a large difference between the two considered signal wavelengths. In the large-signal regime, a signal wavelength at 1530 nm gives a noisier amplifier than the one at 1550 nm. It is therefore most advantageous to move the ASE-suppressing isolator close to the signal input end for the 1530-nm signal, which, consequently, has a shorter optimum location than the one for 1550-nm signals. However, for decreasing signal power, the ASE builds up strongly for the 1550-nm signal, resulting in shorter isolator locations when using this wavelength. In the range where the signal saturation is determinative for the population inversion, the relative isolator location is seen to increase for decreasing signal powers. However, for signals below the 3-dB saturation point, the location remains unchanged at 34% and 24% for 1530 and 1550-nm signals, respectively.

The influence of pump power variations upon the amplifier performance of the 1550-nm signal is investigated for pump wavelengths at both 980 and 1470 nm when using the pump bypass coupling for the former. In Figure 8.10, calculations are presented for a signal input power level at 60 dBm. This very low signal input power level, which may be impractical for actual system applications, is chosen in order to ensure small-signal operation. The results in Figure 8.10 will, however, not deviate from situations of more practical input power levels, which may be 20 to 30 dB higher than the chosen value.

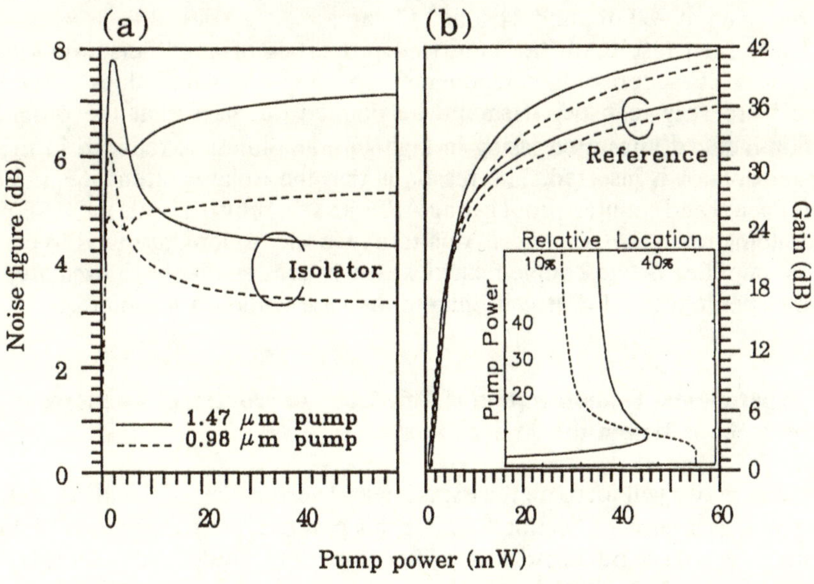

Figure 8.10 Noise figure (a) and corresponding optimum gain (b) versus the pump power to the erbium-doped fiber. Calculations are made both with and without an inserted isolator. Dotted lines show the results for 980-nm pumping, while the solid lines illustrate the results for 1470-nm pumping. Insert: optimum isolator location versus pump power, where λ_s = 1550 nm; P_s = −60 dBm. (Figure from [10].)

Pump power levels sufficient to achieve a 20-dB gain by 980-nm pumping and 24-dB gain by 1470-nm pumping are required before the use of the isolator results in improvements in gain and noise figure performance. However, above this limit, both the gain and the noise figure improvements increase with increasing pump power. For the latter, this appears in the curves in the left part of Figure 8.10 as the reference noise figure increases for increasing pump power due to a faster increase in ASE than in the gain. By inserting the isolator, the course is opposite, since the gain will increase more rapidly than the ASE.

The optimum isolator location as represented in the lower right corner of the figure shows large changes with pump power. In the uninteresting range for small pump powers, where the reference yields a better performance than the use of the isolator does, the optimum location tends towards 50% for 980-nm and towards 0% for 1470-nm pumping. The former is due to equal attenuation of both forward and backward pumping in the bypass coupling. It makes this position optimum, since the smallest amount of pump power is attenuated by the isolator when placed in the middle of the fiber. At 1470-nm pumping, the attenuation of backward pumping is 39.3 dB larger than for forward, making a position in the beginning of the fiber optimal. However, for power levels well above the previously mentioned application limits, it is seen that the optimum position remains unchanged at 15% when pumping at 980 nm and 24% at 1470 nm.

These final results on the improved properties of amplifiers with isolators inserted as ASE-suppressing components also show that gain values in the range of 40 dB are very realistic. It should be pointed out here that the problem of reflection-induced unwanted lasing in high-gain amplifiers is reduced in the case where an isolator is inserted. The reason is that the isolator attenuates the backward-reflected and counter-propagating ASE, and thereby prevents the ASE power from building up in the amplifier. An alternative way to look at this is to consider that the amplifier is two coupled cavities, which are less likely to oscillate, since each are operated at a lower gain than in the case without the isolator.

8.1.4 Experimental Demonstration of Efficiency Improvement by Insertion of an Isolator Within the Active Fiber

In this section, we will focus on the experimental verification of the ideas presented in the previous section of Section 8.1. For this purpose, the experimental setup for evaluating the gain characteristics on a fiber amplifier including an isolator is shown in Figure 8.11. A bi-directional pumping configuration is applied using two InGaAsP/InP lasers, each emitting at 1470 nm and capable of delivering up to a total of 70 mW of launched pump power into the erbium-doped fiber. In each of the pump branches, the laser is followed by a 1:99 coupler that is used for measuring the pump power reference. The output arm of this coupler is spliced to the WDM

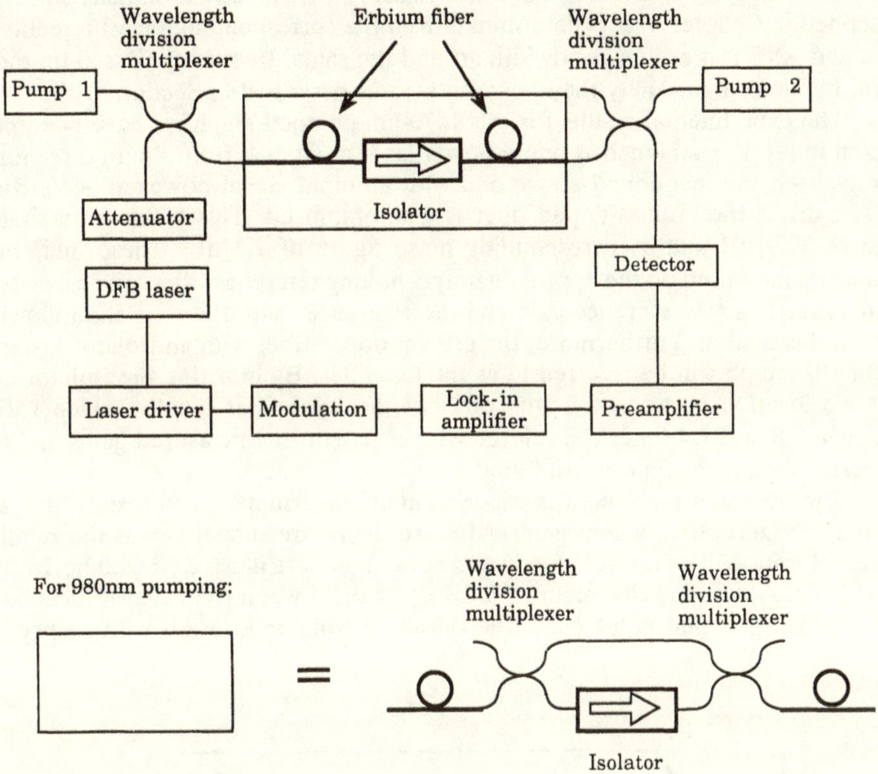

Figure 8.11 Experimental setup for gain measurements on fiber amplifier including an isolator. Note the bypass coupling necessary for efficient 980-nm pumping. DFB: distributed feed back laser. (Figure from [10].)

coupler located at each end of the EDFA. All free fiber ends in these and in the other couplers in the setup are cleaved in angles of 10 deg to prevent unwanted reflections.

The modulated signal light at 1550 nm originates from a temperature-tuned *distributed-feedback* (DFB) laser, and the signal power is adjusted by an attenuator to a level of −40 dBm at the input end of the EDFA. An isolator is inserted to prevent backward-traveling ASE from reaching the signal laser. To allow for a reference measurement of the signal, a 10:90 coupler is inserted into the signal arm before the WDM coupler at the input end of the amplifier. On the output end of the EDFA, the signal is guided through the WDM coupler before it is recorded by the use of a lock-in amplifier with an integrating sphere detector.

The setup for evaluating the noise figure, F, is somewhat similar, since (as described in Chapter 4) it is determined from the corresponding measurements of gain and ASE power in a bandwidth around the signal frequency. Based on these data, the noise figure may then be calculated as described in Section 5.1.4.

The experimental results for this 1470-nm-pumped amplifier have been presented in [9]. For a launched pump power level at 60 mW from the bi-directional pumped scheme mentioned above and with an input signal power at −40 dBm, the length of the erbium-doped fiber is gain-optimized. This yields a maximum gain of 37.9 dB and a corresponding noise figure of 7.4 dB. These numerical quantities measured on the approximately 5-m-long reference fiber are designated, respectively, as the reference gain and the reference noise figure in the following part of this section. Furthermore, the erbium-doped fiber with an isolator inserted within its length will be referred to as the test fiber. By inserting the isolator at a relative location between 0% and 50% of the total fiber length, which varies between 0.8 and 1.4 times the reference fiber length, the measured gains and the corresponding noise figures are found.

The optimum placement is experimentally determined, and results are presented in Figure 8.12, where gain and noise figure are shown versus the relative isolator location. The reference gain and noise figure are illustrated with horizontal solid lines. A gain improvement of 3 dB is obtained when the isolator location is 25% from the signal input end. The optimum isolator location with respect to

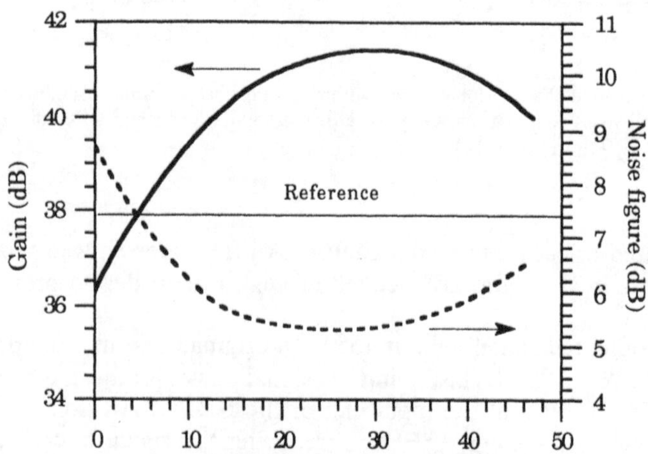

Figure 8.12 Gain and noise figure versus the relative isolator placement. The measured reference gain and noise figure are shown as a horizontal solid line. The total 1470 nm pump power is 60 mW. Measurements are indicated with markers. (Data from [9].)

maximum gain is seen to be within the range of 20% to 35% of the total length. This is in agreement with the predicted optimum isolator location within the first third of the fiber. Related to this, the largest reduction at 2 dB in noise figure is obtained at the same location. The noise figure of 5.3 dB is close to the theoretical lower limit when pumped at 1470 nm. The measurements clearly show how important the position of the isolator is. If we now consider a fiber length similar to the reference, then a gain reduction of 4.9 dB and a corresponding noise figure deterioration of 3.5 dB are caused by moving the isolator from its optimum location to the signal input end.

Comparative measurements employing either 980- or 1470-nm pumping have also been performed, and we will refer to the central results here. It should, however, be noted that for 980-nm pumping, the signal wavelength has been changed to 1530 nm. The reason for this is simply a question of available components at the time of the experiments. In Figure 8.13, the noise figure versus the launched pump power is shown for different positions of the isolator within the EDFA. These positions are given relative to the total fiber length, and in each case the total length of the erbium fiber has been optimized. With 1470-nm pumping and the isolator positioned 25.5% from the signal input end of a 5m-long erbium fiber, the noise figure is decreased by 2.1 to 5.3 dB. With 980-nm pumping, a relative isolator position of 34%, and 2.7m of fiber, the noise figure is decreased

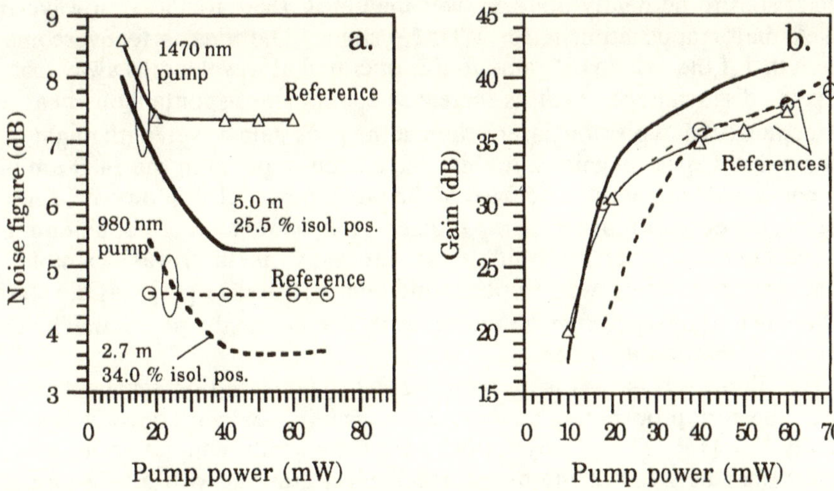

Figure 8.13 (a) Noise figure versus launched pump power. (b) Fiber gain versus launched pump power. Solid lines: bi-directional 1470-nm pumping and 1550-nm signal; dashed lines: co-directional 980-nm pumping and 1530-nm signal; triangles: 1470-nm-pumped reference fiber; circles: 980-nm-pumped reference fiber. (Data from [12].)

by 1.2 to 3.6 dB. Also, the gain versus the launched pump power is shown in Figure 8.13(b). For 1470-nm pumping, an optimum improvement of 4.1 dB is obtained with the isolator positioned 21.8% within a 5.8 m fiber. For 980-nm pumping, a 0.8-dB increase in gain to 40.2 dB is obtained with the isolator positioned 34% within the fiber.

The magnitude and the curve progress for the calculated gain improvements versus the isolator location shows very good agreement with the measurements in the range that has been experimentally investigated. The largest gain difference found between accurate calculations and measurements was approximately 1 dB. A difference of less than 1.5 dB has been observed between the calculated and measured noise figure improvements. Theoretical as well as experimental results determine that the optimum relative isolator location is near 20% of the total fiber length for 980-nm pumping and near 30% to 35% for 1470-nm pumping.

8.1.5 Erbium-Doped Fiber Amplifiers With Flattened Gain Spectrum

The incorporation of ASE-suppressing filters or isolators within the amplifier as described in the previous section had the purpose primarily of simultaneously improving gain and noise properties at a specific operation wavelength of the amplifier. However, the inclusion of internal optical filters can also be used with a somewhat different goal: to modify the overall gain spectrum and the saturation characteristics to be nearly uniform over the entire 1530- to 1560-nm wavelength band. A major application is for WDM systems. The idea is to overcome the problem that if the "standard" amplifier is operated at wavelengths away from the peak gain, disadvantages such as increased spontaneous-spontaneous beat noise (see Chapter 10) and possible laser action at the peak gain wavelength might occur. Other means of spectral gain smoothing have been to pump in the 1480-nm pump band, combined with a careful choice of the pump power [13]. However, since this was at the expense of reduced pump efficiency, a lower value of population inversion, and hence a higher noise figure, an alternative method was desirable. The filtering approach, which was proposed and demonstrated in [14], has the strength of flattening the gain spectrum without severely decreasing the population inversion within the erbium-doped fiber.

The filter used in [14] is an optical notch filter based on resonant coupling between the core propagating mode and the cladding leaky mode, which are subsequently lost [15]. This is physically realized by sandwiching a short length of amplifier fiber between 390-mm-long mechanical gratings with a 0.78-mm pitch. Both the center wavelength and the strength of the filter attenuation may be tuned by changing the operation angle and pressure, respectively. Furthermore, since the coupling band is far away from the 980-nm pump wavelength, the pump power might be chosen so it is not attenuated by the filter.

By placing the filter in the middle of the amplifier and tuning it to suppress the gain spectrum at the peak wavelength, a broadband amplifier with a 3-dB bandwidth of 33 nm and a gain of 27 dB was obtained as illustrated in Figure 8.14. For comparison, it can be seen that the amplifier without a filter has a 3-dB bandwidth of only 5 nm around the maximum gain wavelength of 1532 nm. Figure 8.14 also shows (besides a gain suppression at the peak wavelength) that the gain around 1550 nm is increased. The reason for this is the ASE suppression as described in the previous sections.

As was the case for ASE suppression, the location of the filter in the amplifier is important for optimum pump efficiency. If the filter is located at the output end of the amplifier, it will only shape the gain spectrum by attenuating the frequencies within its stop band. This causes an accompanying reduction of the amplifier saturation power and, hence, a lower pump efficiency. If the filter, on the other hand, is placed within the amplifier, it will improve the population inversion due to ASE suppression. Also, the signal power lost in the middle of the amplifier will be smaller and the output saturation power less affected than in the case of external filtering.

Finally, it should be noted that in [14] it is also shown that the gain profile will remain uniform even when the amplifier gain is saturated. This becomes important for applications in WDM systems, where a further broadening of the gain spectrum would be attractive. Even though this goal has not been accomplished by any single fiber amplifier, broadband amplifiers for operation around 1600 nm

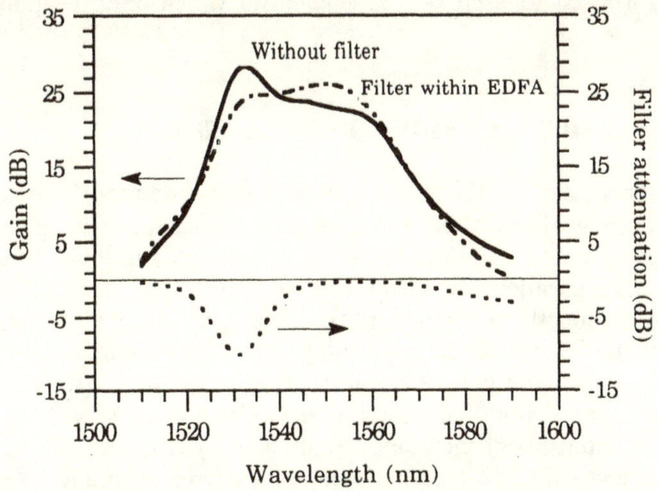

Figure 8.14 Gain spectra of EDFA showing gain flattening due to filtering at different locations. Also shown is the loss spectrum of the applied notch filter. (Data from [14].)

have been demonstrated [16]. Such amplifiers, which are realized by applying fiber lengths much longer than the maximum gain length, would, in parallel with an amplifier as described in this section, provide a total gain bandwidth of approximately 70 nm.

8.2 PUMP POWER REFLECTION

The common pump configurations used for EDFAs apply a one-way signal amplification, and pump light is injected from either one or both ends of the erbium-doped fiber [17]. For these standard configurations, a considerable part of the pump energy is wasted and transmitted towards either the signal source or the detector, causing possible noise problems.

In order to achieve a better utilization of the pump energy and thereby increase the gain coefficient (the ratio between gain and launched pump power as defined in Chapter 5) of the amplifier, new configurations have been proposed [18–20]. These are of the reflective type, where either both signal and pump are reflected from one end of the amplifier [18] or only the pump is reflected [20]. For the former, however, the necessary inclusion of an optical circulator was expected to lead to a noisier performance of the amplifier. In Section 8.2.1, therefore, we will not only address the subject of presenting this method but also analyze the gain and noise properties of amplifiers of the reflective type. Furthermore, an improved configuration of the reflective type of amplifier will be presented, and the result from the experimental verification of the idea is referred to. Secondly, Section 8.2.2 will be focused on the scheme in which only the pump power is reflected.

8.2.1 Configurations Using End-Face Reflective Mirrors

To define what is meant by the term "reflective-type amplifier," it is advantageous to look at a schematic illustration and compare it to a standard pump configuration. For this purpose, we will consider the three pump configurations shown in Figure 8.15, where configuration A, using co-propagating signal and pump (Figure 8.15(a)), is shown together with the reflective-type amplifier B [18] (Figure 8.15(b)). Also shown is the further developed configuration C of Figure 8.15(c), which was proposed [18, 21] for the purpose of improving the noise characteristics. This is done by introducing a 3-nm band-pass filter (with a loss of 1.5 dB) to remove the ASE. To avoid filtering out the pump simultaneously, a dichroic coupler is placed before the bandpass filter to lead the pump to a mirror, where it is reflected. It is important to note that the reflective-type amplifiers shown in Figure 8.15 include an optical circulator to separate in- and outpropagating signals.

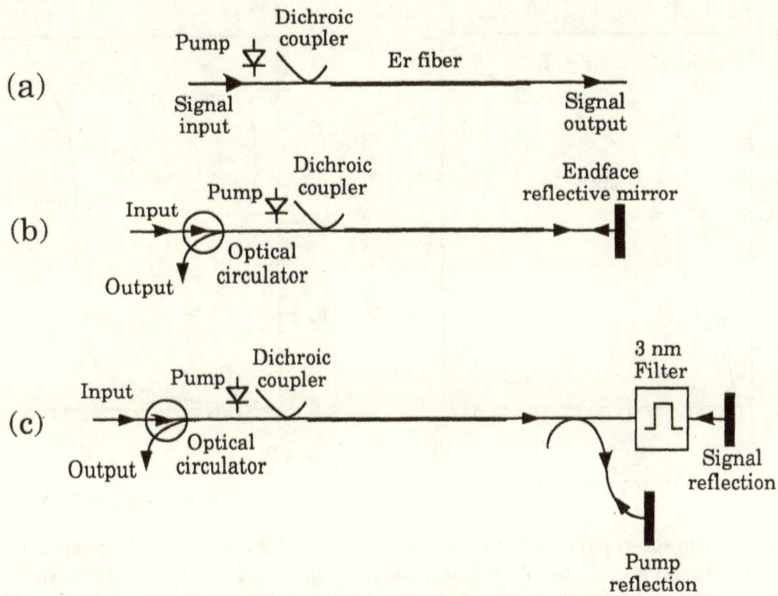

Figure 8.15 (a) Typical EDFA configuration where forward pumping is used. (b) EDFA configuration using the reflective method [18], where a mirror is used to reflect signal, ASE, and pump. (c) EDFA configuration where the ASE is filtered and the pump and signal are reflected. (Figure layout from [21].)

A natural way to evaluate the value of reflective-type amplifiers as illustrated in Figure 8.15 is by choosing realistic and typical data for the components of the EDFAs. Therefore, the fiber considered is an aluminum-co-doped fiber with a core radius of 1.8 μm and an NA of 0.25. The erbium concentration is 2.0×10^{24} m^{-3}. The optical circulator has a loss of about 1.2 dB and an isolation of 25 dB [22], and a fiber coupler loss of 0.3 dB is assumed. The end-face reflection coefficient is 0.99. In the calculations, the signal wavelength is 1530 nm, which corresponds to the emission peak of the specific amplifiers [23].

The three amplifier configurations are analyzed in both the 980- and 1480-nm pump band. The results from using a pump wavelength of 1480 nm are, however, not shown in the following figures of this section, but only referred to in parentheses because of the qualitative similarity to 980-nm pumping.

In Figure 8.16(a), the maximum obtainable gain, G_{opt}, is shown as a function of the pump power for configurations A, B, and C at the 980-nm pump wavelength. Here the fiber lengths L_{opt} are those giving the maximum gain. Figure 8.16(a) shows an improvement of the gain coefficient from 4.8 dB/mW (3.3 dB/mW) in config-

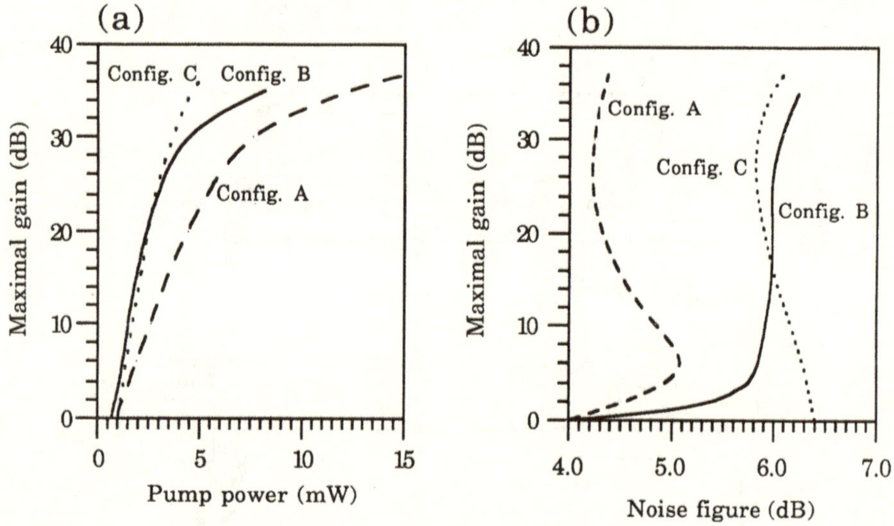

Figure 8.16 (a) Gain versus pump power for all three configurations in Figure 8.15 pumped at 980 nm. The results are calculated for maximum gain lengths L_{opt}. (b) Maximal gain shown versus noise figure for the three configurations pumped at 980 nm. The results are calculated for maximum gain lengths L_{opt}.

uration A to 8.5 dB/mW (5.8 dB/mW) for the reflective type B. With configuration C, a further increase to 8.9 dB/mW (6.4 dB/mW) is obtained. Comparing the gain values of Figure 8.16(a), it is found that configuration C can give a gain value equal to configuration A, but using only 60% of the pump power. Furthermore, an unsaturated gain of 23 dB is obtained for configuration C.

Gain slopes of 13.1 dB/mW (9.1 dB/mW) and 14.7 dB/mW (12.8 dB/mW) for configurations B and C indicate that the reflective type with or without filtering is very sensitive to variations of launched pump power. This demands a stable pump source. The threshold pump power increases for configuration C compared to that for configuration B. This is caused by the attenuation in the band-pass filter (1.5 dB) and in the extra dichroic coupler (0.3 dB). For gain values higher than 21 dB (19 dB), the gain is improved in configuration C.

In Figure 8.16(b), the noise characteristics for the three amplifiers are shown as a function of the gain. The noise figure, F, according to the description in Chapter 5, is calculated as $F = (S_{ASE}(\nu_s)/h\nu_s + 1)/G$. Here, $S_{ASE}(\nu_s)$ is the spectral density of the ASE power co-propagating with the output signal, ν_s is the signal frequency, h is Planck's constant, and G is the signal gain. Figure 8.16(b) shows an increase in the noise figure of 1.5 dB in configuration B compared to configuration A, where the noise levels are about 5.9 dB (7.2 dB) and 4.4 dB (8.7 dB), respectively. Here, the optical circulator is the main reason for the noise level increase, since it adds a 1.2-dB loss to the amplifier input signal. The use of optical circulators with lower loss would therefore improve the noise figure in configuration B. For gain values higher than 18 dB, Figure 8.16(b) shows that configuration C has a better noise performance than configuration B. Since the main part of the ASE is concentrated around the signal wavelength, a narrower bandpass filter would improve the noise characteristics, provided that the loss of such a filter is not increased simultaneously.

In practical situations, the available pump power and the desired gain might be limited, and it is of interest, therefore, to compare the amplifier configurations for a fixed pump power level P_p. Figure 8.17(b) shows the noise figure versus gain using a constant pump power of 4.1 mW for all three configurations. This value is chosen because it gives the maximum pump efficiency with a gain of 20 dB for configuration A (Figure 8.16(a)). For this fixed pump power, the fiber lengths of configurations B and C are adjusted to a gain of 20 dB, and we find a noise figure increase of only 0.5 to 0.6 dB. In configurations B and C (Figure 8.17(a)), the lengths of the EDFAs are L = 12m (conf. B) and L = 13.5m (conf. C), which are almost three times shorter than in configuration A (L = 35 m). Compared to Figure 8.16(b), an improvement in noise figure of about 1 dB is obtained, which is interesting with the use of the EDFA as a preamplifier.

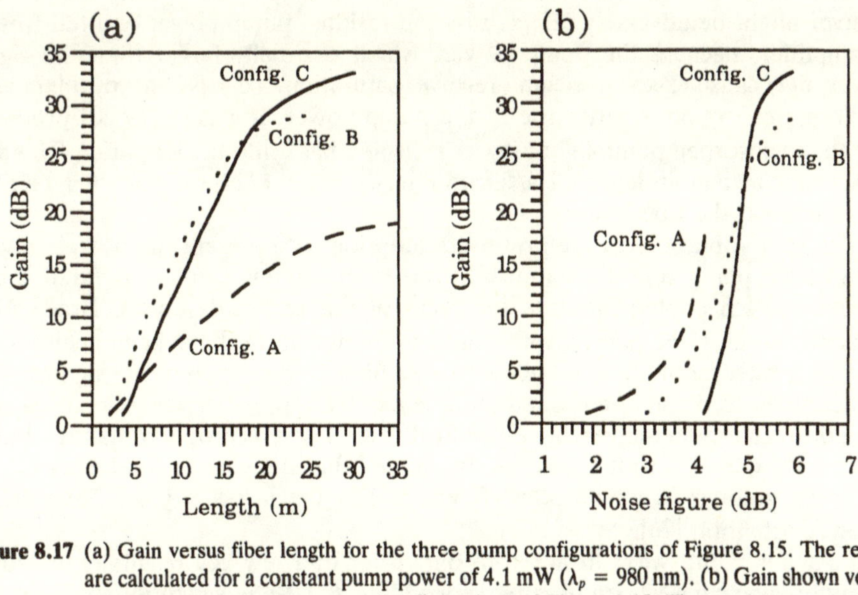

Figure 8.17 (a) Gain versus fiber length for the three pump configurations of Figure 8.15. The results are calculated for a constant pump power of 4.1 mW (λ_p = 980 nm). (b) Gain shown versus noise figure for the three pump configurations of Figure 8.15. The results are calculated for a constant pump power of 4.1 mW (λ_p = 980 nm). (Figures from [21].)

This section has illustrated how improved performance of EDFAs may be achieved using reflectors and optical circulators to double-pass the pump and signal beams through the erbium-doped fiber. However, as pointed out in [19], the reflection of only the pump light is simpler to implement, and it also increases the amplifier gain and saturated output power and lowers the amplifier noise. In Section 8.2.2, therefore, we will turn our attention to amplifiers, which only aim at single-pass signal amplification, but with improved properties compared to "standard" configurations, since they apply double-pass pumping. The somewhat complicating circulator is also avoided.

8.2.2 Pump Light Reflected by Fiber Bragg Filters

In Section 8.1, we have seen how optical filters could be included advantageously within the fiber amplifier with the primary purpose of suppressing the ASE, which builds up within the amplifier. This approach was used in Section 8.2.1 in addition to a signal and pump double-pass amplifier realized by reflection of both these beams. However, in both descriptions the pump power was ideally transmitted unattenuated through the ASE-suppressing components. A significant percentage of pump power will therefore finally be transmitted out through the amplifier output end. This residual pump light emitted from an optical fiber amplifier may be a problem in a telecommunication system. For example, the performance of a receiver might be adversely affected by the residual pump power emitted from a preamplifier, because the pump power, which is usually larger than the signal power, may cause excess noise and receiver saturation. To solve this problem and, furthermore, to find a better use of the pump power, it was in [19,20] proposed that the unabsorbed pump light may be reflected back into the amplifier. This was either done by a multi-layer silica/silicon fiber-edge filter [24] or by applying a fiber Bragg filter at the fiber output end.

Since the purpose of the present description is to present the overall advantages of the pump reflection approach rather than going through a detailed discussion of each implementation, the focus of this section is on the Bragg filter reflector, because the highest reflectance presented so far have been achieved by the Bragg filter method, and it allows for all-fiber transmission of the signals. This is in contrast to the end-deposited multi-layer filter approach which includes coupling through other components. Specifically, the low-loss property makes all fiber devices attractive. It should also be mentioned that the idea of producing photorefractive gratings in glass optical fibers [25] in the last few years have drawn renewed attention [26].

At this point, we will therefore only refer to a few key results obtained by the multi-layer silica/silicon fiber-end edge filter. In [19], it is mentioned that pump reflectors having 40% to 60% reflectance were used with 12- to 13-dB gain amplifiers to increase the gain by 0.9 dB and reduce the noise factor by 0.4 dB. It should, however, be noted that the specific improvements are closely related to the quality of the pump reflectors, and improvements in these are predicted to lead to increases in the gain and saturated output power of up to 2 dB. Furthermore, it is important to note that the end-deposited filters might be very low-cost items. They might, therefore, be incorporated into other amplifier components, such as optical isolators or ASE filters, with which the question of complicating the amplifier further is almost eliminated.

An alternative method is to fabricate efficient fiber-based filters employing Bragg gratings on a polished fiber. This is done by creating a grating in the photoresist with two-beam interference and then etching the grating into the fiber by ion beam milling as described in [27]. The reflection bandwidth of the resulting filters can be determined by the chirp of the grating to be from 0.3 to 22 nm [28]. Furthermore, a high reflection efficiency of the grating is maintained by increasing the interaction length of the grating as the chirp is increased. Such grating filters have been demonstrated and optimized for both the 1480-nm pump band [28] and the 980-nm pumping situation [20].

The filter referred to in [20] is characterized by a peak reflectivity of 99% back into the input fiber and a reflection bandwidth of 12 nm around 980 nm. The reflectivity of more than 75% in a bandwidth of 10 nm is particularly advantageous due to the coverage of the 980-nm pump band. The filter has been demonstrated with a low insertion loss of less than 0.5 dB at all wavelengths between 1100 and 1600 nm.

The resulting amplifier configuration seen in Figure 8.18 has been tested in an experimental setup using 980-nm pumping [29]. Although this does not represent a thorough investigation of the possibility of this configuration, a residual pump power reduction of 30 dB has been demonstrated, leading to a small increase in signal output power.

These results conclude this presentation of fiber amplifiers with improved properties accomplished by the end reflection of pump light. They demonstrate that a few low loss fiber components may improve the gain and noise properties of fiber amplifiers. However, as we shall see in the following section, it will be

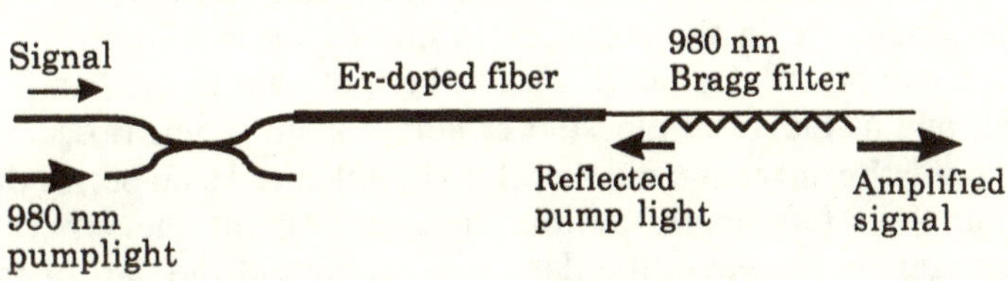

Figure 8.18 Fiber amplifier using a Bragg fiber filter for pump reflection.

possible for certain applications to avoid these additional components by specially designing the erbium-doped fiber.

8.3 TAPERED FIBER AMPLIFIERS

Because EDFAs today have become key components in optical communication systems, their different applications are the subject of thorough experiments and investigations.

One of the main objectives is, as we have already seen, to design a low-noise component while keeping a high gain performance. More traditional ways of achieving this have been a combination of erbium confinement in the core center of the fiber [30] and the use of different pump configurations [17] (co- or counter-propagation or pumping in both directions). The choice of pump wavelength is especially important if noise properties are in focus, and because of the possibility of achieving full population inversion, 980-nm pumping is most favorable [31]. In practical systems, limited pump power consumption is important, and additional means of lowering the noise figure for a fixed pump power are therefore highly relevant.

In addition to the conventional methods of achieving low noise performance are approaches such as those described in Section 8.1. However, as we shall see in this section, a complementary approach to achieving low-noise amplifiers has been suggested [32]. The idea is to apply tapering to the fiber and thereby ensure a better population inversion of the erbium in the signal input end of the fiber. Finally, the impact of the improved noise properties and maintained high gain performance will be evaluated by describing improved receiver sensitivity in communication systems with optical preamplifiers.

8.3.1 Theoretical Basis of Low-Noise Amplifier

To make a low-noise EDFA, the amplifier must be designed according to the fundamental rule, which is that the more rapid the gain increases in the beginning of the EDFA, the lower the noise figure becomes [33]. We therefore have to ensure a high population inversion in the signal input end of the active fiber.

Improved noise properties of fiber amplifiers can therefore be achieved by narrowing the fiber core in the signal input end of the amplifier. Two complementary effects can be obtained in this way: the spot sizes of the pump and signal are increased, and at the same time the erbium dopant radius is decreased. The area of high population inversion can therefore be enhanced by tapering the erbium-doped fiber up to its full length towards the signal output end. The principle is illustrated in Figure 8.19, where the dark areas represent the part of the fiber in which more than 50% of the erbium ions are excited. It should be noted that the

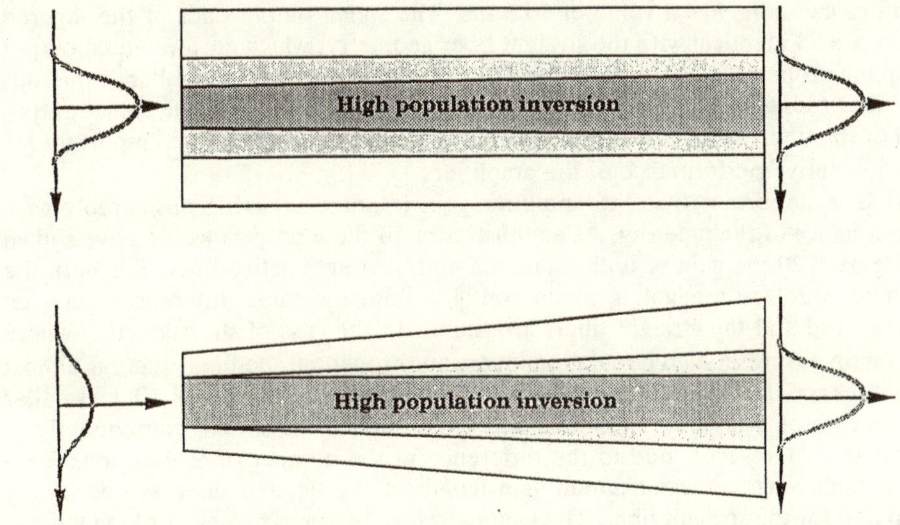

Figure 8.19 Signal spot-size variation in a straight and a tapered active fiber. Counterpropagating pumping is considered in both cases and the dark areas represent parts of the fibers, where more than 50% of the erbium ions are excited. (Figure layout from [32].)

slope of the tapers must be chosen small enough to avoid coupling between different fiber modes.

The tapered fibers are described theoretically as a cascade of very short pieces of active fiber, each with a slightly different radius, but with a constant NA. The mode fields are calculated separately in each section. The erbium distribution in the fiber is assumed to follow the index profile, which is considered a very good approximation as long as aluminum-co-doped active fibers are analyzed (see Section 3.1.2). The overall pump and signal distribution along the amplifier length are calculated, along with the co- and counter-propagating ASE, through an iterative solution of the governing differential equations [34].

8.3.2 Low-Noise Performance of Up-Tapered Erbium-Doped Fibers

In the following examples, all comparisons are made with a straight active fiber with a core radius of 2.0 μm, a fiber radius of 62.5 μm, and a cutoff wavelength at 980 nm, which coincides with the pump wavelength being considered. The signal wavelength is 1530 nm, corresponding to the peak value of the emission and absorption spectra. The erbium concentration is $\rho = 2.0 \times 10^{24}$ m^{-3}, and the absorption cross-section values are the same as those listed in [34]. The fluorescence lifetime in all calculations has a value of 10.5 ms. The signal output ends of the tapered fibers are all identical with the straight fiber geometry, which ensures equal output coupling properties. The taper slope of 1.0 μm/m expresses the number of microns the fiber radius increases per meter of amplifier length towards the signal output end of the fiber. Note that the actual values mentioned above are unimportant for the qualitative performance of the amplifiers.

It is important that the amplifier gain is not decreased considerably as a consequence of the tapering. As an illustration of these properties, we have shown in Figure 8.20 the gain of both a tapered and a straight active fiber. For both the co- and counter-propagating pump configuration, the same differences between the tapered and the straight fibers are found. In the case of short lengths, where the pump ensures a high population inversion throughout the fiber, there is almost no difference. For increased fiber length, the number of erbium ions will be smaller in the case of a tapered amplifier, and the straight fiber shows, consequently, a larger gain. However, due to the difference in the number of erbium ions for a given fiber length, the maximum gain length of the tapered fiber will be longer than that for the straight fiber. The gain therefore becomes highest for long lengths of the tapered fiber.

Since the gain properties are not degraded, we can expect to have almost full advantage of the noise figure improvement, also shown in Figure 8.20. We will

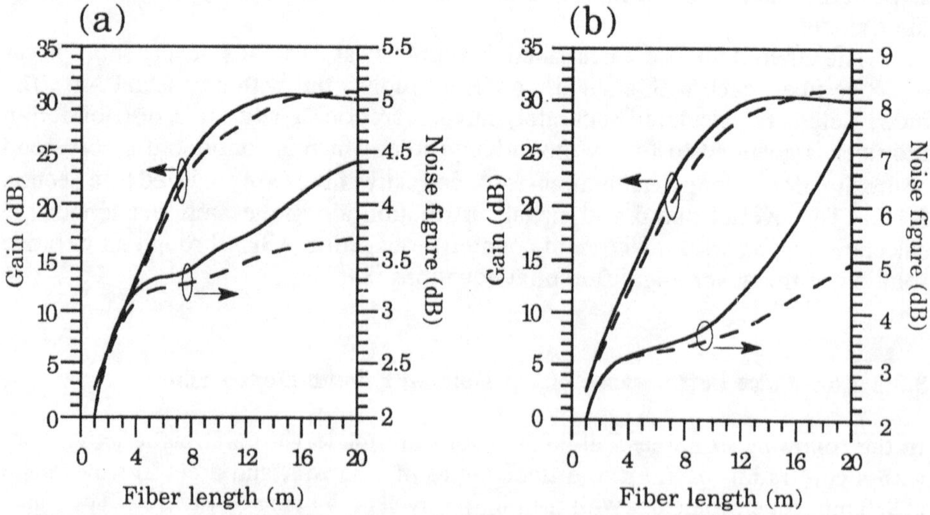

Figure 8.20 Gain and noise figure versus fiber length for a straight (solid curve) and a tapered fiber with slope 1.0 μm/m. 15-mW pump power was used at 980 nm. (a) Results for co-propagating pump. (b) Results for counter-propagating pump. (Figures from [32].)

now, as an example, look at a 20-m-long fiber. In the case of a co-propagating pump configuration, tapering will lead to a noise figure decrease from 4.4 to 3.7 dB. In the case of a counter-propagating pump configuration, the noise figure reduction is much more considerable from 7.6 to 4.9 dB by tapering.

To get a more balanced view of the gain and noise properties of the tapered active fiber, the application as preamplifier in a direct detection system is considered. The theoretical basis and choice of detector parameters of the description given here is based on [35], but for the sake of simplicity, only the counter-propagating pump configuration is considered.

The first of the interesting parameters is the amplifier input coupling efficiency η_{in}. In this particular application, it is calculated from the spot-size mismatch between the active fiber and a passive fiber with an rms spot size of 4.5 μm. It should, however, be pointed out that special care has to be taken with respect to the splicing of standard and tapered fibers, since they can have a different outer fiber diameter. For the amplifier output coupling efficiency, it is considered necessary to include an optical filter; we therefore use the lower value of $\eta_{out} = 0.6$. In agreement with [35], the calculations are carried out for a bit rate of 5 Gbit/s, and consequently the electrical bandwidth $B_e = 2.5$ GHz. Furthermore, the base receiver sensitivity is -25 dBm, and in all cases the receiver sensitivity is given as the power at the output of the transmission fiber necessary for obtaining a BER of 10^{-9}. The infinite extinction ratio is chosen, together with a 50% duty cycle.

The receiver sensitivity in Figure 8.21 is based on the specifications given above and is found for different lengths of the active fibers. For the straight fiber, the minimum receiver sensitivity is -40.95 dBm for an 8m-long active fiber. To define a measurement of the tolerance within which the amplifier length can be varied for a given pump power, the 1-dB penalty points have been marked for the straight fiber. If we now look at the interval for which the tapered amplifier also has a sensitivity better than 39.95 dBm, we find that the interval describing the acceptable fiber length is increased by 70%. It should be noted that the short-length limit is not changed, since this is given by the amplifier shot noise. The upper limit is, however, moved upwards due to the lower noise figure of the tapered amplifier, and the minimum receiver sensitivity has been improved by 0.25 dB and can probably be further improved by optimization.

Note finally that experimental results of tapered fibers indicate that the predicted noise improvements are obtainable [36]. It should also be added, however, that the use of tapered fibers is most realistic in cases where it becomes difficult or impossible to obtain equivalent advantages with isolator insertion.

8.3.3 Other Applications of Tapered Active Fibers

The tapered erbium-doped fiber has, as described in Section 8.3.2, advantages with respect to improved noise properties. However, for other applications such as power

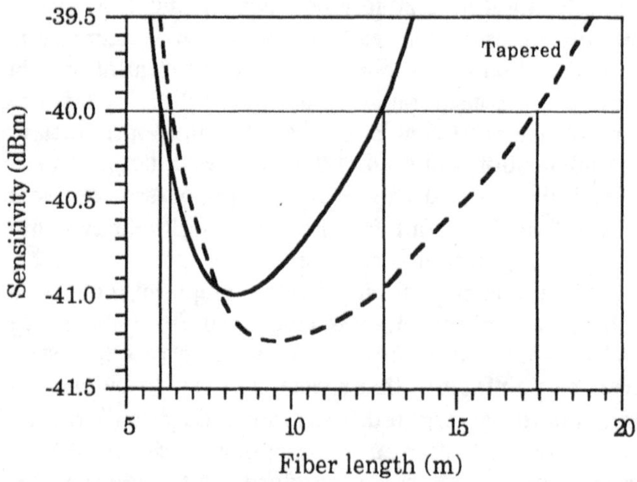

Figure 8.21 Receiver sensitivity versus fiber length for preamplifiers based on a straight (solid curve) and tapered fiber with slope 1.0 μm/m, respectively. The lines indicate the 1-dB penalty points given relative to the minimum receiver sensitivity for the straight amplifier. The receiver parameters are specified in the text. 15-mW counter-propagating pump power was used at 980 nm. (Figure from [32].)

booster amplifiers, the tapering may lead to additional improvements. Among these is the possibility of spot-size matching to the standard communication fibers, resulting in reduced coupling losses. Furthermore, the tapering results in less restrictive tolerance requirements in the choice of optimal fiber length or, correspondingly, in weakened fiber production.

8.4 ADDITIONAL METHODS FOR NOISE FIGURE REDUCTION

In this section, we will present two additional methods for fiber amplifier improvement. The first of these focuses on the problem of optimal coupling between optimal erbium doped fibers and standard fibers for transmission of optical signals. The second method deals with the question of how to build fiber amplifiers with high energy conversion efficiency and simultaneous low-noise properties, achieved by simultaneous pumping at more than one pump wavelength.

8.4.1 Matching Fibers for Low-Loss Coupling into Fiber Amplifiers

As was illustrated in Chapter 6, the efficiency of EDFAs will increase greatly by manufacturing fibers with higher NAs. This means that the spot size in the erbium-doped fiber will be small compared to spot sizes of standard single-mode fibers, which results in a mismatch between the field distributions, leading to a large splice loss. Since this degrades the gain, the signal-to-noise ratio, and the amplifier output power, a solution to the mismatch problem is highly relevant, and we have already touched on the subject in the previous section.

Besides the obvious, but not necessarily optimal solution of fabricating active fibers with spot sizes closer to the values of a standard fibers, diffusion tapering [37] has been used to partly overcome the problem. An alternative method has been proposed [38] in which the idea is to splice one or more (untapered) fibers between the amplifier and the standard fiber.

To illustrate the value of this approach, we will briefly discuss the example presented in [38]. We consider her the theoretical fraction of power transmitted at a splice between a standard fiber with normalized frequency $V = 1.95$ (corresponding to $\lambda_c = 1.25\ \mu m$) and an NA $= 0.13$ ($\Delta = 0.004$), and an erbium-doped fiber with $V = 1.45$ (see description of optimal design in Section 6.3.1). In Figure 8.22(a), this fraction of transmitted power between standard fiber and amplifier is shown. Also, the maximum slope efficiency relative to that obtainable with a perfect splicing is shown together with the net relative slope efficiency, taking the splice loss of the pumping into account. The results show that, although the gain is increased with Δ, much of the potential gain is lost due to a large splice loss. Figure

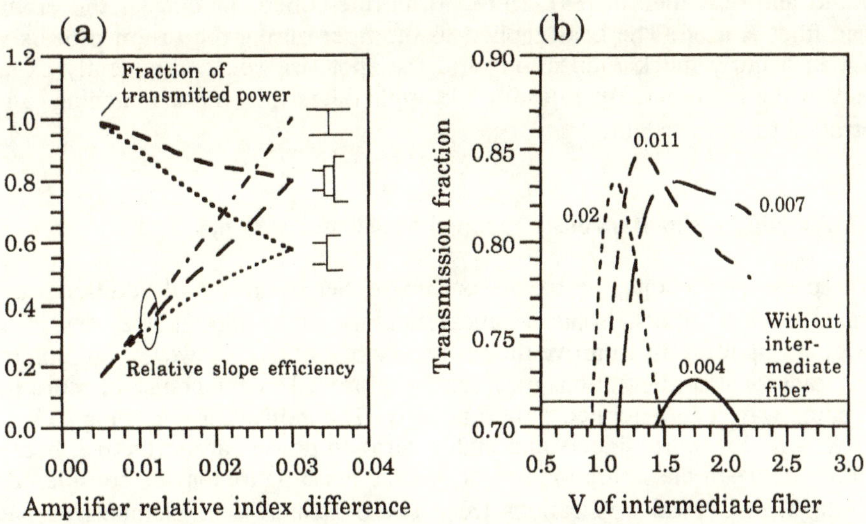

Figure 8.22 (a) Fraction of power transmitted from standard fiber to amplifier and the effect of coupling improvement on amplifier slope efficiency. (b) Fraction of transmitted power as a function of the normalized frequency V of the intermediate fiber. (Data from [38].)

8.22(a) also shows that the use of an intermediate fiber spliced between the amplifier and the standard fiber will improve the relative slope efficiency. In [38], this is reported to yield results as good as those presented for diffusion tapering [37].

In order to get more accurate information on how to choose the intermediate fiber, Holmes et al. [38] investigated how the fraction of transmitted power depends on the normalized frequency V and relative index difference Δ of this fiber. Results similar to these investigations are shown in Figure 8.22(b), and it is shown that more than 82% transmission may be obtained over a wide range of intermediate fiber parameters. The weak sensitivity of the results for the V number of the intermediate fiber suggests that this proposed method for improving coupling loss will work for a wide range of standard fiber and amplifier parameters.

For the sake of generality, it is in [38] also noted that more than one intermediate fiber might be included, resulting in a further improvement of transmission between standard fiber and amplifier. Note, however, that a larger number of intermediate fibers also increases the possibility of introducing splice errors, which will reduce the predicted power transmitted through the interconnection. One should also consider that, in practice, the heat of fusion causes some diffusion of dopants at the joint, increasing the transmission from the theoretical value based on the fields of the undiffused profiles. It may, therefore, not be considered a good strategy to include more than one or two intermediate fibers, because the possible practical value of further additions might only be marginal.

A more practical approach to solving the mode matching problem has been adopted and described in [39]. Here, a fluorine-doped cladding of the erbium-doped fiber is used. The heat applied to the fiber during the fusion process will result in a fairly quick F-diffusion, and the spot size will consequently expand locally at the fiber end. By adjusting the applied heat, the mode matching can be controlled to some extent.

8.4.2 Amplifiers Simultaneously Pumped at 980 and 1480 nm

In the previous sections, we have seen how either by means of additional components, such as filters, isolators, and reflectors, or by special fiber design it is possible to significantly improve the properties of EDFAs. However, amplifiers in more traditional configurations may still be improved by the choice of more than one pump wavelength applied simultaneously. The motivation for doing so is that EDFAs with low-noise figures and high saturation powers are attractive, because the ratio between the saturation power and the noise figure may be considered an advantage for several applications [35]. These include multi-channel receivers, where high sensitivity and high saturation power increase the number of channels to be received, and in-line applications, where the dynamic range is proportional to the advantage mentioned above.

For 980-nm pumping, the noise figure may reach values close to the quantum limit [40], whereas the minimum noise figure obtainable with 1480-nm pumping is approximately 5 dB [41] (see also Chapter 7). Although the noise figure under 1480-nm pumping has been found to be 2 dB worse than for 980-nm pumping, the former offers very high output power or near-quantum-limited conversion efficiency from pump to signal. Reports have therefore been on simultaneous low-noise and high output power operation of the EDFA. One way to achieved these amplifiers is simply by pumping co-directionally at 980 nm and counter-directionally at 1480 nm [42]. Alternatively, a two-stage amplifier that contains two fibers separately pumped at 980 and 1480 nm has been realized [6], which also includes the advantage of internal isolator placement within the amplifier.

We will at this point concentrate on the hybridly pumped amplifier without isolator, because the two-stage amplifier represents a way of combining several methods for improving the amplifier. The effect of simultaneous 980- and 1480-nm pumping might therefore be camouflaged by the isolator.

The results shown in Figure 8.23 are therefore related to a situation in which co- and counter-directional pumping of the active fiber is provided by 980- and 1480-nm wavelength semiconductor lasers, respectively. The data, which relate to the work presented in [42], are measured for a 15m-long Er/Al/Ge-doped fiber

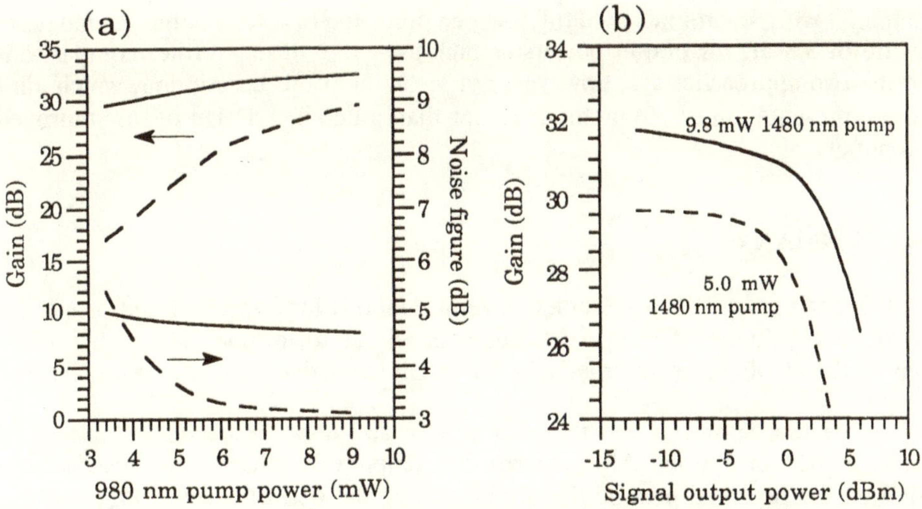

Figure 8.23 (a) Gain and noise figure shown versus the 980-nm pump power with 1480-nm pump power as a parameter. Solid curves: 9.8-mW, 1480-nm pump; dashed curves: 5.0-mW, 1480-nm pump. (b) Gain as a function of signal output power for 9-mW pump power at 980 nm and with 1480-nm pump power as a parameter. (Data from [42].)

with an erbium doping concentration of 5.8×10^{24} m^{-3}, a core radius of 1.85 μm, and a refractive index difference of 0.021. The EDFA is described for a 1552-nm signal.

The gain and noise figure of the EDFA are, as shown in Figure 8.23(a), not surprisingly influenced by the total coupled pump power and the ratio of 980- to 1480-nm pump power. A minimal noise figure of 3.1 dB is measured for a gain of 29.5 dB for pump powers of 9 and 5 mW at 980 and 1480 nm, respectively. This shows that no noise degradation results from the 1480-nm counter-directional pumping. However, as the 980-nm pumping decreases or the 1480-nm pumping increases, the noise figure increases to approximately 5 dB. The latter is cause by the population inversion in the input end of the fiber being forced to a lower level due to the higher intensity of 1480-nm light.

This accounts for the improved noise properties, but in order to see the positive effects on the saturation output power induced by the backward-traveling 1480-nm pumping, Figure 8.23(b) is included. Here, the gain is shown as a function of the signal output power for various pump power levels, illustrating that the saturation output power increases with increased 1480-nm power. This leads to the demonstration of a diode-pumped EDFA with a noise figure of 3.1 dB, a gain of 29.5 dB, and a saturation output power of 2 dBm. This was realized at a signal wavelength of 1552 nm by pumping co-directionally at 980 nm and counter-directionally at 1480 nm, with a total pump power of only 14 mW.

In comparison, it should be noted finally that the hybridly pumped fiber amplifier with isolator included [6] has been predicted to show a nondegraded noise figure of 3.8 dB for output powers as high as +15.5 dBm. Further comparisons of the two approaches are, however, not included in this description, which aims at the presentation of the methods rather than a detailed design of the improved amplifiers.

8.5 SUMMARY

In this chapter, we have presented several methods for improving the properties of fiber amplifiers. The approach has been either to include extra components within the amplifier or, alternatively, to design the active fiber in a more complex way.

The first approach was the idea of gain and noise figure improvements by the inclusion of internal ASE-suppressing components. Numerical investigations showed the potential possibilities for improving the performance of high-NA Er/Ge-doped fiber amplifiers around two possible gain peaks at 1534 and 1554 nm by the implementation of ASE-attenuating components exemplified by either a 1-nm filter or a 30-dB isolator, or a combination of the two. For the considered pump bands of 980 and 1480 nm, gain improvements in the range of 10 to 20 dB was shown, and it was found that the noise figures may be reduced to their theoretical limits. Using a filter as ASE attenuator, only modest improvements were found in noise figures, and the gain improvements are below those found for the case of isolator inclusion. The combination of an isolator and filter seems to be the best choice when the signal wavelength at 1534 nm is considered. For the higher signal wavelength at 1554 nm, it is the insertion of an isolator that gives the best performance when the amplifier is pumped with 980-nm light, while the combination of filter and isolator is slightly better when pumped with a 1480-nm source.

The amplifier improvements by using the isolator exist in the whole input signal power range, where ASE is determinative upon the population inversion. Gain improvements can be obtained for input signal powers below the 3-dB saturation point and noise figure improvements for signal powers below 5 dBm. The optimum isolator location increases for decreasing signal power and becomes unchanged in the small-signal regime.

Also, experimental results showed that the insertion of an isolator within the length of an erbium-doped fiber has been shown to enable near-quantum-limited operation of very-high-gain amplifiers. When pumping at 1470 nm, the isolator was directly installed into the fiber, as only a minor insertion loss was present at this wavelength. For 980-nm pumping, a large insertion loss would cause the pump to deteriorate dramatically; however, a bypass coupling of the pump has been shown to solve this problem efficiently. As in the case of being without an ASE attenuation component, the best pump configuration with respect to high gain and low noise is the bi-directional one. Experiments have yielded gain improvements of up to 4.1 dB and noise figure improvements of 2.0 dB by using the isolator in such a scheme. Still, even larger improvements are predicted for the co-propagating configuration. Depending on the chosen pumping scheme, the optimal isolator location varies within 15% to 37% of the total fiber length. More specifically, the optimum isolator position is theoretically and experimentally determined between 15% and 20% for 980-nm pumping. For 1470-nm pumping the optimal location was determined to be in the range of 20% to 35% of the total erbium fiber length.

The discussion concerning the applications of internal filtering was concluded with a short description of how spectral gain equalization can also be obtained. It was shown that a carefully positioned filter within the amplifier improves the pump efficiency at wavelengths away from the gain peak. At the same time, it does not lower the amplifier saturation output power. A gain of 27 dB and a 3-dB bandwidth of 33 nm were reported for 38 mW of pump power at 980 nm.

Another approach for improving the amplifier is used in the reflective-type amplifiers, in which both pump and signal double-pass through the amplifier. To describe this method, three different configurations are analyzed for Er/Al-doped fiber amplifiers in the 980- and 1480-nm pump bands. The use of a mirror to reflect signal and pump yields two-way amplification and pumping. The pump power necessary to achieve a specified gain can be reduced to about half and the gain coefficient can be doubled. Noise figure investigations show that the noise level at maximum obtainable gain increases about 1.5 dB compared to the usual EDFA configuration. This deterioration in noise figure is mainly set by the loss in the optical circulator, which becomes necessary for signal double-pass amplifiers. When using fiber lengths below the maximum gain length and keeping the gain around 20 dB, the noise figure improves and is only 0.5 dB above that of the usual EDFA configuration.

The other representative of the reflective-type amplifiers is the pump-reflective amplifier, in which the signal only experiences the single-pass gain. The pump reflection is accomplished by a Bragg filter at the signal output end of the amplifier, and a residual pump power reduction of 30 dB was demonstrated.

As a complementary approach to the above methods of gain and noise figure improvement, the tapered active fiber was presented. This represents a new strategy for noise figure reduction, in which variation along the fiber length has been used to improve the amplifier properties. It has been theoretically demonstrated that the tapered fiber amplifier has considerably improved noise properties. Also, the introduction of an extra free parameter gives the possibility of spot-size matching and thereby lower coupling losses in and out of the fiber amplifier.

Finally, this chapter dealt with two complementary approaches, of which the first was a method for reducing coupling losses between standard single-mode transmission fibers and erbium-doped fibers, and it is based on the strategy of including intermediate fibers at the input and output ends of the fiber amplifier. The second complementary approach does not deal directly with a principal change in the setup for a bi-directionally pumped EDFA. The improvements are achieved by simultaneous pumping with 980-nm light co-propagating with the signal and 1480-nm light counter-propagating with the signal. A non-degraded noise figure of 3.8 dB for output powers as high as +15.5 dBm is thereby achieved. A total pump power of only 14 mW also resulted in a noise figure of 3.1 dB and a gain of 29.5 dB for a 1552-nm signal.

REFERENCES

[1] Shimizu, M., M. Yamada, M. Horiguchi, T. Takeshita, and M. Okayasu, "Erbium-Doped Fibre Amplifiers With an Extremely High Gain Coefficient of 11.0 dB/mW," *IEE Electronics Letters*, Vol. 26, No. 20, 1990, pp. 1641–1643.

[2] Hansen, S. L., K. Dybdal, and C. C. Larsen, "Upper Gain Limit in Er-Doped Fiber Amplifiers Due to Internal Rayleigh Backscattering," *Optical Fiber Communications OFC '92,* San Jose, 2–7 February 1992, p. 68.

[3] Olshansky, R., "Noise Figure for Erbium-Doped Optical Fibre Amplifiers," *IEE Electronics Letters*, Vol. 24, No. 22, 1988, pp. 1363–1365.

[4] Laming, R. I., and D. N. Payne, "Noise Characteristics of Erbium-Doped Fiber Amplifier Pumped at 980 nm," *IEEE Photonics Technology Letters*, Vol. 2, No. 6, 1990, pp. 418–421.

[5] Povlsen, J. H., A. Bjarklev, O. Lumholt, H. Vendeltorp-Pommer, and K. Rottwitt, "Optimizing Gain and Noise Performance of EDFAs With Insertion of a Filter or an Isolator." *Proceedings of Society of Photo-Optical Instrumentation Engineers Conference, OE/Fibers '91*, part: Fiber Laser Sources and Amplifiers III, Boston, September 1991, Vol. 1581, pp. 107–113.

[6] Aoki, Y., T. Saito, K. Fukagai, Y. Sunohara, S. Ishikawa, and S. Fujita, "Low Noise and High Saturation Output Power Erbium-Doped Fiber Amplifiers Pumped With 0.98 μm and 1.48 μm LDs for Long-Distance Optical Communication," *European Conference on Optical Communication ECOC '90*, Paris, 9–12 Sept. 1991, Proc. Vol. II, paper WeC9-2, pp. 585–588.

[7] Giles, C. R., and D. J. DiGiovanni, "Dynamic Gain Equalization in Two-Stage Fiber Amplifiers," *IEEE Photonics Technology Letters*, Vol. 2, No. 12, 1990, pp. 866–868.

[8] Lumholt, O., K. Dybdal, C. C. Larsen, S. Dahl-Petersen, K. Schüsler, A. Bjarklev, J. H. Povlsen, T. Rasmussen, and K. Rottwitt, "Er-La Doped Fibre Amplifier, Pumped at 1.47 μm," *Proc. of 17th European Conference on Optical Communication ECOC '91*, Paris, 9–12 Sept. 1991, paper TuPS-1, pp. 285–288.

[9] Lumholt, O., K. Schüsler, A. Bjarklev, S. Dahl-Petersen, J. H. Povlsen, T. Rasmussen, and K. Rottwitt, "Efficiency Improvement of EDFAs by Insertion of an Isolator Within the Active Fibre," *Conference on Lasers and Electro-Optics CLEO '92*, Anaheim, 10–15 May 1991, pp. 354–355.

[10] Lumholt, O., J. H. Povlsen, K. Schüsler, A. Bjarklev, S. Dahl-Petersen, T. P. Rasmussen, and K. Rottwitt, "Quantum Limited Noise Figure Operation of High Gain Erbium Doped Fiber Amplifiers," *IEEE Journal of Lightwave Technology*, in press.

[11] Pedersen, B., J. Chirravuri, and W. J. Miniscalco, "Gain and Noise Penalty for Detuned 980-nm Pumping of Erbium-Doped Fiber Power Amplifiers," *IEEE Photonics Technology Letters*, Vol. 4, No. 4, 1992, pp. 351–353.

[12] Lumholt, O., K. Schüsler, A. Grunnet-Jepsen, A. Bjarklev, S. Dahl-Petersen, J. H. Povlsen, T. P. Rasmussen, K. Rottwitt, and C. C. Larsen, "Low Noise High Gain Preamplifier Using an Isolator Within the Erbium Doped Fibre," *European Conference on Optical Communication ECOC '92*, Berlin, 27 Sept.–1 Oct. 1992, Proc. Vol. I, paper MoA3.5, pp. 93–96.

[13] Atkins, C. G., J. F. Massicott, J. R. Armitage, R. Wyatt, B. J. Ainslie, and S. P. Craig-Ryan, "High-Gain, Broad Spectral Bandwidth Erbium-Doped Fiber Amplifier Pumped Near 1.5 μm," *IEE Electronics Letters*, Vol. 25, No. 14, 1989, pp. 910–911.

[14] Tachibana, M., R. I. Laming, P. R. Morkel, and D. N. Payne, "Erbium-Doped Fiber Amplifier With Flattened Gain Spectrum," *IEEE Photonics Technology Letters*, Vol. 3, No. 2, 1991, pp. 118–120.

[15] Unger, H. G., *Planar Optical Waveguides and Fibres*, Oxford: Clarendon Press, 1977.

[16] Massicott, J. F., J. R. Armitage, R. Wyatt, B. J. Ainslie, and S. P. Craig-Ryan, "High Gain, Broadband, 1.6 μm Er^{3+} Doped Silica Fibre Amplifier," *IEE Electronics Letters*, Vol. 26, No. 20, 1990, pp. 1645–1646.

[17] Pedersen, B., A. Bjarklev, H. Vendeltorp-Pommer, and J. H. Povlsen, "Erbium Doped Fibre Amplifier: Efficient Pumping at 807 nm" *Optics Communications*, Vol. 81, No. 1-2, 1991, pp. 23–26.

[18] Nishi, S., K. Aida, and K. Nakagawa, "Highly Efficient Configuration of Erbium-Doped Fibre Amplifier," *European Conference on Optical Communication ECOC '90*, Amsterdam, 1990, Proc. Vol. I, pp. 99–102.

[19] Giles, C. R., J. Stone, L. W. Stulz, K. Walker, and C. A. Burns, "Gain Enhancement in Reflected-Pump Erbium-Doped Fiber Amplifiers," *Techn. Dig. of Optical Amplifiers and Their Applications OAA '91,* Snowmass, CO, 24–26 July 1991, paper ThD2, pp. 148–151.

[20] Farries, M. C., C. M. Ragdale, and D. C. Reid, "Fibre Bragg Filters for Pump Rejection and Recycling in Erbium Doped Fibre Amplifiers," *Techn. Dig. of Optical Amplifiers and Their Applications OAA '91,* Snowmass, CO, 24–26 July 1991, paper ThD1, pp. 144–147.

[21] Lauridsen, V., R. Tadayoni, A. Bjarklev, J. H. Povlsen, and B. Pedersen, "Gain and Noise Performance of Fibre Amplifiers Operating in New Pump Configurations," *IEE Electronics Letters*, Vol. 27, No. 4, 1991, pp. 329–331.

[22] Iwamura, H., H. Iwasaki, K. Kubodera, Y. Torii, and J. Noda, "Simple Polarisation-Independent Optical Circulator for Optical Transmission Systems," *IEE Electronics Letters*, Vol. 15, 1979, pp. 830–831.

[23] Dybdal, K., A. Bjarklev, C. Dam-Hansen, B. Pedersen, J. H. Povlsen, H. Vendeltorp-Pommer, and C. C. Larsen, "Detailed Experimental and Theoretical Investigation of High-Gain Er-Doped Fibre Amplifier," *European Conference on Optical Communication ECOC '90*, Amsterdam, 1990, Proc. Vol. I, pp. 443–446.

[24] Stone, J., and L. W. Stulz, "FiEnd Filters: Passive Multilayer Thin-Film Optical Filters Deposited on Fibre Ends," *IEE Electronics Letters*, Vol. 26, 1990, pp. 1290–1291.

[25] Hill, K., Y. Fujii, D. C. Johnson, and B. S. Kawasaki, "Photosensitivity in Optical Fiber Waveguide: Applications to Reflective Filter Fabrication," *Applied Physics Letters*, Vol. 32, No. 10, 1978, pp. 647–649.

[26] Meltz, G., W. W. Morsey, and W. H. Gleen, "Formation of Bragg Grating in Optical Fibers by a Transverse Holographic Method," *Optics Letters*, Vol. 14, No. 15, 1989, pp. 823–825.

[27] Bennion, I., D. C. J. Reid, C. J. Rowe, and W. J. Stewart, "High-Reflective Monomode-Fibre Grating Filters," *IEE Electronics Letters*, Vol. 22, 1986, pp. 341–343.

[28] Farries, M. C., C. M. Ragdale, and D. C. J. Reid, "Chirped Bragg Optical Filter With 22 nm Bandwidth," *Workshop on Active and Passive Fibre Components*, Monterey, 1991, paper MC2.

[29] Farries, M. C., C. M. Ragdale, and D. C. Reid, "Broadband Chirped Fibre Bragg Filters for Pump Rejection and Recycling in Erbium Doped Fibre Amplifiers," *IEE Electronics Letters*, Vol. 28, No. 5, 1992, pp. 487–489.

[30] Desurvire, E., J. L. Zyskind, and C. R. Giles, "Design Optimization for Efficient Erbium-Doped Fiber Amplifiers" *IEEE Journal of Lightwave Technology*, Vol. 8, No. 11, 1990, pp. 1730–1741.

[31] Payne, D. N., and R. I. Laming, "Optical Fibre Amplifiers," *Optical Fiber Communications OFC '90*, San Francisco, 1990, tutorial, ThF1, pp. 331–354.

[32] Bjarklev, A., O. Lumholt, J. H. Povlsen, T. Rasmussen, K. Rottwitt, S. Dahl-Petersen, and C. C. Larsen, "New Tapered Active Fibres for Very Low Noise Amplifier Applications," *IEE Electronics Letters*, Vol. 28, No. 1, 1992, pp. 34–35.

[33] Pedersen, B., A. Bjarklev, J. H. Povlsen, K. Dybdal, and C. C. Larsen, "The Design of Erbium-Doped Fiber Amplifiers," *IEEE Journal of Lightwave Technology*, Vol. 9, No. 9, 1991, pp. 1105–1112.

[34] Bjarklev, A., S. L. Hansen, and J. H. Povlsen, "Large Signal Modelling of an Erbium Doped Fiber Amplifier," *Proceedings of Society of Photo-Optical Instrumentation Engineers Conference, OE/Fibers '89*, part: Fiber Laser Sources and Amplifiers, Boston, 6–8 Sept. 1989, SPIE Vol. 1171, pp. 118–129.

[35] Olsson, N. A., "Lightwave Systems With Optical Amplifiers," *IEEE Journal of Lightwave Technology*, Vol. 7, No. 7, 1989, pp. 1071–1082.

[36] Grunnet-Jepsen, A., K. Schüsler, S. Dahl-Petersen, B. Palsdottir, A. Bjarklev, K. Rottwitt, J. H. Povlsen, and T. P. Rasmussen, "Experimental Investigation of Tapered Erbium-Doped Fiber Amplifiers," *Proceedings of Society of Photo-Optical Instrumentation Engineers Conference, OE/Fibers '92*, part: Fiber Laser Sources and Amplifiers IV, Boston, September 1992, paper 1789-16.

[37] Mortimore, D. B., and J. V. Wright, "Low-Loss Joints Between Dissimilar Fibres by Tapering Fusion Splices," *IEE Electronics Letters*, Vol. 22, 1986, pp. 318–319.

[38] Holmes, M. J., F. P. Payne, and D. M. Spirit, "Matching Fibres for Low Loss Coupling Into Fibre Amplifiers," *IEE Electronics Letters*, Vol. 26, No. 25, 1990, pp. 2102–2104.

[39] Kashiwada, T., M. Shigematsu, T. Kougo, H. Kanamori, and M. Nishimura, "Characteristics of 1.48 μm Pumped Erbium-Doped Fiber Amplifier With High Efficiency," *Techn. Dig. of Optical*

Amplifiers and Their Applications OAA '91, Snowmass, CO, 24–26 July 1991, ThE2-1, pp. 166–169.

[40] Laming, R. I., and D. N. Payne, "Noise Characteristics of Erbium-Doped Fiber Amplifier Pumped at 980 nm," *IEEE Photonics Technology Letters*, Vol. 2, 1990, pp. 418–421.

[41] Giles, C. R., and E. Desurvire, "Propagation of Signal and Noise in Concatenated Erbium-Doped Fiber Optical Amplifier," *IEEE Journal of Lightwave Technology* , Vol. 9, No. 2, 1991, pp. 147–154.

[42] Mikkelsen, B., C. G. Joergensen, J. A. van den Berk, C. F. Petersen, C. C. Larsen, T. Durhuus, and K. E. Stubkjaer, "High Receiver Sensitivity at 2.5 Gb/s Obtained With a Highly Efficient Low Noise Diode Pumped Erbium Doped Fiber Amplifier," *Techn. Dig. of Optical Amplifiers and Their Applications OAA '91,* Snowmass, CO, 24–26 July 1991, paper FA4-1, pp. 192–195.

Chapter 9
The 1300-nm Fiber Amplifier

Most of the interest in rare-earth-doped fiber amplifiers has been in the Er^{3+}-doped silica fiber because of its potential as a high-gain optical amplifier for 1550-nm signals. For this reason, by far the most advanced results have been achieved for amplifiers operating in this wavelength band, and 1550-nm amplifiers consequently play a significant role in this book. However, the majority of the installed optical communication systems operate at 1300 nm. Therefore, the development of high-gain optical amplifiers for this communication window is of great interest.

It was hoped that Nd^{3+}-doped SiO_2 fibers, as the first candidate, would fulfill this requirement, but they have been shown to reveal ESA near 1300 nm rather than the expected gain. The ESA spectrum is sensitive to glass composition, and in fluoride glasses the spectrum is shifted downward in wavelength and strength [1], opening up the possibility of 1300-nm amplification. The last basic obstacle for the 1300nm amplifier, which was the fabrication of single-mode ZBLANP fibers, has also been solved, and gain of up to 7 dB at 1340 nm has been reported [2]. As this gain value clearly indicates, the neodymium-doped ZBLANP fiber does show results far below the values obtained by erbium-doped 1550-nm amplifiers. Furthermore, the gain overlap with the 1300-nm telecommunication window is relatively poor [1,2], since the gain maximum is located around 1340 nm and the 3-dB bandwidth (FWHM) is on the order of 40 nm. Among the major reasons for the limited gain of these amplifiers are the ASE around 900 nm and 1050 nm [2,3]. Of these two, the long-wavelength emission is known from Nd-SiO_2 lasers, where, due to the differences in host material, it is spectrally located at slightly higher wavelengths. Thus, one has to be careful to avoid lasing in the fiber amplifier at 1050 nm. What is more important is the strong ESA band at 1300 nm due to the $^4F_{3/2} \rightarrow {}^4G_{7/2}$ transition [4]. All these properties will be further clarified in Section 9.1, where the model and the basic characteristics of the neodymium-doped ZBLANP fiber amplifier are presented. However, methods of improving neodymium-doped fiber amplifiers have been proposed [5], and as we shall see in Section

9.2 and 9.3, such considerable improvements of the gain may be achieved by filtering out the 1050-nm ASE in Nd^{3+}-doped fiber amplifiers. The discussion of neodymium-doped fiber amplifiers will be rounded out in Section 9.4, where a new design for a neodymium-doped fluoride fiber for 1300-nm small-signal amplification is presented. The idea is to place the active material so that the 1050-nm ASE is discriminated [6], whereby a gain improvement of up to 9 dB is predicted.

Another and maybe even more promising candidate for the 1300 nm fiber amplifier is represented by the Pr^{3+}-doped fluorozirconate fiber amplifier. Such amplifiers have recently been demonstrated to provide gain in a broad wavelength band centered in the 1300-nm window [7–9]. Even though these praseodymium fibers still suffer from a low quantum efficiency and very high pump requirements [10,11], net gains of around 30 dB have been achieved. We will, therefore, in Section 9.5 discuss the latest results from modeling and optimization of praseodymium-doped fluorozirconate fibers. Finally, the chapter includes a short discussion in Section 9.6 on the question of how to make approximate gain estimates for idealized neodymium- and praseodymium-doped fiber amplifiers.

9.1 THEORY OF NEODYMIUM-DOPED FIBER

For the purpose of giving a quantitative description of the properties and problems of neodymium-doped fiber amplifiers, the focus of this section will be the elements of a detailed model. This was developed as a modification of the model for Er^{3+}-doped silica fibers [12] so that Nd^{3+}-doped ZBLANP fibers could be analyzed [4]. The primary difference between the original model and the neodymium-doped fiber model is that the latter may be described by a four-level laser system. The basic equations describing the population inversion in the fiber will therefore be different, as we shall see in Section 9.1.1.

9.1.1 Population Inversion and Power Distribution in Fluoride Glass Fibers

The Nd^{3+}-ion is described by the energy level diagram shown in Figure 9.1. The transitions labeled *fast phonons* are assumed to be instantaneous decays yielding a complete depopulation of the levels 2, 3, 5, and 6. Note that the excitation to level 6, caused by ESA, in agreement with [4], is assumed to be followed by an instantaneous decay to level 1. It must be pointed out here that this contrasts with the assumptions of other authors [1], who have suggested a decay to level 4. In addition, we may add that, at this time, it is still unclear [13] whether or not a significant part of the decays are from levels 6 to 4. However, the approach chosen here ensures that the gain is not overestimated, since it represents the worst-case situation, where a signal photon is absorbed at the same time as the population inversion is decreased. Note, therefore, that even though ESA in Nd^{3+}-doped fibers

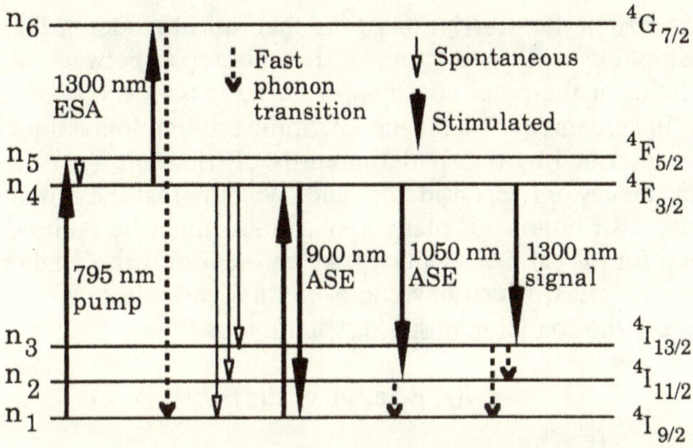

Figure 9.1 Energy level diagram for the Nd^{3+} ion. (Figure layout from [5].)

causes a loss of signal photons and not, as in Er^{3+}-fibers, a loss of pump photons, the present model includes an indirect need for pump photon absorption. It should be emphasized that as long as the neodymium fiber amplifier is operated in the small-signal regime, the specific process of decay from level 6 will be insignificant for the gain. This may be quantified by the fact that detailed calculations have shown no more than 5% differences in gain (in the wavelength range of 1330 to 1340 nm) when either one or the other decay assumption is made.

The model describes the ASE spectrally in the three most significant wavelength bands around 880, 1050, and 1300 nm. As seen from Figure 9.1, both the emission and absorption cross sections must be included for the 900-nm band. However, due to the assumption of instantaneous decay from levels 2 to 1, no absorption is present for the 1050-nm band, and only the emission cross section has to be applied. Also, in the 1300-nm band, the absorption from the lower laser level is very unlikely, but the emission cross section is not sufficient for an accurate description in this band, where the ESA (from levels 4 to 6) should also be included.

Besides the ASE spectra in the three wavelength bands, the signal and pump are represented as light with discrete wavelengths. It must be pointed out here that the signal also suffers from ESA in the 1300-nm band.

In the present description, the ASE is assumed to be guided solely in LP_{01}-modes. This assumption might at first seem somewhat strict, since the published results for single-mode ZBLAN fibers have shown cutoff wavelengths higher than 1200 nm [2,14]. This leaves the fibers multimoded in the 900- and 1050-nm ASE band. However, calculations for erbium-doped fibers have shown a relatively weak mode excitation dependence on the amplifier characteristics, which indicates that only a minor error is introduced by the LP_{01}-mode simplification (see Section 5.3).

The next step in the description of the theoretical model of the neodymium-doped fiber amplifier will be to focus on the interaction between atom and electromagnetic fields at the polar coordinates (r, ϕ, z) with the z-axis placed along the fiber axis. In agreement with the model for the erbium-doped fiber (see Chapter 5) the light is described by the spatial intensity distribution $I(r, \phi, z, \nu)$, where ν denotes the frequency of the considered field. We note that the azimuthal variation of the intensity distribution for many applications might be omitted. This is due to the fact that, for the sake of simplicity, we consider only the fundamental pump, signal, and ASE-modes, which have no azimuthal variations. We may, however, generally express the spatial intensity distribution as follows:

$$I(r, \phi, z, \nu) = P(z) \cdot |E(r, \phi, \nu)|^2$$
$$\int_0^{2\pi} \int_0^b |E(r,\phi,\nu)|^2 r dr d\phi = 1 \tag{9.1}$$

where $P(z)$ is the power distribution along the fiber, $E(r, \phi, \nu)$ is the normalized transverse electrical field distribution in a given LP-mode, and b is the radius of the fiber.

For the calculation of the LP-modes, we have chosen to use the multi-layer method [15] in the preceding results in this chapter, which may be used for accurate mode field calculations for arbitrary index profiles. It should be noted, however, that Gaussian approximations for LP_{01}-modes might also be applied with good results for fibers with step index profiles.

The ionic population in energy level number i at a place given by the polar coordinates (r, ϕ, z) is denoted by $n_i(r, \phi, z)$. Furthermore, the probability of photon absorption and stimulated emission are given by the respective cross-section values σ_{ij}, corresponding to a transition from levels i to j.

The general rate-equations for the six energy levels may, based on the description of [16], be written as

$$\frac{dn_6(r, \phi, z)}{dt} = n_4(W_{SESA} + W_{ESA}) - \frac{n_6}{\tau_{61}} \tag{9.2}$$

$$\frac{dn_5(r, \phi, z)}{dt} = n_1 W_p - \frac{n_5}{\tau_{54}} \tag{9.3}$$

$$\begin{aligned} \frac{dn_4(r, \phi, z)}{dt} = {} & \frac{n_5}{\tau_{54}} + n_1 W_{abs} \\ & - n_4(W_{SSE} + W_{SESA} + W_{ESA} + W_{900} + W_{1050} + W_{1300}) \\ & - n_4\left(\frac{1}{\tau_{41}} + \frac{1}{\tau_{42}} + \frac{1}{\tau_{43}}\right) \end{aligned} \tag{9.4}$$

$$\frac{dn_3(r, \phi, z)}{dt} = n_4\left(W_{SSE} + W_{1300} + \frac{1}{\tau_{43}}\right) - n_3\left(\frac{1}{\tau_{31}} + \frac{1}{\tau_{32}}\right) \tag{9.5}$$

$$\frac{dn_2(r, \phi, z)}{dt} = n_4\left(W_{1050} + \frac{1}{\tau_{42}}\right) + \frac{n_3}{\tau_{32}} - \frac{n_2}{\tau_{21}} \tag{9.6}$$

$$\frac{dn_1(r, \phi, z)}{dt} = \frac{n_6}{\tau_{61}} + n_4\left(W_{900} + \frac{1}{\tau_{41}}\right) + \frac{n_3}{\tau_{31}} + \frac{n_2}{\tau_{21}} - n_1(W_{abs} + W_p) \tag{9.7}$$

In order to understand (9.2) to (9.7), a number of parameters need to be defined. W_p is the pump efficiency and W_{SSE} is the stimulated emission efficiency for the signal. W_{SESA} and W_{ESA} are ESA efficiencies for the signal and the ASE in the 1300-nm wavelength band, respectively. W_{900}, W_{1050}, and W_{1300} are the stimulated emission efficiencies for the ASE in the respective wavelength bands around 900, 1050, and 1300 nm. Finally, W_{abs} denotes the absorption efficiency for the 900-nm ASE band. We will return to the exact definitions of these efficiencies after a few simplifications of the equations.

In (9.2) to (9.7), the parameter τ_{ij} denotes the lifetime related to the transition between levels i and j. However, for real systems, it is usually necessary first to make a number of simplifying assumptions regarding lifetimes and branching ratios in order to limit the otherwise excessively complex system of coupled equations to deal with a few states only. Therefore, we may with good approximation relate immediate decays [13] to the transitions $6 \rightarrow 1$, $5 \rightarrow 4$, $3 \rightarrow 2$, $3 \rightarrow 1$, and $2 \rightarrow 1$. This means that $\tau_{61} = \tau_{54} = \tau_{32} = \tau_{31} = \tau_{21} = 0$. As a consequence, the population in all states other than the ground state and the $^4F_{3/2}$ (level 4 in Figure 9.1) may be assumed to be negligible.

The number of coupled rate equations is therefore reduced to the following:

$$\begin{aligned}\frac{dn_4(r, \phi, z)}{dt} &= n_1(W_p + W_{abs}) \\ &\quad - n_4\left(W_{SSE} + W_{SESA} + W_{ESA} + W_{900} + W_{1050} + W_{1300} + \frac{1}{\tau_4}\right)\end{aligned} \tag{9.8}$$

$$\begin{aligned}\frac{dn_1(r, \phi, z)}{dt} &= n_4\left(W_{SSE} + W_{SESA} + W_{ESA} + W_{900} + W_{1050} + W_{1300} + \frac{1}{\tau_4}\right) \\ &\quad - n_1(W_p + W_{abs})\end{aligned} \tag{9.9}$$

where

$$\frac{1}{\tau_4} = \frac{1}{\tau_{41}} + \frac{1}{\tau_{42}} + \frac{1}{\tau_{43}} \tag{9.10}$$

In the case of a traveling wave optical amplifier, the next natural step of the model development is to treat the case of steady state. This means that all the first derivatives of the population densities with respect to time are identically zero, and the steady-state situation describes very well the situation of normal operation conditions. More specifically, this means that the modulation rate should be very much faster than the characteristic time constants (gain recovery time and fluorescence lifetime) of the system. Then the optical power levels in the amplifier will only depend on the time average of the signal and pump powers.

Under this steady-state assumption, (9.8) and (9.9) may be written as follows:

$$0 = n_1(W_p + W_{abs}) - n_4\left(W_{SSE} + W_{SESA} + W_{ESA} + W_{900} + W_{1050} + W_{1300} + \frac{1}{\tau_4}\right) \quad (9.11)$$

Now we only need to combine this with the radial distribution $\rho(r)$ of the Nd^{3+} ions, which is given by the fiber manufacturing process. This is done by using the fact that the sum of the densities in the different states equals the total density of ions:

$$\rho(r) = n_1(r, \phi, z) + n_4(r, \phi, z) \quad (9.12)$$

From (9.11) and (9.12), we reach the following expressions for the populations in the ground state $^4I_{9/2}$ (n_1) and in the upper state of the gain transition $^4F_{3/2}$ (n_4), respectively:

$$n_1(r, \phi, z) = \frac{W_{SSE} + W_{SESA} + W_{ESA} + W_{900} + W_{1050} + W_{1300} + (1/\tau_4)}{(W_p + W_{abs} + W_{SSE} + W_{SESA} + W_{ESA} + W_{900} + W_{1050} + W_{1300} + 1/\tau_4)} \cdot \rho(r) \quad (9.13)$$

$$n_4(r, \phi, z) = \frac{W_p + W_{abs}}{(W_p + W_{abs} + W_{SSE} + W_{SESA} + W_{ESA} + W_{900} + W_{1050} + W_{1300} + 1/\tau_4)} \cdot \rho(r) \quad (9.14)$$

Under the assumption that only one pump source and one signal are present, the efficiency rates may be expressed in terms of cross sections, photon energies, and intensities:

$$W_p(r, \phi, z) = \frac{\sigma_{15}}{h\nu_p} \cdot I_p(r, \phi, z) = \frac{\sigma_{15}}{h\nu_p} \cdot P(z) \cdot |E(r, \phi, \nu_p)|^2 \tag{9.15}$$

$$W_{SSE}(r, z) = \frac{\sigma_{43}(\nu_s)}{h\nu_s} \cdot P_s(z) \cdot |E(r, \nu_s)|^2 \tag{9.16}$$

$$W_{SESA}(r, z) = \frac{\sigma_{46}(\nu_s)}{h\nu_s} \cdot P_s(z) \cdot |E(r, \nu_s)|^2 \tag{9.17}$$

$$W_{900}(r, z) = \int_0^\infty \frac{\sigma_{41}(\nu)}{h\nu} (P_{900}^+(z, \nu) + P_{900}^-(z, \nu))|E(r, \nu)|^2 d\nu \tag{9.18}$$

$$W_{abs}(r, z) = \int_0^\infty \frac{\sigma_{14}(\nu)}{h\nu} (P_{900}^+(z, \nu) + P_{900}^-(z, \nu))|E(r, \nu)|^2 d\nu \tag{9.19}$$

$$W_{1050}(r, z) = \int_0^\infty \frac{\sigma_{42}(\nu)}{h\nu} (P_{1050}^+(z, \nu) + P_{1050}^-(z, \nu))|E(r, \nu)|^2 d\nu \tag{9.20}$$

$$W_{1300}(r, z) = \int_0^\infty \frac{\sigma_{43}(\nu)}{h\nu} (P_{1300}^+(z, \nu) + P_{1300}^-(z, \nu))|E(r, \nu)|^2 d\nu \tag{9.21}$$

$$W_{ESA}(r, z) = \int_0^\infty \frac{\sigma_{46}(\nu)}{h\nu} (P_{1300}^+(z, \nu) + P_{1300}^-(z, \nu))|E(r, \nu)|^2 d\nu \tag{9.22}$$

In (9.15) to (9.22), ν_s denotes the signal frequency, ν_p is the frequency of the pump light, and h is Planck's constant. P_s and P_p are pump and signal powers, respectively, and $P_{900}^{\pm}$ are the spectrally dependent powers (per Hz) of forward (+) and backward (−) propagating ASE in the 900-nm band. The forward direction is defined as the propagation direction of the signal. Similarly, $P_{1050}^{\pm}$ and $P_{1300}^{\pm}$ are the powers (per Hz) of the forward- and backward-propagating ASE in the 1050- and 1300-nm wavelength bands. It should also be noted that the possibility of considering higher order pump modes is included through the azimuthal dependence of W_p.

The local population inversions dependent on the pump, signal, and ASE powers have been found, and it is now straightforward to establish the set of coupled differential equations describing the evolution of significant powers along the fiber.

For the pump light we find the equation

$$\frac{dP_p^{\pm}(z)}{dz} = \pm g_p(z) \cdot P_p^{\pm}(z) \tag{9.23}$$

where the attenuation factor $g_p(z)$ is given by

$$g_p(z) = \sigma_{15} \int_0^{a_d} \int_0^{2\pi} |E(r, \phi, \nu_p)|^2 n_1(r, \phi, z) r dr d\phi \tag{9.24}$$

a_d denotes the dopant radius, and the $\pm$ in (9.23) indicates whether the pump is propagating in the positive direction of the z-axis (upper sign) or in the negative direction (lower sign). Note that the integration, which for the mode normalization in (9.1) was performed over the full fiber cross section, here only needs to be done within the neodymium-doped part of the fiber. In the preceding equations, the same significance of the upper or lower signs is implied.

The equation prescribing the signal variation along the fiber may be written as follows:

$$\frac{dP_s^{\pm}(z)}{dz} = \pm g_s(z) \cdot P_s^{\pm}(z) \tag{9.25}$$

The gain factor $g_s(z)$ is defined by the following relation:

$$g_s(z) = (\sigma_{43}(\nu_s) - \sigma_{46}(\nu_s)) \int_0^{a_d} \int_0^{2\pi} |E(r, \nu_s)|^2 n_4(r, \phi, z) r dr d\phi \tag{9.26}$$

which includes both the emission and the absorption cross section at the signal frequency. To make the system of coupled differential equations complete, it becomes necessary also to describe ASE in the 900-, 1050-, and 1300-nm bands.

For the 900-nm ASE band, the equations become

$$\begin{aligned} \frac{dP_{900}^{\pm}(z, \nu)}{dz} &= \pm g_{900}(z, \nu) \cdot P_{900}^{\pm}(z) \pm 2h\nu \cdot a_{900}(z, \nu) \\ g_{900}(z, \nu) &= \int_0^{a_d} \int_0^{2\pi} |E(r, \nu)|^2 \\ &\quad (n_4(r, \phi, z) \cdot \sigma_{41}(\nu) - n_1(r, \phi, z) \cdot \sigma_{14}(\nu)) r dr d\phi \\ a_{900}(z, \nu) &= \sigma_{41}(\nu) \int_0^{a_d} \int_0^{2\pi} |E(r, \nu)|^2 n_4(r, \phi, z) r dr d\phi \end{aligned} \tag{9.27}$$

The 1050-nm ASE is described spectrally by the expressions

$$\frac{dP_{1050}^{\pm}(z, \nu)}{dz} = \pm g_{1050}(z, \nu) \cdot P_{1050}^{\pm}(z) \pm 2h\nu \cdot a_{1050}(z, \nu)$$

$$g_{1050}(z, \nu) = \sigma_{42}(\nu) \int_0^{a_d} \int_0^{2\pi} |E(r, \nu)|^2 n_4(r, \phi, z) r dr d\phi \qquad (9.28)$$

$$a_{1050}(z, \nu) = \sigma_{42}(\nu) \int_0^{a_d} \int_0^{2\pi} |E(r, \nu)|^2 n_4(r, \phi, z) r dr d\phi$$

Finally, the 1300-nm ASE develops along the fiber according to

$$\frac{dP_{1300}^{\pm}(z, \nu)}{dz} = \pm g_{1300}(z, \nu) \cdot P_{1300}^{\pm}(z) \pm 2h\nu \cdot a_{1300}(z, \nu)$$

$$g_{1300}(z, \nu) = (\sigma_{43}(\nu) - \sigma_{46}(\nu)) \int_0^{a_d} \int_0^{2\pi} |E(r, \nu)|^2 n_4(r, \phi, z) r dr d\phi \qquad (9.29)$$

$$a_{1300}(z, \nu) = \sigma_{43}(\nu) \int_0^{a_d} \int_0^{2\pi} |E(r, \nu)|^2 n_4(r, \phi, z) r dr d\phi$$

The first term in each of the differential equations (9.27) to (9.29) for the ASE spectral power density represents amplification. For the first two bands, the amplification term reflects the absence of ESA and the three- and four-level nature of the 900- and 1050-nm transitions, respectively. For ASE in the signal band, this term is proportional to the difference between the stimulated emission and ESA cross sections and to the ASE intensity and density of ions in the upper state, n_4.

The second and remaining terms in the differential equations (9.27) to (9.29) represent the addition of new photons into the two bound orthogonal polarization modes due to spontaneous decay [13] (see also the similar derivations in Chapter 5). It should in this connection be noted that an extra term might be added to the differential equation (9.29) if the ESA caused excitation to level 6 is later assumed to be followed by a decay to level 4. This is described in [13], but is not included here, since the influence on the final results is considered insignificant based on the information available today.

The equations derived in this section account for the propagation of signal, pump, and ASE light through the neodymium-doped ZBLAN fiber. Due to the different modal structure of the terms in the denominators, these equations may only be solved numerically by iterative integration back and forth along the fiber. It has, however, been shown in [13] that the similarity between the propagation equations for the signal (equations (9.25) to (9.26)) and the ASE intensity spectral density in the signal band (equation (9.29)) makes it possible to express the corresponding power spectral density in a semi-analytical form in terms of the total gain G at the frequency ν:

$$P_{1300}^{-}(0, \nu) = P_{1300}^{+}(L, \nu) = 2h\nu\sigma'(G - 1) \qquad \sigma' = \frac{\sigma_{43}}{\sigma_{43} - \sigma_{46}} \tag{9.30}$$

From (9.30), two important theoretical predictions follow. First, the expression shows that the generated ASE is the same in both directions relative to the pump direction. Second, the presence of ESA ($\sigma_{46}(\nu) > 0$) will cause an enhanced production of ASE for a given value of the gain. This naturally has implications for the noise properties of the amplifier, as we will see in Section 9.1.4.

9.1.2 Derivation of Fundamental Parameters for the Neodymium Fiber

A number of properties may become immediately evident from the equations derived in Section 9.1.1. However, before this becomes clear, it is necessary to find the respective magnitudes of the relevant fundamental parameters of the neodymium-doped fiber amplifier. Furthermore, it is a simple demand for the applications of the model to achieve knowledge of the fundamental parameters governing the processes of gain and noise generation in the fiber amplifier. We will, therefore, in this section focus on the determination of these. This includes the cross sections for any possible stimulated processes in the Nd^{3+}/ZBLAN system, the spontaneous lifetimes and branching ratios of any level involved, the neodymium concentration and distribution, and the index profile geometry. However, the assumptions applied in the text above reduces the number of lifetimes and branching ratios considerably. Therefore, we only have to determine one lifetime: the lifetime $\tau = \tau_4$ of the upper laser level $^4F_{3/2}$. This may be measured by monitoring the decaying fluorescence from a length of fiber pumped with a chopped laser beam at 795 nm and short enough to ignore ASE effects. The $1/e$ decay time of the fluorescence is found to be 0.72 ms in [13].

Another important parameter is the cross section for pump absorption σ_{15} at 795 nm. According to the procedure in [13], this may be found to be $\sigma_{15}(795\ \text{nm}) = 20.5 \times 10^{-25}\ \text{m}^2$. If we investigate the energy level diagram in Figure 9.1 further, it becomes obvious that the absorption cross-section spectrum in the $^4I_{9/2} \rightarrow {}^4F_{3/2}$ band, σ_{14}, is of relevance also. This spectrum may be derived from absorption measurements, and available data on bulk glass samples have been published in [17]. According to these and [13], the peak value of σ_{14} is $5.5 \times 10^{-25}\ \text{m}^2$ measured at 860 nm.

Just as important as the pump absorption strength is the cross-section spectrum for stimulated emission from the upper laser level $^4F_{3/2}$. Figure 9.2 shows the emission cross-section spectrum used by the model. This spectrum was derived from measurements [2,13] of the fluorescence spectrum from a short length of fiber

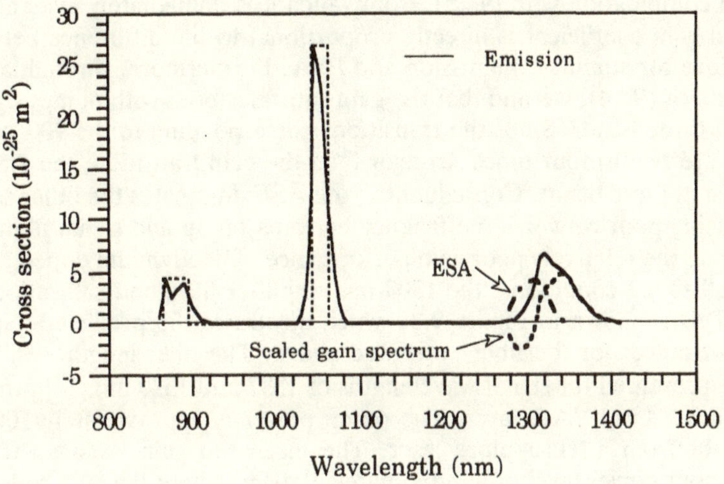

Figure 9.2 Emission cross-section spectrum for neodymium-doped ZBLAN fibers (solid) with 3-dB bandwidth (dashed). Also shown is the ESA cross-section spectrum in the 1300-nm band for Nd^{3+} in ZBLANP glass. The resulting small signal gain is also included. (Data from [2,13].)

pumped to full inversion and from the measured spontaneous lifetime of level 4 ($^4F_{3/2}$). It should especially be noted that the ASE around 880 and 1050 nm may be evaluated with good approximation in an equivalent 3-dB bandwidth, as shown dashed in Figure 9.2.

Finally, the ESA spectrum was evaluated as the difference between the emission spectrum (see Figure 9.2) and the scaled measured gain spectrum [2]. The relevant ESA cross sections were evaluated to be 0.5×10^{-25} m^2 at 1340 and 1337 nm, and 3.4×10^{-25} m^2 at 1320 nm. The more specific spectral dependence of the ESA is also illustrated in Figure 9.2, together with the small-signal gain, which is proportional to ($\sigma_{43} - \sigma_{46}$). Lasing with near unity quantum efficiency was measured at 1345 nm [18]. This implies the absence of ESA at this wavelength. σ_{46} may then be calculated as the difference between the stimulated emission cross-section spectrum and the scaled small-signal gain spectrum [13].

In the following calculations, a ZBLANP fiber [14] doped with 500 ppm by weight of Nd^{3+} ions is used. The fiber has a core diameter of 5.8 μm and a cutoff wavelength of 1200 nm.

9.1.3 Output Signal Saturation in Neodymium-Doped Fluoride Glass Fibers

The limiting effects on the gain of neodymium-doped fiber amplifiers become clearer after the parameter specification given in the previous section. This may be seen in combination with (9.25), from which it is immediately evident that the differential gain coefficient is directly proportional to the difference between the cross sections for stimulated emission and ESA. Furthermore, through additional examination of (9.14), we find that the gain saturates both with pump, signal, and ASE in all three bands. Since the transitions corresponding to the 900- and 1050-nm bands are two to four times stronger than the gain transition, the ASE builds up very fast in these bands. Consequently, the ASE dominates the other saturation terms, causing poor conversion efficiency between pump and signal photons.

Despite the relatively poor gain performance of neodymium doped amplifiers under small-signal conditions, the 1300-nm amplifier has good saturation characteristics. This is shown in Figure 9.3, which illustrates the predicted gain versus signal input power for the fiber presented above. The fiber length is 0.75 m and results are presented for signal wavelengths of 1337 and 1320 nm [14] provided by a diode pumped Nd:YAG laser. The pump power is approximately 100 mW of 795-nm light from a Ti:sapphire laser. The measured gain is constant up to a launched input power level of approximately 0 dBm, where the gain rolls off. The theoretical results are seen to match the experimental results well on the flat parts of the curves and roll off at only slightly higher signal powers.

The reason for the high saturated output power may be seen from (9.14) and (9.25), where we find that the gain saturates with the signal through the rates W_{SSE}

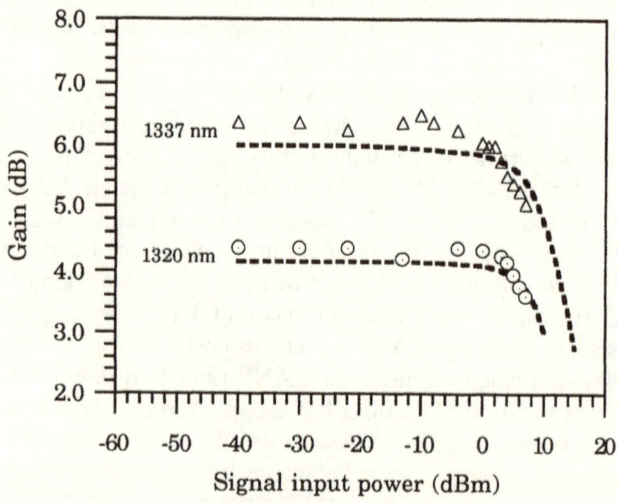

Figure 9.3 Measured and calculated gain values as a function of the total signal input power at the two wavelengths 1337 and 1320 nm. The dashed curves indicate the ±20% pump power intervals. (Data from [5].)

and W_{SESA}. From (9.16) to (9.17), it then follows that the signal saturation intensity (i.e., the signal intensity at which the density of ions in the upper state of the gain transition is pulled down to half its unsaturated value) is given by [13]:

$$I_s^{sat} = \frac{h\nu}{(\sigma_{43} + \sigma_{46})\tau} \tag{9.31}$$

The sum of σ_{43} and σ_{46} is on the order of 10^{-24} m^2 and the lifetime is 0.72 ms. This leads to a signal saturation intensity of 2×10^{11} mW/m^2, which in a plane wave approximation corresponds to a saturating signal power of +7.4 dBm.

Another measure of the validity of the neodymium-doped fiber amplifier as a power booster is the photon conversion efficiency (i.e., how many pump photons are converted to signal photons). Since Nd^{3+} is a four-level system at the 1300-nm signal wavelength, a conversion efficiency of up to 100% may be expected [19], provided that ESA could be avoided. Further calculations employing the above model shows that a conversion efficiency of 80% is realistic if 100-mW pump power and 10-mW signal power are launched into an amplifier consisting of an optimized fiber and filters for ASE suppression [20].

9.1.4 Noise Figure for the Neodymium-Doped 1300-nm Amplifier

Chapters 5 and 7 discussed how the re-absorption of the signal in the three-level Er^{3+} system will normally lead to incomplete inversion, resulting in increased spontaneous emission per unit gain. While the fraction of ions in the excited state does not affect the signal-to-noise ratio in four-level systems, the presence of ESA at the signal wavelength will cause an enhanced generation of spontaneous emission in the signal band. This will result in an increased noise level compared to the situation without ESA.

To examine this effect further, we will look at (9.30), which relates the amplifier gain and the signal absorption and ESA cross section values with the ASE spectral density. Under the assumption that the launched signal is coherent, the noise at the input end of the amplifier is quantum-limited (i.e., only shot noise would be detected). The noise figure, which is defined as the degradation in the signal-to-noise ratio from the input to the output of the amplifier, may be written as [19,21] (see also Chapter 5):

$$F = \frac{P_{1300}^{+}(L, \nu)}{h\nu_s} + 1 \cdot \frac{1}{G} \tag{9.32}$$

By combining equations (9.30) and (9.32), the resulting expression for the noise figure is

$$F = \left(2\frac{\sigma_{43}(\nu_s)}{\sigma_{43}(\nu_s) - \sigma_{46}(\nu_s)}(G-1)+1\right)\cdot\frac{1}{G}$$
$$\approx 2\frac{\sigma_{43}(\nu_s)}{\sigma_{43}(\nu_s) - \sigma_{46}(\nu_s)} \quad \text{for} \quad G \gg 1 \tag{9.33}$$

Several conclusions may be drawn from (9.33). First of all, in a four-level system with no ESA, the quantum-limited noise figure would always be 3 dB ($\sigma_{46}(\nu_s) = 0$). However, for the proposed neodymium-doped fibers, ESA is present and the noise figure will consequently be degraded by a spectrally dependent factor $\sigma' = \sigma_{43}(\nu_s)/(\sigma_{43}(\nu_s) - \sigma_{46}(\nu_s))$, as defined in (9.30). Second, the noise figure defined in (9.33) is independent of the inversion, pump power level, and pump direction (provided that the gain $G \gg 1$). This contrasts with the case of a three-level system (e.g., Er^{3+}), where the noise figure may be brought close to the quantum limit (given that the pump band is different from the upper band of the gain transition). Third, the noise figure expression is independent of the 900- and 1050-nm ASE because the signal and ASE are only detected within a small bandwidth $\Delta\nu$ around the signal frequency. Note also that the independence of pump power direction is a direct consequence of the fact that the generated ASE is the same in both directions, as shown in (9.30).

The first noise spectra of neodymium-doped fibers were published in [21]. In Figure 9.4, the noise figure is shown as a function of the signal wavelength, which

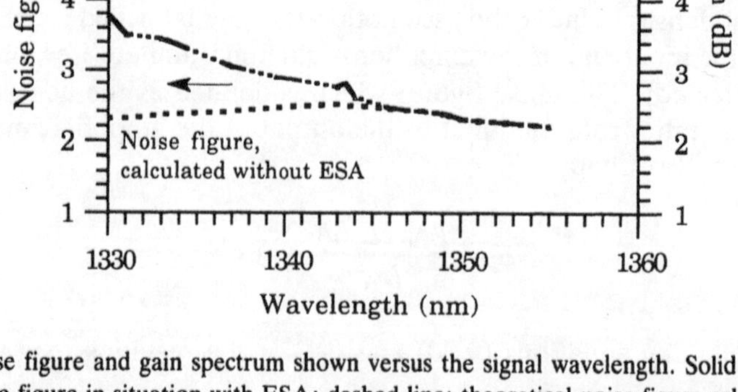

Figure 9.4 Noise figure and gain spectrum shown versus the signal wavelength. Solid line: measured noise figure in situation with ESA; dashed line: theoretical noise figure calculated without ESA. (Data from [21].)

corresponds to a variation of the signal absorption and ESA cross-section ratio. Measured results are shown, along with theoretical noise figure estimates based on (9.33), calculated under the assumption that ESA is no longer present and that the same gain values apply. It should be noted that good agreement between measured and predicted gain and noise figure values are found in [21].

For wavelengths below 1345 nm, where ESA becomes significant, the noise figure increases from approximately 2.5 to 3.8 dB at 1330 nm. This clearly demonstrates the tendency of increasing noise figure with increasing ESA cross section values. It is also noteworthy that the noise figure is below 3 dB for wavelengths above 1337 nm. This is because the 3-dB limit only holds under optimal conditions and not for gain values as low as those in the case of Figure 9.4.

In conclusion, we have found that the noise figure increases with increasing ESA cross section. Consequently, the improvement in system performance achieved by the amplifier is highest in the long-wavelength region of the gain spectrum.

9.2 METHODS FOR IMPROVING NEODYMIUM-DOPED FIBER AMPLIFIERS

As indicated in the previous section, Nd^{3+}-doped ZBLANP fibers have only shown modest gain in the wavelength interval between 1300 and 1400 nm. The difficulty in obtaining high gain is for a secondarily due to ASE at 880 nm, but primarily the 1050-nm ASE limits the gain performance of the neodymium amplifiers. In this section, we will discuss how the ASE saturated neodymium-doped fiber amplifiers may be improved by filtering the 1050-nm ASE, and we will furthermore present the optimum standard fiber design for such ZBLAN fibers.

9.2.1 Incorporation of ASE-Suppressing Filters

As previously suggested, a possible route to higher gain is filtering the ASE powers. This approach is principally the same as that described in Chapter 8 for erbium fibers, except that the realization of the ASE filters is less critical in the present case because the signal and dominating ASE bands are separated in wavelength. To investigate the point of filtering, further configurations consisting of equal-length fibers cascaded with filters in between are analyzed. The filters are assumed to completely absorb the ASE in the 1050-nm wavelength band and leave all other wavelengths unaffected.

In order to understand the influence of a filter, Figure 9.5 shows pump, signal, and 1050-nm ASE power versus the distance from the input fiber end. The basic fiber parameters presented above are used in the calculations. The solid curves indicate the situation without the filter, and the dotted curves represent the situation in which a single filter is inserted. The filter is placed so that the amounts of

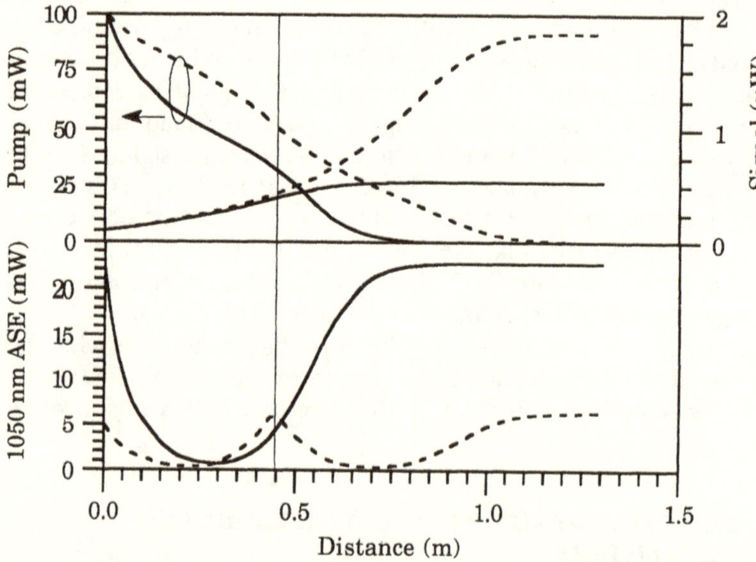

Figure 9.5 The pump power, the signal power, and the total 1050-nm ASE power in watts as a function of the distance from the input fiber end. The vertical lines in both figures represent the placement of the filter. Solid curves are for the situation without a filter. (Figure from [22].)

absorbed pump power are equal in the two fiber sections. In Section 9.2.2, this is shown to be the optimum placement for a filter in Nd^{3+}-doped fibers.

In the case without filters, the signal saturates shortly after propagating through the first fiber section, as most of the pump photons are converted into useless 1050-nm ASE photons. However, the inclusion of the filter reduces the pump attenuation and thereby increases the absorbing fiber length by 50%. This is because the 1050-nm ASE (as can be seen on the lower part of Figure 9.5) is not allowed to saturate the amplifier completely. In the extra absorbing length, the signal power increases by more than a factor of three.

This illustrates how the inclusion of a single filter may improve the properties of the neodymium-doped fiber amplifier. However, before we study the optimum placement of a single filter, a more generalized approach is discussed. In this approach, the filters are equally spaced along the fiber length and the number of filters is increased, pointing to the ideal situation of distributed ASE suppression.

In Figure 9.6(a), the small-signal gain is shown versus the total fiber length. Curves are presented for configurations using zero to nine filters and using an infinite number of filters. The signal wavelength is 1340 nm and the pump power

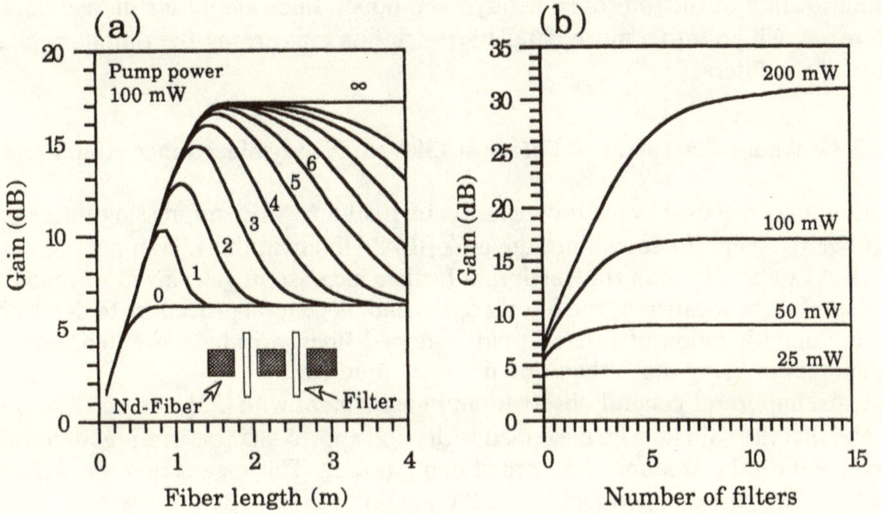

Figure 9.6 (a) Small-signal gain versus total fiber length for zero to nine filters and for an infinite number of filters. (b) Maximum gain versus number of filters for four pump powers. (Data from [5].)

is 100 mW. The pump is co-propagating with the signal. The curve for zero filters and the curve for an infinite number of filters enclose the remaining curves. When the ASE at 1050 nm is suppressed, the maximum obtainable gain increases. However, for a very long total fiber length, there will be an inversion in the first amplifier only. Consequently, this first amplifier completely determines the gain for the whole amplifier, yielding the same gain as that for an amplifier with no filters. Therefore, when using filters, there will be a maximum gain at an optimum fiber length. As can be seen in Figure 9.6(a), the optimum fiber length and the maximum gain increase when the number of filters increases.

Figure 9.6(b) shows the maximum gain versus the number of filters for four pump power values (25, 50, 100, and 200 mW). It can be seen that the gain improvement increases with the number of filters. It should be noted that for a larger number of filters the ratio between the maximum gain (dB) and the pump power (mW) is almost constant (0.16 dB/mW).

These results show, together with the reports in [2,5], that filtering of the ASE at 1050 nm is required for obtaining a high gain in the amplifier. Lately, this was experimentally demonstrated, when it was reported [3] that a 10-dB gain at 1343 nm used a two-section amplifier with a filter placed in the middle of the fiber to suppress the 1050nm ASE. We will return to the subject of the experimental demonstration of the improved neodymium-doped fiber amplifiers in Section 9.3, where we will go into a more detailed description concerning the optimum placement of the filters.

9.2.2 Optimum Placement of Filters in 1300-nm Neodymium Fiber Amplifiers

Section 9.2.1 discussed how the inclusion of 1050-nm ASE-suppressing filters may improve the gain of neodymium-doped ZBLAN fibers in the 1300-nm wavelength band. As we shall see in this section, a further increase in gain may, however, be predicted as the location of the filter is optimized. A general procedure to determine the optimum location of filters in Nd^{3+} -doped fibers will be presented, and the advantages of using more than one filter are predicted.

An important general observation, in agreement with [23], for a Nd^{3+}-doped fiber is that the gain may be described with good approximation as a function solely dependent on the amount of absorbed pump power. This means that the different fiber sections are independent when the amplifier is operated in the small-signal regime, and the dominant 1050-nm ASE is suppressed by filters between each fiber section. The total gain of an amplifier, separated in $(N + 1)$ sections by the insertion of N filters, may then be expressed as

$$G = \sum_{i=1}^{N+1} g_i(P_{ai}) = \sum_{i=1}^{N} g_i(P_{ai}) + g_{N+1}\left(P_{in} - \sum_{i=1}^{N} P_{ai}\right) \tag{9.34}$$

where g_i is the gain in decibels and P_{ai} is the absorbed pump power in fiber section number i. P_{in} is the total available pump power. Symmetry with regard to the separate fiber sections means that the maximum gain is obtained by choosing equal gain values and, therefore, an equal amount of absorbed pump power for each fiber section.

The independent fiber section assumption yields that the optimum placement of the filters can be determined from a few measured parameters. Figure 9.7(a) illustrates a general procedure that is valid for an arbitrary Nd^{3+}-doped ZBLANP fiber. The left part of the figure shows the gain as a function of the absorbed pump power. The signal wavelength is 1340 nm. In the right part of the figure, the absorbed pump power is shown as a function of the fiber length. Curves are shown for launched power levels ranging from 20 to 200 mW. The procedure for placing a filter is as follows: a total gain of 12 dB is required and one filter is to be used. The required gain in each section is 6 dB and, as can be seen from the left part of Figure 9.7(a), 60 mW of pump power have to be absorbed in each section. From

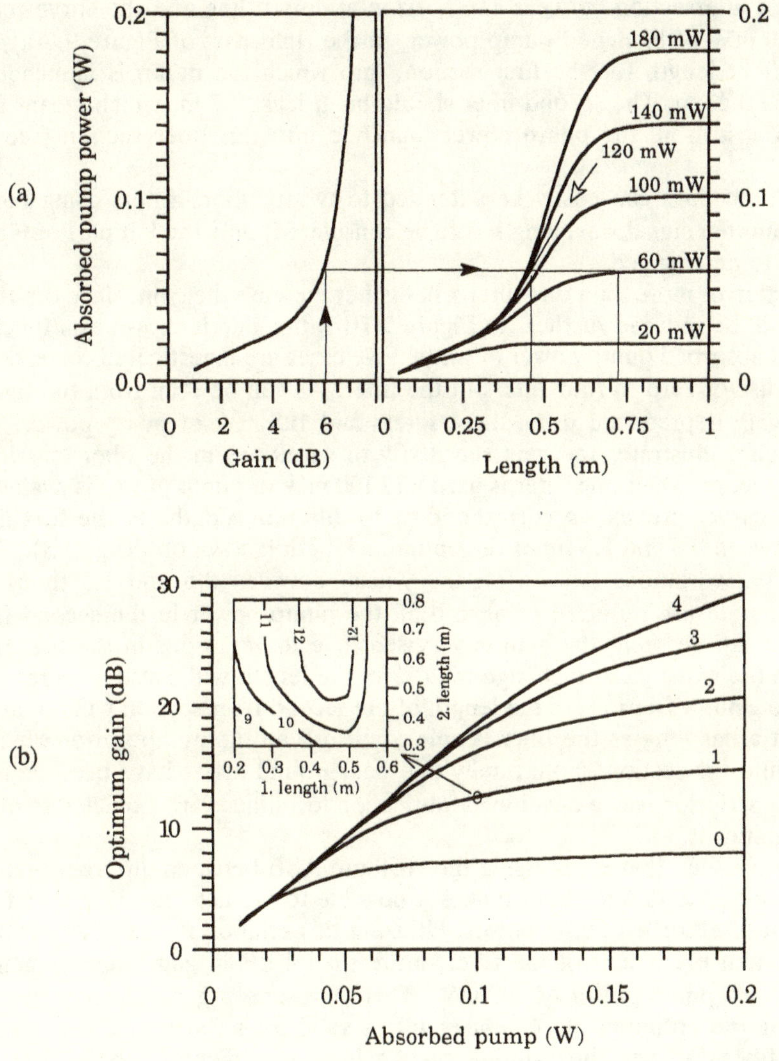

Figure 9.7 (a) Left side: absorbed pump power in watts required to obtain the gain shown on the x-axis. Right side: absorbed pump power in watts as a function of the fiber length. The different curves correspond to different input pump powers. The thin lines on the figures illustrate a design example. (b) The gain in decibels as a function of the total amount of absorbed pump power in watts. Five curves are shown, corresponding to the cases with zero to four filters. Insert: gain-level curves for variation of the fiber length before the filter (horizontal) and after the filter (vertical) for the case in which only one filter is used. (Data from [22].)

the point of intersection between the horizontal dotted line and the curve representing 120 mW of launched pump power, in the right part of Figure 9.7(a), the optimum fiber length for the first section, into which the pump is launched, is found to be 0.47 m. The second fiber should be at least 0.7 m, which means long enough to absorb all the pump power launched into this fiber section (see the figure).

The procedure can easily be extended to two or more filters using Figure 9.7(a). If another signal wavelength is to be considered, only the left part of Figure 9.7(a) has to be changed.

Insertion of more than one filter will further increase the gain, since the 1050-nm ASE will be reduced further. In Figure 9.7(b), the gain is shown as a function of the total absorbed pump power in watts. Five cases are investigated corresponding to the use of zero to four filters in the fiber. As can be seen from the figure, 20 dB of gain is predicted using three filters and 125 mW of pump power. The inserted figure illustrates the gain sensitivity to variations in the fiber lengths of each fiber section when one filter is used and 100 mW of pump power is available. Gain-level curves are shown corresponding to different lengths of the first fiber section between 0.5 and 1.3 times the optimum length (shown on the *x*-axis). They illustrate that variations (shown on the *y*-axis) between 0.5 and 1.5 times the minimum length are required to absorb all the pump power in the second fiber section. As can be seen, the gain is very sensitive to variations in the first fiber section. In the worst case, a change of 20% in the length will cause a decrease of 3 dB in the gain. Variations in the length of the second fiber section will not affect the gain at all as long as the fiber is able to absorb all the pump power which is launched into the section. Note finally that background losses have been omitted from this description, since developed fabrication techniques are expected to make these insignificant.

We have seen that suppressing the 1050-nm ASE between different sections of a Nd^{3+}-doped ZBLANP fiber makes it possible to predict the gain solely from the amount of absorbed pump power. Utilizing this relation in the determination of an optimum placement of the filter, more than 5 dB in gain improvement is predicted for a pump power of 100 mW. This represents a general procedure for determining the optimum placement of filters valid for an arbitrary Nd^{3+}-doped ZBLANP fiber. Further, the examination of gain improvements by the use of more than one filter yields an average gain improvement of up to 5 dB for each filter. A 20-dB gain is predicted using three filters and 125 mW of pump power. In addition, the gain sensitivity of the fiber length optimization is investigated for the case in which one filter is used. This shows a worst case of 3-dB gain penalty if the first fiber length is 20% larger than its optimum value.

In the next section, we turn to improvements to be gained in the design of the fiber itself.

9.2.3 Optimum Design of Neodymium-Doped Fiber-Optical Amplifiers

In the previous predictions of significant improvement of the neodymium-doped fiber by effectively filtering the 1050-nm ASE, the design of the fiber itself has not been the major task. The reason has been that, due to the manufacturing difficulties of single-mode ZBLAN fibers, the focus of the work has been to improve amplifiers, which could be realized using already manufactured fibers. However, as we will show here, the efficiency for both small-signal and large-signal operations may be improved by optimizing the waveguide geometry. The analysis, which is based on the full-scale numerical model presented in Section 9.1, is performed both with and without a 1050-nm filter. Optimizations of EDFAs have shown that confinement of the active ions increases the gain [4]. As a result, we have included this in the analysis of the neodymium amplifier. The neodymium-doped fiber has a step-index profile, and the pump and signal wavelengths are 795 and 1340 nm, respectively.

The optimum cutoff wavelength versus pump power (P_p) is shown in Figure 9.8(a) for two NAs. The dashed curves refer to a situation where an optical filter has been inserted at the optimum distance from the input end (as prescribed in Section 9.2.2). This filter is assumed to absorb the 1050-nm ASE totally and leave all other wavelength bands unaffected. The insets show the small-signal gain versus cutoff wavelength (λ_c) for NA = 0.1 and NA = 0.4. The figure shows that one should choose the NA as high as possible. For small pump powers, the cutoff wavelength should be chosen to 800 nm, but for larger pump powers, the cutoff wavelength is a function of both NA and the pump power. To ensure single-mode operation of the signal, λ_c should be chosen less than or at 1300 nm. The figure also shows that if a filter is inserted, the optimum cutoff wavelength is lowered towards 800 nm. A further analysis has shown that if the ASE at 1050 nm was totally removed throughout the fiber, the optimum cutoff wavelength would be 800 nm for all pump powers.

Figure 9.8(b) shows the gain as a function of input signal power (P_s) for two different designs ((a): NA = 0.4, λ_c = 0.8 μm; (b): NA = 0.4, λ_c = 1.3 μm). The pump power is 100 mW. In the small-signal region (the flat part of the curves), there is a significant difference between the two designs, as discussed above, but in the booster region (the decreasing part of the curves), no significant difference can be seen. This is also evident in the inset in Figure 9.8(b), which shows the QCE defined as $(G - 1)P_s\lambda_s/P_p\lambda_p$, against λ_c for NA = 0.1 and NA = 0.4. G is the gain and λ_s and λ_p are the signal and pump wavelengths, respectively. There is a maximum around 800 nm, but this is not significant. Notice that with one filter inserted, a QCE of nearly 80% can be reached with an input power of 10 dBm and a pump power of 100 mW.

In order to investigate the influence of Nd^{3+} ion confinement on small-signal gain and booster QCE, investigations were made, and the results are reported in

(a)

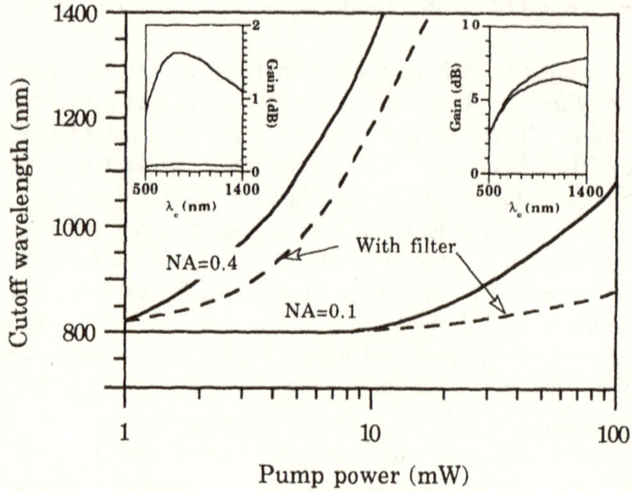

(b)

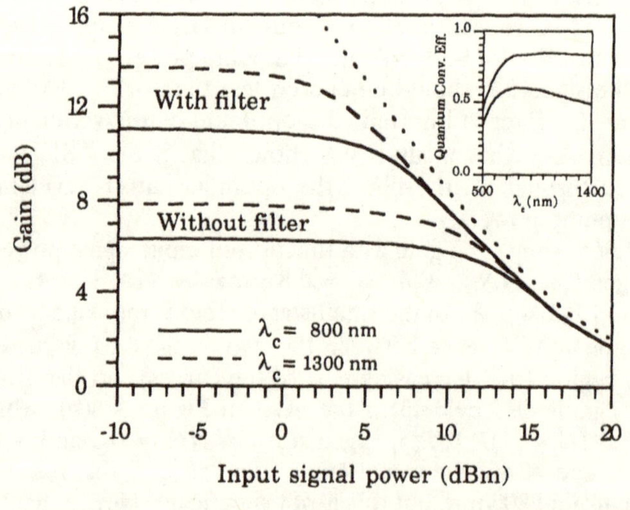

Figure 9.8 (a) Optimum cutoff wavelength versus pump power. The input signal power is -40 dBm. The left inset shows gain versus cutoff wavelength for a pump power of 1 mW, and the right for 100 mW. The upper lines are for NA = 0.4 and the lower for NA = 0.1. (b) Gain versus signal power for a pump power of 100 mW. The dotted line to the right is the theoretical limit when the QCE is 100%. The inset shows the QCE versus cutoff wavelength. (Data from [20].)

[20]. It was shown that a confinement of 50% improves the small-signal gain when the pump power is modest, but degrades the gain for larger pump powers. To explain this, the radial distributions of the involved intensities (all LP_{01} modes) have to be considered. Since the signal wavelength is 1340 nm and the dominant ASE wavelength is 1050 nm, the signal has the broadest field distribution. Therefore, confining the active ions will be more favorable to the ASE than the signal. When the pump power level is small, the ASE power is also small, and confinement increases the signal gain.

In a booster, the ASE is negligible, since the signal power is large. Thus, confining should increase the QCE, which is also the case for small NAs [20]. However, for fairly large values of the NA of 0.4, the signal is well confined in the core, so there is no significant improvement by confining the active ions in this case.

The waveguide parameters for a neodymium ZBLANP optical-fiber amplifier have been optimized for the small-signal and booster operation. In conclusion, the NA should always be chosen as high as possible, and the optimum cutoff wavelength should be chosen to 800 nm, provided that the 1050-nm ASE can be removed in the small-signal case. A booster QCE of nearly 80% may be reached for an input power of 10 dBm and pump power of 100 mW with the use of one filter. Finally, in a standard fiber design, the neodymium dopant radius should be chosen equal to the core radius for the small-signal case. In the booster case, there is an improvement in QCE of 2.5% and 0.1% for NAs of 0.1 and 0.4, respectively, when the Nd^{3+} ion distribution is confined to 50% of the core radius. The reason is that neodymium confinement leads to a better mode overlap with the signal. In a booster amplifier where the 1050-nm ASE becomes less significant, since the signal and pump dominate the population inversion, confinement is slightly advantageous because shorter fiber lengths and therefore lower background losses may be obtained. However, as we shall see in Section 9.4, this explanation does not hold for small-signal amplifiers that are dominated by 1050 nm ASE.

9.3 EXPERIMENTAL DEMONSTRATIONS OF 1300-nm GAIN IMPROVEMENT

We have seen that the performance of neodymium in ZBLAN is limited by ESA in the signal band and by competition between the gain transition ${}^4F_{3/2} \rightarrow {}^4I_{13/2}$ and the stronger transition ${}^4F_{3/2} \rightarrow {}^4I_{11/2}$ around 1050 nm. Whereas the first results in a shift towards longer wavelengths (a red shift) of the gain spectrum away from the desired wavelength for telecommunication applications, it may only be overcome by changing the host glass composition. The second problem causes a strong buildup of ASE around 1050 nm, which saturates the amplifier and locks the gain at a low value (around 7 dB). In the previous section, however, we have already seen one solution to the latter problem: suppressing the 1050-nm ASE by filtering it out along the Nd^{3+}-doped ZBLAN fiber.

This filtering may be done at discrete points, and, as shown in Section 9.2.2, the filters should in this situation be positioned so that equal amounts of pump power are absorbed in each section. This principle has been demonstrated in practice [3] by placing a 1050/1300-nm dichroic fiber coupler between two separately pumped fiber amplifiers. In this way, saturation caused by mutual 1050-nm ASE of the two amplifier sections was avoided, and a total gain of 10 dB at 1343 nm was reported.

The approach of discrete filtering might, however, become less attractive if large gains and therefore many filters have to be applied. The reason is that we must also consider in this case the coupling losses between the neodymium ZBLAN fiber and the filters, and that a large number of filters might increase the price of the amplifier considerably. Therefore, it is advantageous if an alternative filtering approach is applied. An answer to this may be found in the use of distributed mode coupling between the guided mode of the 1050-nm ASE and leaky cladding modes. This may be accomplished through the application of a mechanical grating notch filter, which is tunable, movable, and nondestructive to the fiber. In the following two sections, we will discuss the principle of such filters and the experimental results of their application [24].

9.3.1 Physical Realization of 1050-nm Mode Coupling Filter

For laboratory purposes, a very attractive way of producing a 1050-nm ASE attenuating filter is to press the amplifying fiber between mechanical gratings. This introduces periodic microbends that introduce perturbations of the refractive index profile of the fiber and result in wavelength-dependent coupling between the guided LP_{01} mode and leaky discrete LP_{1p}-modes [25]. The suffix p indicates the radial mode number of the specific cladding mode.

The microbending loss due to coupling to a given cladding mode at a given wavelength depends partly on the coupling coefficient between core and cladding mode and partly on the magnitude of the spatial deformation spectrum at the spatial frequency $\Delta\beta_{1p}$ introduced by the grating. Here, $\Delta\beta_{1p}$ denotes the propagation constant difference between the LP_{01}-mode and the cladding LP_{1p}-mode found at the respective wavelength.

When the microbends appear by sandwiching the fiber between a mechanical grating with a period Λ_μ, the spatial deformation spectrum peaks at frequencies in multiples of $\Delta\beta_\mu = 2\pi/\Lambda_\mu$. However, due to the elasticity of the fiber coating, the peaks at higher spatial frequencies are suppressed [26] and only one peak at $\Delta\beta_\mu$ becomes significant. Maximum attenuation will therefore occur at wavelengths where the spatial frequency $\Delta\beta_\mu$ of the microbends equals the difference in propagation constants between the guided LP_{01}-mode and the cladding LP_{1p}-modes [25]. Note, however, that the spectral positions of these mode coupling losses are highly dependent on the index profile.

From this description, it is also evident that the spectral position of the coupling loss peaks may be changed by varying the grating pitch (i.e., by changing $\Delta\beta_\mu$). Furthermore, the filter attenuation increases with pressure and the length of the filter. It should also be mentioned that it is important to ensure a uniform pressure distribution along the fiber/filter interaction path and by, appropriate handling, suppress the pure bending losses at the ends of the filter. Otherwise, the performance of such filters is severely limited.

9.3.2 Measurements on Gain-Improved Amplifiers

The use of mode coupling filters for suppression of 1050-nm ASE in neodymium-doped ZBLAN fiber amplifiers was first demonstrated in [24]. In this section we will review the central results of this work.

The fiber used for the demonstration [24] was manufactured from ZBLAN and doped with 500 ppm Nd^{3+} ions. The core diameter was 5.5 μm, the refractive core index was 1.51, and the cutoff wavelength was approximately 1030 nm. A signal wavelength of 1328 nm was chosen, and the input signal power was kept below -10 dBm to ensure small-signal operation. The pump light at 795 nm (from a Ti:sapphire laser) was launched co-directionally with the signal. Furthermore, the neodymium ZBLAN fibers were butt-coupled with SiO_2 fibers that were index-matched to minimize Fresnel reflections. This becomes especially important for avoiding 1050-nm lasing, since the stimulated emission at 1050 nm is around eight times stronger than the effective emission strength of the signal. For practical reasons, it was not possible to distribute the filter along the full fiber length, since the filters were 130 mm long and had a pitch of 0.258 mm.

In Figure 9.9(b), the gain versus the position of the filter is shown. This is done for a situation in which high pressure is applied to the filter, and, for reference, also included are the gain value measured without pressing the filter against the fiber. The highest gain of 9.2 dB was obtained by placing the filter 39 cm from the input end of the fiber. This corresponds to 11-dB gain at the peak of the gain spectrum. The gain for the filter that is not pressured is not constant for different positions, which is presumably due to variations in the reflections.

This demonstration has shown that it is possible to improve the gain (in decibels) by a factor of approximately 1.8 by introducing one mechanical notch filter. This approach may be extended to more filters in future experiments, and other means of realizing the notch filters may come into consideration in order to

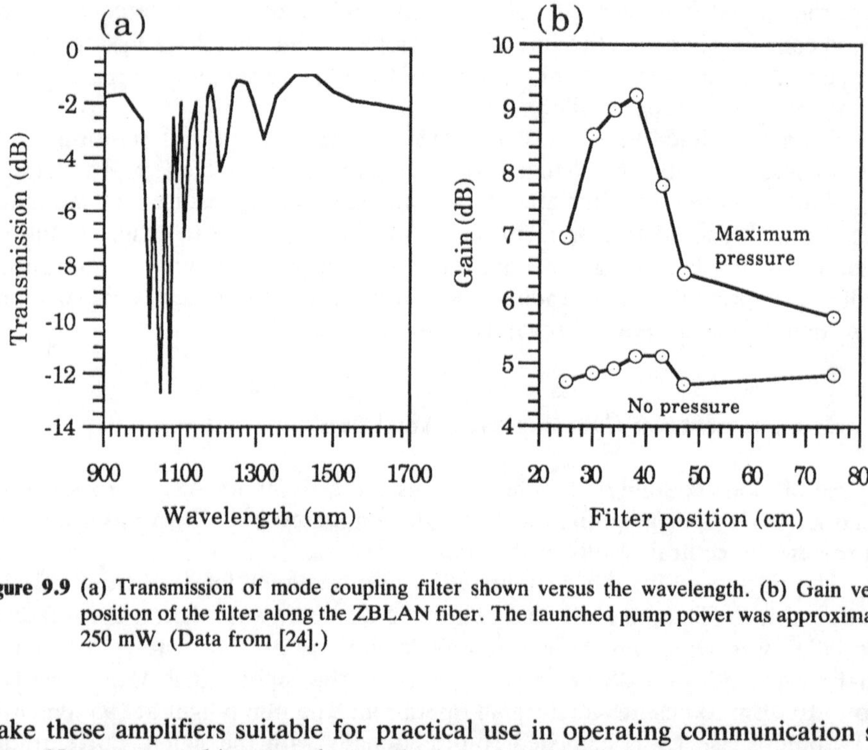

Figure 9.9 (a) Transmission of mode coupling filter shown versus the wavelength. (b) Gain versus position of the filter along the ZBLAN fiber. The launched pump power was approximately 250 mW. (Data from [24].)

make these amplifiers suitable for practical use in operating communication systems. However, at this stage the results have been included in the description for the primary purpose of demonstrating the method and thereby illustrating the physical problems and possibilities of neodymium-doped fiber amplifiers.

9.4 1300-nm GAIN IMPROVEMENT BY AN ASE-SUPPRESSING FIBER DESIGN

In this chapter we have already discussed how the problem of the signal ESA was addressed by choosing ZBLAN glasses [1] as host material for the Nd^{3+} ions. However, the suggested solution of the problem of the 1050-nm ASE has so far only been limited to the use of filters [3–5] in small-signal neodymium-doped amplifiers.

Since the realization of filters might be shown to be a problem, it has been suggested that an alternative and new fiber design be used [6]. It is based on the difference in radial mode distribution between the 1050-nm ASE and the 1340 nm signal, so it in itself limits the 1050-nm ASE power buildup. This leads to a simple but significantly improved optimum fiber design which will be discussed in the following sections.

9.4.1 Basic Method for Gain Improvement by Advanced Fiber Design

An established way of improving erbium-doped optical amplifiers [27] is by confining the rare-earth dopant around the fiber axis in a radius smaller than the fiber core. This ensures a higher population inversion in the fiber and thereby a better power conversion from pump to signal. However, when neodymium-doped fibers are designed, two basic differences must be considered. First, the Nd^{3+} ions act as a four-level laser system, and a low population inversion does not therefore lead to signal absorption, as is the case for the Er^{3+} ion, which is described by a three-level system. Second, the dominant 1050-nm ASE in neodymium-doped fibers has a different fundamental LP_{01} mode distribution than the 1340-nm signal, which is described by a broader radial intensity distribution than the ASE.

The first difference suggests that the Nd^{3+} ions can be placed at a larger distance from the fiber axis, where the signal intensity, as shown in Figure 9.10(a), is larger than the ASE power intensity. Thus increasing the gain, further improvements can be achieved by discriminating the 1050-nm ASE from the signal by removing the active ions from the center of the fiber. The optimum placement of the Nd^{3+} ions is also dependent on the radial pump power distribution. Therefore, additional improvements may be obtained by launching the pump power into the LP_{11} mode to achieve a better overlap with the neodymium dopant profile. It should in this connection be noted that the possible use of LP_{11} pumping causes two basic problems: (1) how to excite the LP_{11} mode efficiently and (2) designing a dichroic coupler to the propagation constant of the LP_{11} mode. We will not, however, go into further details of this problem here, but merely present results for the case in which LP_{11} pump modes could be obtained.

In order to quantify the obtainable improvements, the point of reference can be found in the optimum fiber designs (i.e., λ_c = 800 nm in the case where the total ASE power is suppressed) as described in Section 9.2.3. All calculations are made for a 5 m-long ZBLANP fiber with a neodymium dopant concentration of $\rho = 9.5 \times 10^{24}$ m^{-3}.

Figure 9.10(b) shows the gain as a function of the inner and outer radii, a_1 and a_2, respectively, for the Nd^{3+} dopant distribution. The gain values should be compared with the standard fiber design ($a_1/a_{core} = 0.0$, $a_2/a_{core} = 1.0$). For a fiber with NA = 0.3 and a launched pump power of 200 mW, we find a gain improvement from 6.7 dB for the standard design to a maximum gain G_{opt} = 13.3 dB (for $a_1/a_{core} = 1.4$ and $a_2/a_{core} = 3.0$).

The existence of this maximum gain point may be understood from the following explanation. By removing Nd^{3+} ions from the area around the fiber axis (i.e., increasing a_1), the 1050-nm ASE buildup is discriminated from the signal gain. This effect works positively until a ratio $a_1/a_{core} = 1.4$ is reached, and a further increase of a_1 above this value will result in a disadvantageous decrease in the overlap between the Nd^{3+} dopant profile and the pump mode.

(a)

(b)

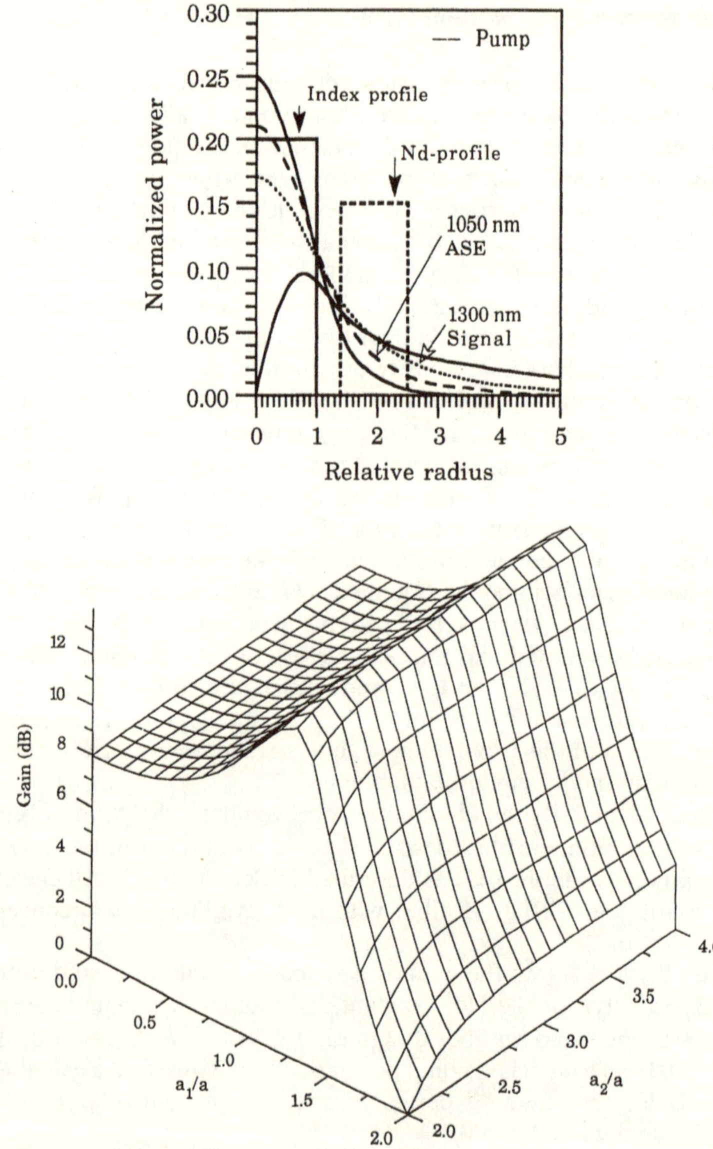

Figure 9.10 (a) Normalized radial intensity distributions of the 1340-nm signal, the 1050-nm ASE, and 795 nm pump LP_{01} modes. Also shown is the 795-nm LP_{11} pump mode. (b) The maximum obtainable gain at 200-mW pump power as a function of both the inner Nd^{3+} dopant radius a_1 and the outer Nd^{3+}-dopant radius a_2. Both radii are given relative to the fiber core radius, and the fiber is specified by NA = 0.3 and the cutoff wavelength λ_c = 800 nm. (Figure from [6].)

The outer Nd^{3+} dopant radius a_2 can be increased considerably with a positive effect, since it favors the signal gain compared to the ASE-gain (as seen in Figure 9.10(a)). However, for a value larger than a_2/a_{core} = 3.0, a further increase will result in decreased gain. This is because the Nd^{3+} ions placed at a large distance from the core axis will only result in pump mode attenuation, whereas the signal intensity is too small to result in significant stimulated emission contribution.

9.4.2 Optimum Distribution of Nd^{3+} Ions

The gain improvement described in the previous section make use of different radial intensity distributions of signal, ASE, and pump power. It is therefore not only dependent on the radial shape of the pump mode, but also on the power level of the launched pump. This is illustrated in Figure 9.11(a), where the optimum gain of a neodymium-doped fiber (NA = 0.3 and λ_c = 800 nm) is given as a function of pump power. We find that for LP_{01} pumping, a direct gain improvement of up to 6.6 dB (for P_p = 200 mW) is achieved.

The improvement obtained by LP_{11} pumping is an additional 2.1 dB at 200-mW pumping, and the positive effect of the higher order pump mode appears for pump power levels above 15 dBm. Note also that the optimum choice of a_1 and a_2

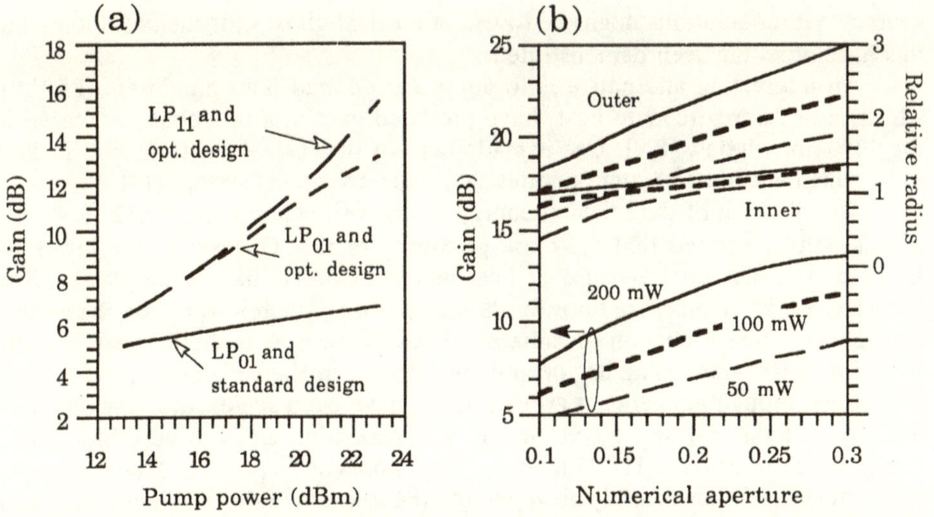

Figure 9.11 (a) The maximum gain value versus pump power launched into the fiber for NA = 0.30. (b) The optimum choices of the inner and outer Nd^{3+} dopant relative radii, a_1/a_{core} and a_2/a_{core}, respectively, as a function of NA and LP_{01} pump power. Also shown are the corresponding gain values. (Data from [6].)

in the case of LP_{11} pumping is different from the case of LP_{01} pumping. The following description will, however, concentrate on pumping in the fundamental mode, since the gain advantages of higher order pumping are considered to be too small compared to the problems of handling the LP_{11} pump situation.

The optimum placement of the neodymium dopant as a function of NA is given in the upper part of Figure 9.11(b) for pump powers of 50, 100, and 200 mW launched into LP_{01}. The lower part of Figure 9.11(b) shows the corresponding gain. As an example, we consider a pump power of 100 mW and an NA of 0.20. The thick dotted lines in the figure should be used. As can be seen, the optimum placement of the neodymium dopant is from an inner radius of $1.1 \cdot a_{core}$ to an outer radius of $1.7 \cdot a_{core}$. This design will result in a gain of 9.2 dB.

9.5 PRASEODYMIUM-DOPED FIBER AMPLIFIERS

As demonstrated in the previous sections of this chapter, amplification in neodymium-doped ZBLAN fibers is limited not only by the ASE generation around 1050 nm, but also because ESA prevents amplification below 1310 nm and realistic gain values are only obtained above 1330 nm. This is a fundamental problem, since the upgrading of already existing 1300-nm systems by means of amplifiers is severely hampered if it is also necessary to change the center wavelength of the signal sources. Other solutions might be to use other host glasses for the Nd^{3+} ions, but this has not so far been demonstrated.

An interesting alternative is to apply Pr^{3+}-doped fiber amplifiers (PDFA), which, in a fluorozirconate host, have provided gain in a broadband centered in the 1300-nm window [8,9]. Use is made here of the $^1G_4 \rightarrow {}^3H_5$ transition in Pr^{3+} ions, which around 1300 nm operates as a quasi-four-level system [17].

The diagram of the relevant energy levels is shown in Figure 9.12. It should at this point be noted that the close proximity of the $^1G_4$ level to the lower $^3F_4$ levels makes necessary the use of host materials other than oxide glasses. The known candidates today are fluoride glasses because of their lower phonon energies and the expected reduction in radiative decay. However, it should also be mentioned that 1080-nm lasing has been demonstrated in silica fibers [28].

Even though net gains of around 30 dB have been achieved, a low quantum efficiency of the $^1G_4$ upper level in the gain transition leads to very high pump power requirements [10,11]. Due to the large impact of a possible 1300-nm amplifier, a huge effort has recently been put into the solution of the problem of reducing the pump power requirements. For clarifying the nature of this problem and discuss the possible solutions, modeling becomes a relevant tool. In Section 9.5.1, therefore, we will describe how to develop an amplifier model for the Pr^{3+} ion system in fluorozirconate glass and use these results to prescribe how optimal Pr^{3+}-doped

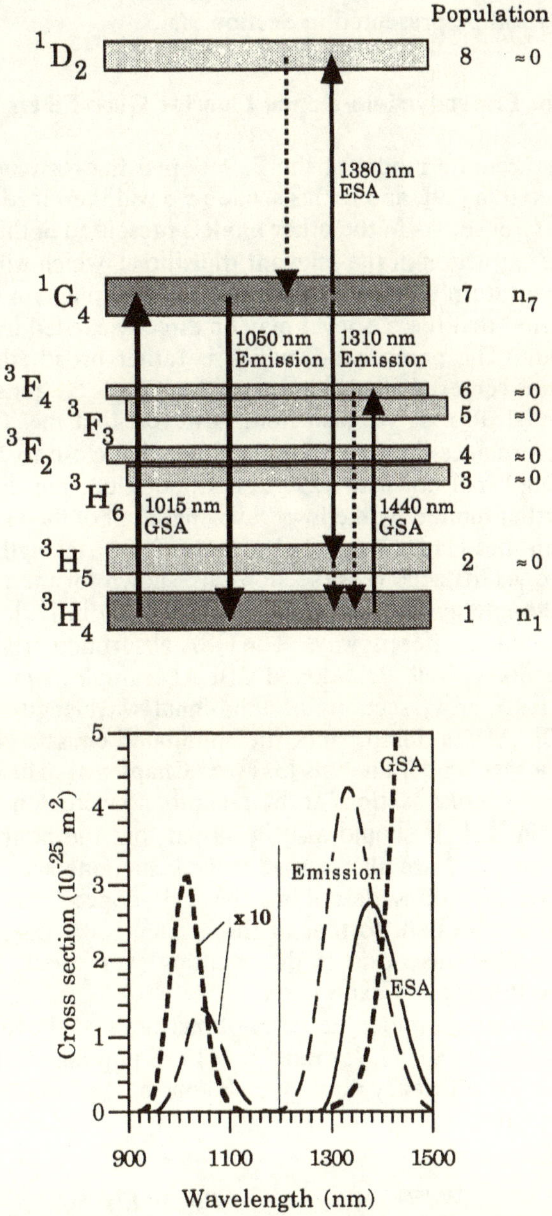

Figure 9.12 (a) Energy level diagram for Pr^{3+} with relevant transitions illustrated by arrows. Also shown are identification numbers of the different levels used in the following model description. (b) Cross-section spectra for the transitions included in the model. (Data from [29].)

fibers should be designed (see Section 9.5.2). Finally, the latest results of Pr^{3+} doped fiber amplifiers will be presented in Section 9.5.3.

9.5.1 Model for Praseodymium-Doped Fluoride Glass Fibers

A qualitatively accurate model of the Pr^{3+}-doped fluorozirconate fiber amplifier was first presented in [29], and in this section we will therefore use this description as a point of reference. As in the other models presented in this book, it is natural to start with an extraction of the relevant transitions, which will be included in the calculations. We return therefore to the energy level diagram in Figure 9.12 from which we first find that the $^1G_4$ level may be directly excited by pumping from the $^3H_4$ ground state. The pump band, which is rather broad (the FWHM width is around 50 nm), is centered at 1015 nm, and the $^1G_4 \rightarrow {}^3H_5$ transition produces gain centered around 1310 nm. We also find, however, that the $^1G_4 \rightarrow {}^1D_2$ transition produces a performance-limiting ESA band peaking close to 1380 nm, and it has a short-wavelength tail down to 1290 nm. In addition, the high-wavelength performance is further limited by the low-wavelength tail of the GSA band that peaks around 1440 nm, but leads to signal absorption for wavelengths $\lambda > 1290$ nm.

On Figure 9.12(b), the cross sections are shown for the transitions indicated by arrows in the energy level diagram. As described in [29], these cross-section values are obtained in different ways. The GSA absorption originates from absorption measurements on bulk Pr^{3+}-doped ZBLAN samples, whereas the stimulated emission and ESA cross sections at the signal wavelength are determined as reported in [30]. At the pump band, the stimulated emission cross sections were obtained from a McCumber analysis [31] (see Chapter 3). The final scalings of the stimulated emission cross sections at the pump band were done using the radiative rate presented in [32]. It should also be noted that the nonradiative transitions indicated in Figure 9.12 are all assumed to be instantaneous.

The central equations describing the Pr^{3+}-doped fiber amplifier may be derived in a way equivalent to that of the erbium- or neodymium-doped fibers. However, some differences exist in the resulting expressions, and we will, consequently, outline the primary expressions.

First we notice that, under the assumptions outlined above, only the $^1G_4$ (level 7 in Figure 9.12) and the $^3H_4$ (ground state) are populated. Therefore, we only need to consider the efficiency rates given below in terms of cross sections, photon energies, and intensities.

$$W_p^{abs}(r, z) = \frac{\sigma_{17}(\nu_p)}{h\nu_p} \cdot P_p(z) \cdot |E(r, \nu_p)|^2 \tag{9.35}$$

$$W_p^{ems}(r, z) = \frac{\sigma_{71}(\nu_p)}{h\nu_p} \cdot P_p(z) \cdot |E(r, \nu_p)|^2 \tag{9.36}$$

$$W_{\mathrm{SSE}}(r, z) = \frac{\sigma_{72}(\nu_s)}{h\nu_s} \cdot P_s(z, \nu_s) \cdot |E(r, \nu_s)|^2 \tag{9.37}$$

$$W_{1050}^{\mathrm{abs}}(r, z) = \int_0^\infty \frac{\sigma_{17}(\nu)}{h\nu} \cdot (P_{1050}^+(z, \nu) + P_{1050}^-(z, \nu)) \cdot |E(r, \nu)|^2 d\nu \tag{9.38}$$

$$W_{1050}^{\mathrm{ems}}(r, z) = \int_0^\infty \frac{\sigma_{71}(\nu)}{h\nu} \cdot (P_{1050}^+(z, \nu) + P_{1050}^-(z, \nu)) \cdot |E(r, \nu)|^2 d\nu \tag{9.39}$$

$$W_{1310}^{\mathrm{ems}}(r, z) = \int_0^\infty \frac{\sigma_{72}(\nu)}{h\nu} \cdot (P_{1310}^+(z, \nu) + P_{1310}^-(z, \nu)) \cdot |E(r, \nu)|^2 d\nu \tag{9.40}$$

All the parameters have the same significance as those in Section 9.1.1, and the frequency integrations should be performed only over the relevant intervals given by the physical limits of the ASE bands.

It is here important to note that the model includes the wavelength-dependent ASE around 1050 and 1310 nm (equations (9.38) to (9.40)). Furthermore, the signal ESA is considered together with the GSA of the signal, but these properties do not appear in the expressions for the efficiency rates given above. The reason is that ions excited by either of the two absorption mechanisms are assumed to relax immediately to the original state. Therefore, both ESA and GSA of the signal will only appear in the equations describing the power distribution along the fiber and will not appear in the expression for the population inversion.

Equivalent to the derivation in Section 9.1.1, the rate equations lead in the steady-state solution to an expression for the population in the upper laser level. In this case we find that

$$n_7 = \frac{W_p^{\mathrm{abs}} + W_{1050}^{\mathrm{abs}}}{W_p^{\mathrm{abs}} + W_p^{\mathrm{ems}} + W_{1050}^{\mathrm{abs}} + W_{1050}^{\mathrm{ems}} + W_{\mathrm{SSE}} + W_{1310}^{\mathrm{ems}} + (1/\tau)} \cdot \rho(r) \tag{9.41}$$

and the population in the ground state may be found from the local Pr^{3+} ion concentration $\rho(r)$ according to

$$n_1(r, z) + n_7(r, z) = \rho(r) \tag{9.42}$$

What remains now is to express the coupled differential equations describing the power distribution along the fiber amplifier. For the pump, the power may be expressed by (9.23), where in this case we only have to use another attenuation factor $g_p(z)$; namely,

$$g_p(z) = -2\pi \int_0^{a_d} |E(r, \nu_p)|^2 \cdot (n_1(r, z) \cdot \sigma_{17}(\nu_p) - n_7(r, z) \cdot \sigma_{71}(\nu_p)) r dr \quad (9.43)$$

Also, (9.25), prescribing the signal variation, may be used here, only the gain factor must now be calculated as

$$g_p(z) = 2\pi \int_0^{a_d} |E(r, \nu_s)|^2 (n_7(r, z)(\sigma_{72}(\nu_s) - \sigma_{78}(\nu_s)) - n_1(r, z) \cdot \sigma_{16}(\nu_s)) r dr \quad (9.44)$$

Note that this includes the stimulated emission, the ESA, and the GSA of the signal.

For the low-wavelength ASE band around 1050 nm, the equations become

$$\begin{aligned} \frac{dP^{\pm}_{1050}(z, \nu)}{dz} &= \pm g_{1050}(z, \nu) \cdot P^{\pm}_{1050}(z, \nu) \pm 2h\nu \cdot a_{1050}(z, \nu) \\ g_{1050}(z, \nu) &= 2\pi \int_0^{a_d} |E(r, \nu)|^2 \cdot (n_7(r, z) \cdot \sigma_{71}(\nu) - n_1(r, z) \cdot \sigma_{17}(\nu)) r dr \\ a_{1050}(z, \nu) &= 2\pi \cdot \sigma_{71}(\nu) \int_0^{a_d} |E(r, \nu)|^2 \cdot n_7(r, z) r dr \end{aligned} \quad (9.45)$$

Finally, the 1310-nm ASE band may be included through the following equations

$$\begin{aligned} \frac{dP^{\pm}_{1310}(z, \nu)}{dz} &= \pm g_{1310}(z, \nu) \cdot P^{\pm}_{1310}(z, \nu) \pm 2h\nu \cdot a_{1310}(z, \nu) \\ g_{1310}(z, \nu) &= 2\pi \int_0^{a_d} |E(r, \nu)|^2 (n_7(r, z) \cdot (\sigma_{72}(\nu) - \sigma_{78}(\nu)) \\ &\quad - n_1(r, z) \cdot \sigma_{16}(\nu)) r dr \\ a_{1310}(z, \nu) &= 2\pi \cdot \sigma_{72}(\nu) \int_0^{a_d} |E(r, \nu)|^2 \cdot n_7(r, z) r dr \end{aligned} \quad (9.46)$$

This completes the description of the model for the Pr^{3+}-doped fiber amplifier. It should be noted that this model was verified in [29] by obtaining good agreement between calculated results and measurements presented in [11]. Note also that a model similar to the one described in this chapter has recently been published in [33].

9.5.2 Optimum Design of Praseodymium-Doped Fibers

In [29] the model for Pr^{3+}-doped ZBLAN fibers was used to describe how not only waveguide design but also the $^1G_4$ lifetime influences the amplifier performance. The latter is related to the question of how to choose a host glass composition, which may enable us to overcome the pump power requirement problems of Pr^{3+} amplifiers. The efficiency of these 1300-nm amplifiers is limited by nonradiative decay, and there is, consequently, strong motivation for finding a means to reduce these nonradiative decays, which will show through an increase in the $^1G_4$ excited-state lifetime. The way of tailoring a new glass host that will result in the largest lifetime increase is at this point unknown. We will therefore in this section not try to tie the description to a specific host composition, but instead apply a more overall procedure where only the potentials of a lifetime increase are investigated. Due to the resemblance with the Er^{3+} system (see previous section), we have furthermore limited the discussion to step-index fibers, since they are the most efficient according to Chapter 6.

For the sake of simplicity, only co-directionally pumped fibers have been considered, with a pump wavelength of 1017 nm. In [29] it was reported that the small-signal gain had an optimum for a cutoff wavelength of approximately 800 nm, which is almost independent of both NA (in a range of 0.2 to 0.5) and pump power. This optimum cutoff wavelength is therefore chosen in the following. It should also be noted that the scattering loss of the fiber has been ignored, since, according to [29], it is typically around 0.1 dB/m, with the potential of further improvement, due to the predicted low-loss properties of ZBLAN glass [34].

If we look initially at the small-signal properties, Figure 9.13(a) shows the small-signal gain versus the NA for co-directional pumping with the pump power as a parameter. Each point of the curve corresponds to the fiber length at which the small-signal gain is maximal. Like previous observations of Er^{3+}-doped fibers, the gain increases with increasing NA due to the improved overlap between the fiber core, in which the Pr^{3+} ions are homogeneously distributed, and the pump and signal modes.

For a modest pump power of 50 mW, the NA requirement is very high, since a gain of only 10 dB necessitates an NA > 0.4. For higher pump powers, more reasonable NA requirements may be set, but we also find that for a pump power of 200 mW the largest improvement with NA is for NA < 0.35. The same effect is seen for 400-mW pump power, where the gain curve, however, already bends over for an NA of around 0.25. The reason is partly that the higher NA also increases the pump absorption due to the better overlap mentioned earlier. However, when the pump absorption rate becomes comparable to the total relaxation rate of the $^1G_4$ level ($1/\tau$), an additional increase in pump absorption rate (see (9.41)) will not lead to as strong an improvement, and the gain increase starts to

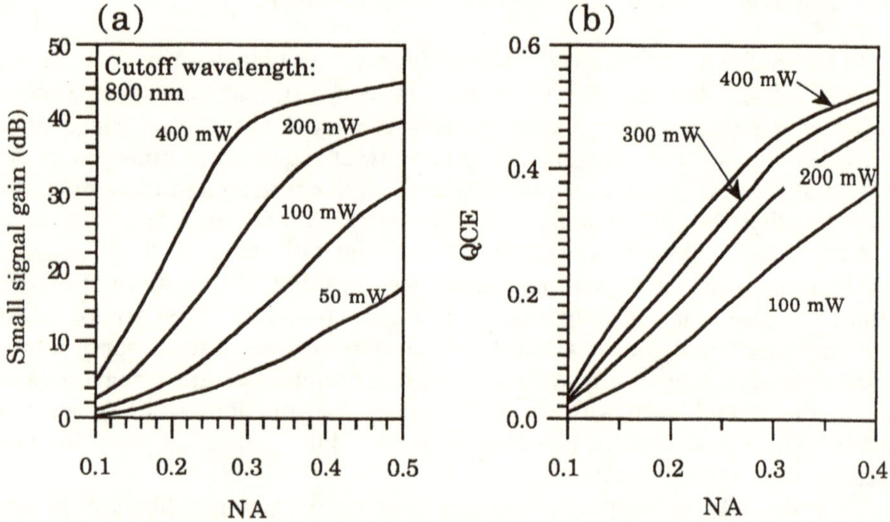

Figure 9.13 (a) Small-signal gain versus NA for step index praseodymium-doped ZBLAN fibers pumped co-directionally. The signal wavelength is 1310 nm and the pump wavelength is 1017 nm. (Data from [29]). (b) QCE shown as a function of NA with pump power as parameter. The signal input power is 5 mW. (Data from [35].)

roll off. Another effect, which contributes to the gain rolloff, is that the amplifier starts to saturate due to ASE buildup.

To describe the large-signal performance of the Pr^{3+}-doped fiber amplifiers, the QCE is used, which in Section 6.2.2 is defined as the ratio of the number of signal photons added to the signal, to the total number of launched pump photons. This power amplifier performance has been studied in [35], and based on this description, the primary results are shown in Figure 9.13(b). Here, the QCE is plotted as a function of NA with pump power as the parameter. We find that the QCE is an increasing function of NA, but it is also evident that for a Pr^{3+}-doped ZBLAN fiber like the one considered here, we may require more than 300 mW to obtain a QCE higher than 0.5, even for an NA as high as 0.4.

It is evident from the above description that the praseodymium-doped ZBLAN fiber suffers from serious problems of low quantum efficiency of the metastable $^1G_4$ state, and these limit the overall amplifier performance. As mentioned earlier, the $^1G_4$ level experiences a rapid nonradiative decay to the $^3F_4$ level through multiphonon emission, leading to an observed lifetime of 110 μs compared to a radiative lifetime of 3.2 ms [32]. The nonradiative relaxation rate of the $^1G_4$ level may be decreased through the use of other glass hosts, which then show either a lower effective phonon energy or a weaker electron-phonon coupling [29,36].

In order to illustrate the potential of such an increase in the lifetime τ of the metastable state, we may start by considering the case in which ASE saturation is ignored. In this case, the available small-signal gain for a given pump power simply scales linearly with this value, or, alternatively, the power necessary to achieve a given gain is inversely proportional to the lifetime. Results of a more detailed examination are shown in Figure 9.14(a), where the small-signal gain is illustrated as a function of τ for pump powers of 100 and 250 mW. The results, which originate from [29], have been calculated using cross sections for Pr^{3+} ZBLAN, and they therefore represent solely the effect of changing the nonradiative relaxation rate. When this is done in praxis through the choice of another host glass composition, measurements of the relevant cross sections must naturally be performed to obtain accurate results for the specific glass composition. Figure 9.14(a) shows that the gain is very sensitive to variations in the relaxation lifetime τ in the interval given by 20 $\mu s < \tau <$ 400 μs. Under the condition that the cross-section values would remain unchanged, we find, for example, that a doubling of the relaxation lifetime will result in a gain increase from 13 to 25 dB for a pump power of only 100 mW. Furthermore, an increase of three times the τ-value to 330 μs means that only 100-mW pump power would be required to obtain 30-dB gain compared to 250 mW for the Pr^{3+} ZBLAN fiber.

The other aspect of the results in Figure 9.14 is that we may evaluate the consequences of reducing the relaxation lifetime. As reported in [37], this will

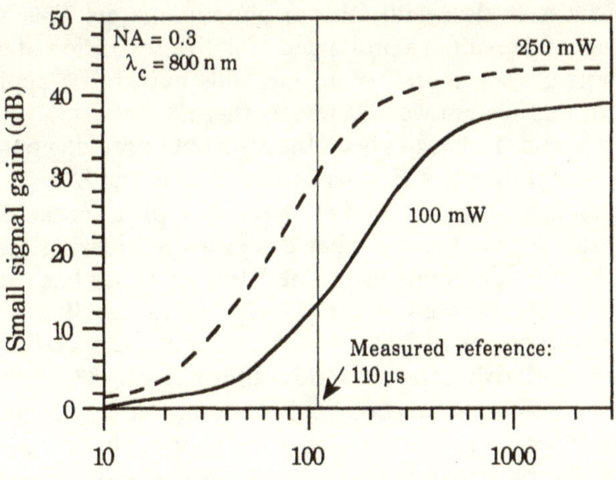

Figure 9.14 Small-signal gain shown as a function of the effective relaxation lifetime τ for the $^1G_4$ level. Cross-section spectra are the same as for ZBLAN glass for all values of τ. (Data from [29].)

happen because of cross relaxation if the Pr^{3+} concentration is increased above 0.1 wt% (or 1000 ppm-wt). As explained in Chapter 3, cross relaxation may occur between two closely located Pr^{3+} ions, and the possible channels are the ($^1G_4$-$^3H_5$) to ($^3H_4$-$^3F_4$) and the ($^1G_4$-$^3H_6$) to ($^3H_4$-$^3F_3$) transfers. The obvious reason for wishing to increase the Pr^{3+} ion concentration above the 1000 ppm-wt limit is that the fiber length and associated scattering loss should become lower. However, as we find from [37], a 10-times higher Pr^{3+} concentration will result in a τ-value below 50 μs. According to Figure 9.14, this will lead to gain below 5 dB for 100-mW pump power. In other words, this means that the gain in decibels is reduced by a factor of two.

Finally, it should be mentioned that the large-signal performance is also strongly influenced by the relaxation lifetime. It is reported in [35] that if the lifetime can be doubled, then the QCE will increase by 0.15 to 0.20 to reach levels just below 0.6 for very high NA fibers (NA $\approx$ 0.4).

9.5.3 Present Results of Praseodymium-Doped Fluoride Glass Fibers

As has become clear from previous sections, the Pr^{3+} doped ZBLAN fiber amplifiers are still far from the development level of Er^{3+}-doped fibers for 1550-nm amplification. The primary difficulties are the fabrication properties (see Chapter 2) and the fundamental problem of a very low quantum efficiency of about 1.5% [38] for demonstrated amplifiers. We found in Section 9.5.2 that an attractive solution would be to find a host glass with a lower phonon energy. This would lower the phonon decay rate and result in a prolonged effective relaxation lifetime. However, before such a host glass is found, other methods have to be applied, and in the remaining part of this section we will review these.

Among these approaches has been the effort of improving pumping efficiency by co-doping with ytterbium, which has absorption in the 850- to 1050-nm wavelength range [39,40]. Another method has been to apply a special fiber design [41], or, in other words, to optimize the fiber design by minimizing the core diameter and increasing the NA simultaneously. The latter approach has led to some of the most promising results reported so far [11,42]; namely, a 38.2-dB gain obtained for a launched pump power of 300 mW. The data of the specific fiber are a core radius of 1.15 μm, a relative refractive index difference of 3.8%, a cutoff wavelength of 1260 nm, a Pr^{3+} ion concentration of 2000 ppm, and a background loss of 0.1 dB/m. It is important to note here that the reported fiber has a higher cutoff wavelength than the optimum value discussed in the previous section. Since the actual fiber makes a gain of 20 dB possible for a pump power of 100 mW, comparable or even better results are within reach for Pr^{3+}-doped fiber amplifiers pumped by laser diodes emitting light at 1017 nm [43].

This has been demonstrated in [44] by the development of a Pr^{3+}-doped fluoride fiber amplifier module pumped by laser diodes. The configuration was a bi-directional pumped amplifier in which the two 1017-nm diodes each emitted 160 mW, and an overall signal gain of 11.0 dB was obtained at 1300 nm. The results of [44] are also interesting in another aspect, since a low loss (<0.5 dB) and stable splice technique is demonstrated between the praseodymium-doped fluoride fiber and a silica fiber. The method is to use a V-groove connection technique in which each fiber is aligned and their end faces are mechanically polished. Thereafter, they were carefully aligned and connected with UV-curable adhesives. This technique shows a way to overcome the problems of the very different melting temperatures of ZBLAN and SiO_2 glass (see Chapter 2), which makes normal fusion splicing impossible.

Also, the large-signal performance of Pr^{3+}-doped fiber amplifiers has recently shown good progress, since an output power of 212 mW has been obtained [45] in a counter pumped configuration where a 900-mW pump at 1010 nm was launched into the praseodymium-doped fiber. The power slope efficiency was reported to be 30%, and the fiber is specified by a refractive index difference of $\Delta n = 0.04$, a cutoff wavelength of 650 nm, a praseodymium concentration of 500 ppm, and a background loss of 0.14 dB/m. Lately, large-signal performance has also been demonstrated with a Pr^{3+}-doped fiber amplifier by using the high-power output from a Nd:YLF laser as a pump source emitting at 1047 nm [46]. With a saturating input of +1 dBm, a maximum output power of 250 mW was obtained for 1.8W pump power with a power slope efficiency of 18%. The advantage of this approach is that the Nd:YLF pump lasers, because they are compact and reliable at high output powers, represent a realistic alternative to semiconductor laser pumping. Therefore, this result is considered important as a step towards the realization of a practical 1300-nm amplifier.

On the subject of noise performance, the Pr^{3+}-doped fiber amplifiers exhibit good properties, since noise figures between 3 and 4 dB have been obtained in the wavelength interval from 1290 to 1320 nm [47]. At wavelengths above 1320 nm, the PDFA degrades with respect to noise properties because of the increased GSA (the ${}^3H_4 \rightarrow {}^3F_4$ transition as shown in Figure 9.12). In relation to these results, it should also be noted that a PDFA has been used as preamplifier in conjunction with a pin-FET receiver, leading to a sensitivity of 37.3 dBm achieved for a 2.5-Gbit/s system [48].

9.6 FIBER AMPLIFIER GAIN EXPRESSIONS FOR AN IDEALIZED FOUR-LEVEL SYSTEM

In Sections 9.1.1 and 9.5.1, respectively, the question of how accurate modeling of a neodymium- and a praseodymium-doped ZBLAN fiber amplifier could be performed was discussed. The major drawback of these approaches is that they are time consuming and require numerical solutions of coupled differential equations. However, this is necessary to obtain accurate results, but for certain applications, where only the approximate magnitude of the gain is required, alternative and approximate solutions might be of interest.

In order to fulfill this need for approximate and fast solution methods, work has been done on idealized four-level systems [23]. What is meant by an idealized four-level system is that only the upper laser level and the ground-state level (different from the lower laser level) are populated. Furthermore, no ASE is considered in the model, which is an important property to keep in mind when the model is applied to a given situation. It rules out application on neodymium-doped fiber amplifiers, which are performance-dominated by 1050-nm ASE. Also, situations in which ASE is assumed to be significant in praseodymium-doped fibers must be considered with great caution. An example of the latter is the application on praseodymium-doped fiber amplifiers at wavelengths above 1320 nm. For the neodymium-doped fiber amplifiers using distributed filtering of 1050-nm ASE, however, such as those described in Sections 9.3.1 and 9.3.2, the approximate formulas might also be applied.

In favor of the idealized model is, of course, its simplicity, and it [23] has recently been further developed [49] and shown to agree with low gain (< 10 dB) experimental results for praseodymium-doped fibers. With this in mind, the compact, approximate expressions will be of value for the fast evaluation of many problems, and the final results of [49] are therefore included in this description.

The parameters that are needed to apply the approximate gain expressions for the idealized four-level system are, first of all, the spot size φ. If this is calculated according to the normal Marcuse polynomial spot-size formula [50], inaccuracies may be found in small-signal gain calculations. According to the oral presentation of [49], the error increases with decreasing cutoff wavelength. However, this problem has been eliminated by the authors of [49] by applying the modified expression, which is as follows:

$$\frac{\omega}{a} = 0.61 + \frac{1.68}{V^{3/2}} + \frac{0.97}{V^6} \tag{9.47}$$

where a is the fiber core radius and $V = ka \cdot \mathrm{NA}$ is the normalized frequency of the fiber. Using these spot sizes, three parameters may be defined:

$$\begin{aligned} F &= \left\{1 - \exp\left[-2a^2\left(\frac{\omega_s^2 + \omega_p^2}{\omega_s^2 \cdot \omega_p^2}\right)\right]\right\} \\ F_p &= \left[1 - \exp\left(-\frac{2a^2}{\omega_p^2}\right)\right] \\ F_s &= \left[1 - \exp\left(-\frac{2a^2}{\omega_s^2}\right)\right] \end{aligned} \tag{9.48}$$

Here, F is related to the three-way overlap between pump mode, signal mode, and fiber core, whereas F_p and F_s are related to the respective overlaps between fiber core and either pump mode or signal mode.

The approach of [49] is now to divide the solution into two limiting cases. In the first case, which may be described as the low pump solution, it is assumed that the pump power is insufficient to depopulate the ground state significantly. For fibers in which the dopant is uniformly distributed throughout the core of the fiber, the above assumption of negligible ground-state depopulation may be shown [49] to lead to the following expressions:

$$G_{Lp} = 4.343 \cdot \frac{\sigma_s \tau}{h\nu_p} \cdot \frac{1}{A} \cdot \frac{2a^2}{\omega_s^2 + \omega_p^2} \cdot \frac{F}{F_p} \cdot P_p^{abs} \tag{9.49}$$

Here, G_{Lp} is the total small-signal gain in decibels which is obtained for an absorbed pump power of P_p^{abs}, and the parameter A is the fiber core area. σ_s is the emission cross-section value at the signal wavelength, and τ is the lifetime in the upper laser level.

The other case is the high-pump solution in which it is assumed that the pump is so large that the population in the upper laser level is approximately equal to unity. Under this assumption, the following expression for the high pump small-signal gain (in fibers with a confinement factor of one) appears:

$$G_{Hp} = 4.343 \cdot \frac{\sigma_s \tau}{h\nu_p} \cdot \frac{F_s}{A} \cdot P_p^{abs} \tag{9.50}$$

The two equations (9.49) and (9.50) differ in that the low-pump equation is related to the overlap of pump and signal with the core region, whereas the high-pump equation is only related to the overlap of the signal mode with the fiber core. The latter is the case because the pump intensity is considered high enough to excite all the rare-earth ions at any location within the doped area.

It should finally be noted that for practical pump powers (<1W), the low pumping limit represented by G_{Lp} gives a good gain approximation [49]. Gain comparisons with experimental results are also shown to give good agreement with (9.49). This further emphasizes that for the praseodymium-doped fiber amplifiers, the assumption of negligible ground-state depletion is a good one.

9.7 SUMMARY AND PERSPECTIVES

In this chapter, a thorough description of the status of fiber amplifiers for 1300-nm operation has been presented. One of the primary points is that it is not possible at present to realize an amplifier based on silica fibers for 1300 nm. ZBLAN or similar heavy-metal fluoride glass fibers offer, however, a choice of transitions due to shifted energy bands and much lower phonon energies. Therefore, the description has a point of reference in amplifiers realized in such host materials. The chapter includes a detailed description of the theoretical basis of the relevant neodymium- and praseodymium-doped fiber amplifiers, as well as a summary of the latest experimental results.

To keep the presentation in a chronological order, we first described a model for the Nd^{3+}-doped fiber amplifier, which was shown to be in good agreement with signal saturation measurements on well-characterized fibers. This description also illustrated that the serious performance-limiting properties of neodymium-doped fiber amplifiers are the strong signal ESA, which forces the amplifier operation to wavelengths above 1320 nm, and the very strong 1050-nm ASE that saturates the amplifier and locks the gain at values around 5 to 7 dB. With regard to noise description, the properties of the four-level system become important, and for an ideal system without ESA, the quantum-limited noise figure would always be 3 dB, provided that the gain was large. However, it was demonstrated that the noise figure increases with increasing ESA cross section. Consequently, the improvement in system performance achieved by the amplifier is highest in the long-wavelength region of the gain spectrum. This also indicates that operation of NDFAs at wavelengths lower than 1330 to 1340 nm will only be possible by the development of host glass compositions different from the ZBLAN glasses applied today. The goal should be to find a host material in which the neodymium ions will exhibit an ESA spectrum which is moved towards shorter wavelengths.

Under all circumstances, methods of additional suppression of 1050-nm ASE will be a central point for NDFAs, and so the chapter also discusses different means of obtaining this ASE suppression. One of these approaches is to use a configuration consisting of fibers cascaded with 1050-nm ASE filters in between. This would ideally make it possible to more than triple the gain in decibels. For a larger number of filters, the ratio between maximum gain in decibels and launched pump power in milliwatts is shown to be a constant of 0.16 dB/mW for a pump power level below 200 mW. Optimal positions of these filters were found and experimental results on partly distributed ASE filtering by 1050-nm mode coupling demonstrated the usefulness of the ASE suppression approach. Another method was also discussed. Here, a special 1050-nm ASE-suppressing design for neodymium-doped fiber optical amplifiers was presented and investigated with an accurate model. It was shown that a gain of up to 13.3 dB may be obtained for a pump power of 200 mW, an NA of 0.30, and a ring-shaped neodymium dopant area located within the inner and outer radii of 1.4 and 3.0 relative to the core radius. If the same amount of pump power is launched into the LP_{11} pump mode, even 15.4 dB of gain could be reached.

The waveguide parameters for a neodymium ZBLANP optical-fiber amplifier was furthermore optimized for the small-signal and booster operation. In conclusion, the NA should always be chosen as high as possible, and the optimum cutoff wavelength should be chosen to 800 nm, provided that the 1050-nm ASE can be removed in the small-signal case. A booster QCE of nearly 80% may be reached for an input power of 10 dBm and pump power of 100 mW with the use of one filter.

The other candidate for rare-earth dopant is praseodymium. The amplifiers are still at an early stage of development, but intense work has been carried out lately to investigate the properties of praseodymium-doped fibers. As a consequence of this, it was lately shown that gain values as high as 38.2 dB may be obtained for a pump power of 300 mW. These numbers, however, also illustrate the basic problem of praseodymium-doped fiber amplifiers; namely, the very low quantum efficiency. This is due to a large percentage of nonradiative decays from the upper laser level.

In order to understand the nature of this problem and its possible solutions, a numerical model has been shown to be a useful tool. We have therefore in this chapter included a description of how to establish such a model. From this it is not only derived that an optimal cutoff wavelength for praseodymium-doped fibers should be 800 nm, but also that gains larger than 10 dB are only practically obtainable for pump powers >50 mW. For pump powers ≤200 mW and NAs <0.35, the small-signal gain increases linearly with the NA. Another very important result is that by increasing the metastable lifetime to 300 μs, the pump power required to obtain 30-dB gain would be reduced from 250 to 100 mW.

Among the most positive reports of the development of 1300-nm fiber amplifiers are the present investigation results on new host glass materials [36,41]. The reason for this optimism is that we find clear indications that it seems within reach to increase the radiation lifetime in praseodymium-doped fibers. Furthermore, development of host glass materials might also be advantageous for neodymium-doped amplifiers, which with optimal designs such as the proposed examples, can then be expected to supplement praseodymium-doped fibers in the long wavelength range of the 1300-nm window. For the immediate future, however, praseodymium-doped fiber amplifiers offer the best prospects for 1300-nm operation, but future development might show new combinations of both host glass and rare-earth compositions.

REFERENCES

[1] Dakss, M. L., and W. J. Miniscalco, "Fundamental Limits on Nd^{3+}-Doped Fiber Amplifier Performance at 1.3 μm," *IEEE Photonics Technology Letters*, Vol. 2, No. 9, 1990, pp. 650–652.

[2] Pedersen, J. E., M. C. Brierley, S. F. Carter, and P. W. France, "Amplification in the 1300nm Telecommunications Window in a Nd-Doped Fluoride Fibre," *Electronics Letters*, Vol. 26, No. 5, 1990, pp. 329–330.

[3] Sugawa, T., Y. Miyajima, and T. Komukai, "10 dB Gain and High Saturation Power in a Nd^{3+}-Doped Fluoro-zinate Fibre Amplifier," *Electronics Letters*, Vol. 26, No. 24, 1990, pp. 2042–2044.

[4] Miniscalco, W. J., L. J. Andrews, B. A. Thompson, R. S. Quimby, L. J. B. Vacha, and M. G. Drexhage, "1.3 μm Fluoride Fiber Laser," *Electronics Letters*, Vol. 24, No. 1, 1988, pp. 28–29.

[5] Øbro, M., B. Pedersen, A. Bjarklev, J. H. Povlsen, and J. E. Pedersen, "Highly Improved Fibre Amplifier for Operation Around 1300 nm," *Electronics Letters*, Vol. 27, No. 5, 1991, pp. 470–472.

[6] Bjarklev, A., T. Rasmussen, J. H. Povlsen, O. Lumholt, K. Rottwitt, S. Dahl-Petersen, and C. C. Larsen, "9 dB Gain Improvement of 1300 nm Optical Amplifier by Amplified Spontaneous Emission Suppressing Fibre Design," *Electronics Letters*, Vol. 27, No. 19, 1991, pp. 1701–1702.

[7] Ohishi, Y., T. Kanamori, T. Kitagawa, S. Takahashi, E. Snitzer, and G. H. Sigel, "Pr^{3+}-Doped Fluoride Fiber Amplifier Operating at 1.3 μm," *Proc. of Conference of Optical Fiber Communication OFC '91*, San Diego, postdeadline paper PD2, 1991.

[8] Durteste, Y., M. Monerie, J. Y. Allain, and H. Poignant, "Amplification and Lasing at 1.3 μm in Praseodymium-Doped Fluorozirconate Fibres," *Electronics Letters*, Vol. 27, No. 8, 1991, pp. 626–628.

[9] Carter, S. F., D. Szebesta, S. T. Davey, R. Wyatt, M. C. Brierley, and P. W. France, "Amplification at 1.3 μm in a Pr^{3+}-Doped Single-Mode Fluorozirconate Fibre," *Electronics Letters*, Vol. 27, No. 8, 1991, pp. 628–629.

[10] Ohishi, Y., T. Kanamori, T. Nishi, and S. Takahashi, "A High Gain, High Output Saturation Power Pr^{3+}-Doped Fluoride Fiber Amplifier Operating at 1.3 μm," *IEEE Photonics Technology Letters*, Vol. 3, No. 8, 1991, pp. 715–717.

[11] Miyajima, Y., T. Sugawa, and Y. Fukasaku, "38.2 dB Amplification at 1.31 μm and the Possibility of 0.98 μm Pumping in Pr^{3+}-Doped Fluoride Fiber," *Techn. Dig. of Optical Amplifiers and Their Applications OAA '91*, Snowmass, CO, 24–26 July 1991, postdeadline paper PD1.

[12] Pedersen, B., K. Dybdal, C. D. Hansen, A. Bjarklev, J. H. Povlsen, H. Vendeltorp-Pommer, and C. C. Larsen, "Detailed Theoretical and Experimental Investigation of High-Gain Erbium Doped Fiber Amplifier," *IEEE Photonics Technology Letters*, Vol. 2, No. 12, 1990, pp. 863–865.

[13] Pedersen, J. E., "Rare Earth Doped Fibre Optical Amplifiers," Ph.D. Thesis EF249, Institute of Physics, University of Aarhus, March 1991.

[14] Pedersen, J. E., and M. Brierley, "High Saturation Output Power From a Neodymium-Doped Fluoride Fibre Amplifier Operating in the 1300 nm Telecommunications Window," *Electronics Letters*, Vol. 26, No. 12, 1990, pp. 819–820.

[15] Danielsen, P. L., "Analytical Expressions for Group Delay in the Far Field From an Optical Fiber Having an Arbitrary Index Profile," *IEEE Journal of Quantum Electronics*, Vol. 17, No. 6, 1981, pp. 850–853.

[16] Øbro, M. O., "Characterization of Fiber Amplifiers," Master's Thesis E453 (in Danish), Electromagnetics Institute, Technical University of Denmark, 1991.

[17] Davey, S. T., and P. W. France, "Rare Earth Doped Fluorozirconate Glasses for Fibre Devices," *British Telecom Technical Journal*, Vol. 7, 1989, pp. 58–68.

[18] Brierley, M. C., and M. Hunt, "Efficient Semiconductor Pumped Fluoride Fibre Lasers," *Proceedings of SPIE OE/FIBERS '89, Fiber Laser Sources and Amplifiers* Boston, Vol. 1171, paper 15, 1989, pp. 157–159.

[19] Dakss, M. L., and W. J. Miniscalco, "A Large Signal Model and Signal/Noise Ratio Analysis for Nd^{3+}-Doped Fiber Amplifiers at 1.3 μm," *Proceedings of SPIE OE/Fibers '90,* part: Fiber Laser Sources and Amplifiers II, San Jose, Vol. 1373, paper 11, 1990, pp. 111–124.

[20] Rasmussen, T., A. Bjarklev, O. Lumholt, M. Øbro, B. Pedersen, J. H. Povlsen, and K. Rottwitt, "Optimum Design of Nd-doped Fibre Optical Amplifiers," *IEEE Photonics Technology Letters*, Vol. 4, No. 1, 1992, pp. 49–51.

[21] Pedersen, J. E., M. Brierley, and R. A. Lobbett, "Noise Characterization of a Neodymium-Doped Fluoride Fiber Amplifier and Its Performance in a 2.4 Gb/s System," *IEEE Photonics Technology Letters*, Vol. 2, No. 10, 1990, pp. 750–752.

[22] Lumholt, O., M. Øbro, A. Bjarklev, T. Rasmussen, B. Pedersen, J. E. Pedersen, J. H. Povlsen, and K. Rottwitt, "Optimum Placement of Filters in 1300 nm Nd-Fibre Amplifiers," *Optics Communications*, Vol. 89, No. 1, 1992, pp. 30–32.

[23] Digonnet, M. J. F., "Closed-Formed Expressions for the Gain in Three- and Four-Level Laser Fibers," *IEEE Journal of Quantum Electronics*, Vol. 26, No. 10, 1990, pp. 1788–1796.

[24] Øbro, M., J. E. Pedersen, and M. C. Brierley, "Gain Enhancement in Nd^{3+} Doped ZBLAN Fibre Amplifier Using Mode Coupling Filter," *Electronics Letters*, Vol. 28, No. 1, 1992, pp. 99–100.

[25] Probst, C. B., A. Bjarklev, and S. B. Andreasen, "Experimental Verification of Microbending Theory Using Mode Coupling to Discrete Cladding Modes," *IEEE Journal of Lightwave Technology*, Vol. 7, No. 1, 1989, pp. 55–61.

[26] Bjarklev, A., and S. B. Andreasen, "Microbending Characterisation of Optical Fibres From Artificially Induced Deformation," *Electronics Letters*, Vol. 25, No. 6, 1989, pp. 417–419.

[27] Pedersen, B., A. Bjarklev, O. Lumholt, and J. H. Povlsen, "Detailed Design Analysis of Erbium Doped Fiber Amplifiers," *IEEE Photonics Technology Letters*, Vol. 3, No. 6, 1991, pp. 548–550.

[28] Percival, R. M., M. W. Phillips, D. C. Hanna, and A. C. Tropper, "Characterization of Spontaneous and Stimulated Emission From Praseodymium (Pr^{3+}) Ions Doped Into a Silica-Based Monomode Optical Fiber," *IEEE Journal of Quantum Electronics*, Vol. 25, No. 10, 1989, pp. 2119–2123.

[29] Pedersen, B., W. J. Miniscalco, and R. S. Quimby, "Optimization of Pr^{3+}-ZBLAN Fiber Amplifiers," *IEEE Photonics Technology Letters*, Vol. 4, No. 5, 1992, pp. 446–448.

[30] Quimby, R. S., and B. Zheng, Excited State Absorption in Pr^{3+}-Doped Fluorozirconate Glass," to be published.

[31] Miniscalco, W. J., and R. S. Quimby, "General Procedure for the Analysis of Er^{3+} Cross Sections," *Optics Letters*, Vol. 16, No. 4, 1991, pp. 258–260.

[32] Ohishi, Y., T. Kanamori, T. Kitagawa, S. Takahashi, E. Snitzer, and G. H. Sigel, "Pr^{3+}-Doped Fluoride Fiber Amplifier Operating at 1.3 μm," *Proc. of Conference of Optical Fiber Communication OFC '91* San Diego, 1991, postdeadline paper PD2.

[33] Karasek, M., "Numerical Analysis of Pr^{3+}-Doped Fluoride Fiber Amplifier," *IEEE Photonics Technology Letters* Vol. 4, No. 11, 1992, pp. 1266–1269.

[34] France, P. W., M. G. Drexhage, J. M. Parker, M. W. Moore, S. F. Carter, and J. V. Wright, "Fluoride Glass Optical Fibres," Blackie and Son, 1990.

[35] Pedersen, B., W. J. Miniscalco, S. Zemon, and R. S. Quimby, "Neodynium- and Praseodymium-Doped Fiber Power Amplifiers," *Techn. Dig. of Optical Amplifiers and Their Applications OAA '92*, Santa Fe, 24–26 June 1992, paper WB4, pp. 16–19.

[36] Newhouse, M. A., R. F. Bartholomew, B. G. Aitken, A. L. Sadd, and N. F. Borrelli, "309 μs Pr^{3+} Excited State Lifetime Observed in a Mixed Halide Glass," *Techn. Dig. of Optical Amplifiers and Their Applications OAA '92*, Santa Fe, 24–26 June 1992, postdeadline paper PD16.

[37] Ohishi, Y., T. Kanamori, and S. Takahashi, "Concentration Quenching in Pr^{3+}-Doped Fluoride Fiber Amplifier Operating at 1.3 μm," *Proc. of 17th European Conference on Optical Communication ECOC '91*, Paris, 9–12 Sept. 1991, paper MOA2-2, pp. 17–20.

[38] Carter, S. F., R. Wyatt, D. Szebesta, and S. T. Davey, "Quantum Efficiency and Amplification at 1.3 μm in a Pr^{3+} Doped Fluorozirconate Single-Mode-Fibre," *Proc. of 17th European Conference on Optical Communication ECOC '91*, Paris, Sept. 9–12, 1991, paper MOA2-3, pp. 21–24.

[39] Allain, J. Y., M. Monerie, and H. Poignant, "Energy Transfer in Pr^{3+}/Yb^{3+}-Doped Fluorozirconate Fibres," *Electronics Letters*, Vol. 27, No. 12, 1991, pp. 1012–1014.

[40] Ohishi, Y., T. Kanamori, T. Nishi, S. Takahashi, and E. Snitzer, "Gain Characteristics of Pr^{3+}-Yb^{3+} Codoped Fluoride Fiber for 1.3 μm Amplification," *IEEE Photonics Technology Letters* Vol. 3, No. 11, 1991, pp. 990–992.

[41] Miyajima, Y., "Progress in 1.3 μm Fluoride Fiber Amplifiers," *Techn. Dig. of Optical Amplifiers and Their Applications OAA '92*, Santa Fe, 24–26 June 1992, invited paper WB1, pp. 4–7.

[42] Miyajima, Y., T. Sugawa, and Y. Fukasaku, "38.2 dB Amplification at 1.31 μm and the Possibility of 0.98 μm Pumping in Pr^{3+}-Doped Fluoride Fibre," *Electronics Letters* Vol. 27, No. 19, 1991, pp. 1706–1707.

[43] Zarem, H., J. Ungar, M. Mittelstein, J. Paslaski, and I. Ury, "High-Power Single-Mode InGaAs Lasers for Pumping Praseodymium-Doped Fiber Amplifiers," *Proc. of Conference of Optical Fiber Communication OFC '92*, San Jose, 2–7 Feb. 1992, poster paper WL7, pp. 160.

[44] Shimuzu, M., M. Yamada, F. Hanawa, Y. Ohishi, J. Temmyo, M. Wada, Y. Terunuma, T. Kanamori, and S. Sudo, "1.3-μm Band Pr-Doped Fluoride Fiber Amplifier Module Pumped by Laser Diodes," *Techn. Dig. of Optical Amplifiers and Their Applications OAA '92* Santa Fe, 24–26 June 1992, postdeadline paper PD3.

[45] Whitley, T., R. Wyatt, D. Szebesta, and S. Davey, "High Output Power From an Efficient Praseodymium Doped Fluoride Fibre Amplifier," *Techn. Dig. of Optical Amplifiers and Their Applications OAA '92* Santa Fe, 24–26 June 1992, paper WB2, pp. 8–11.

[46] Whitley, T., R. Wyatt, D. Szebesta, S. Davey, and J. R. Williams, "Quarter Watt Output at 1.3 μm From a Praseodymium Doped Fluoride Fibre Amplifier Pumped With a Diode-Pumped Nd:YLF Laser," *Techn. Dig. of Optical Amplifiers and Their Applications OAA '92*, Santa Fe, 24–26 June 1992, postdeadline paper PD4.

[47] Sugawa, T., and Y. Miyajima, "Noise Characteristics of Pr^{3+}-Doped Fluoride Fibre Amplifier," *Electronics Letters*, Vol. 28, No. 3, 1992, pp. 246–247.

[48] Lobbett, R., R. Wyatt, P. Eardley, T. J. Whitley, P. Smyth, D. Szebesta, S. F. Carter, S. T. Davey, C. Millar, and M. C. Brierley, "System Characterization of High Gain and High Saturated Output Power, Pr^{3+}-Doped Fluorozirconate Fibre Amplifier at 1.3 μm," *Electronics Letters* Vol. 27, No. 16, 1991, pp. 1472–1474.

[49] Whitley, T., and R. Wyatt, "Analytic Expression for Gain in an Idealised 4-level Doped Fibre Amplifier," *Techn. Dig. of Optical Amplifiers and Their Applications OAA '92*, Santa Fe, 24–26 June 1992, paper WB2, pp. 20–23.

[50] Marcuse, D., "Loss Analysis of Single-Mode Fiber Splices," *The Bell System Technical Journal*, Vol. 56, No. 5, 1977, pp. 703–718.

Chapter 10
Fiber Amplifiers in Digital Direct Detection Systems

The development of optical-fiber amplifiers has been unique in the sense that it has taken only three years from the first laboratory demonstrations of semiconductor-pumped EDFAs to the commercial availability of these components. This very rapid development has been driven by a number of factors, but one of the principal reasons has been the large industrial interest and resulting effort invested in research and development. It was essential that the applications of the EDFAs in optical direct detection systems were obvious from the very beginning of the development process, and the results have completely fulfilled the expectations. This may, for instance, be seen from important system application demonstrations such as [1,2].

The fiber amplifier has meant that previous loss limitations on digital direct detection systems can be totally eliminated over even transoceanic distances [3], and it has opened up the possibility of new and more complex network structures. The subject of fiber amplifiers in digital direct detection systems is therefore one of the most important in the development of optical communication systems. For this reason, we will describe in this chapter the basic properties of EDFAs in direct detection systems.

However, for the completeness of this description, it should be mentioned that optical amplifiers have also found a use in coherent optical communication systems. Near-future implementation of such techniques in operating communication systems seems, on the other hand, not likely due to the extra complexity of coherent detection techniques. This book will therefore not discuss applications of fiber amplifiers in coherent optical communication systems, but merely point out that detailed information on this subject may be found in [4–6].

The chapter is organized as follows. In Section 10.1 we establish a theoretical basis necessary to evaluate the system consequences of optical amplifiers. From then on we concentrate on direct detection systems with only one amplifier included, and Section 10.2 describes the optimum placement of this. In Section 10.3 the description is focused on properties of systems with cascaded amplifiers, and problems such as mutual ASE buildup, automatic gain control, and gain limitations due to Rayleigh reflections are discussed. In Section 10.4 the possibility of remote pumping of fiber amplifiers is evaluated, and Section 10.5 presents the latest results on long-haul direct detection systems. Finally, some system-related subjects such as optical routing and optical time domain reflectometry measurements on systems with optical amplifiers are discussed.

10.1 THEORETICAL DESCRIPTION OF THE AMPLIFIERS IN DIRECT DETECTION SYSTEM APPLICATIONS

An obvious motivation for devoting time to a theoretical description of the system applications of fiber amplifiers in optical communication is the excessive potential of these in present and future optical systems. Furthermore, a description of the system consequences supplies a clearer understanding of the advantages and limitations of the application of optical amplifiers, since it brings together central parameters, such as gain and noise figure, in a more coherent discussion. Keeping these goals in mind, it is not in any way unnatural to limit the description to intensity-modulated direct detection systems, since these are clearly the most relevant for telecommunication applications for the next decade.

The focus will be on the amplifier applications in direct detection systems. One of the most central descriptions of system consequences of optical amplifiers is presented in [7], which, with [8] as the basis, combines amplifier parameters with receiver models and BER calculations. These fundamental descriptions will consequently also form a reference for the present presentation. In addition, useful knowledge about the detection process may be derived from [9]. This will also be used here to establish expressions describing the different noise contributions and their influence on the system performance.

10.1.1 Detection of Intensity-Modulated Signals

The process of optical signal amplification is stochastic (see Chapter 5), and the signal-dependent noise in the photodetector results from the randomness associated with the rate of arrival of the photons at the detector and the randomness of the release of photoelectrons. To establish the relationship between a given optical signal and the current from the photodetector, we will look at the photodetection process. In accordance with [9], the number of generated electrons, l, in a specific time interval of length T may be expressed by a Poisson probability distribution

$$p_l(l) = \frac{m^l}{l!} e^{-m} \quad \text{where} \quad m = \frac{\eta}{h\nu_s} \int_t^{t+T} |f(t)|^2 \, dt \tag{10.1}$$

Here, $f(t)$ denotes the time-dependent electrical field (normalized so that the unit is $W^{-1/2}$), and η is the quantum efficiency of the photodiode. Since $f(t)$ results from a stochastic process, the parameter m becomes a random variable with a probability density, which we will represent with $p_m(m)$. The counting probability for l, therefore, should more precisely be called a conditional Poisson probability, which can be expressed as

$$p_l(l) = \int_0^\infty \frac{m^l}{l!} e^{-m} \, p_m(m) \, dm \tag{10.2}$$

The next step in the derivation is now to find the relationship between the probability distribution function $p_l(l)$ and the photocurrent $i_p(t)$, and to do so, it is necessary to know the current response due to a single electron $h_e(t)$. For this function we have the relation

$$\int_{-\infty}^{\infty} h_e(t) \, dt = q \tag{10.3}$$

where q is the electron charge. The Fourier transform of $h_e(t)$ is the detector transfer function, which consequently defines the bandwidth of the detector. A detailed discussion of the response function $h_e(t)$ is given in [9], and we shall here only note that $h_e(t)$ cannot generally be handled like a deterministic function. As pointed out in [10], the stochastic nature of $h_e(t)$ must be taken into account when *avalanche photodiodes* (APD) are used for the detection. However, in the case where pin diodes are used, it is acceptable to ignore fluctuations in $h_e(t)$, since a simplified shape (step or δ-function) is chosen anyway under normal circumstances. We will, therefore, for the sake of simplicity and the fact that pin diodes are the preferred choice for optical amplifiers [7], limit the description to detectors with pin diodes. It should at this point be noted that APD receivers may still be a viable alternative to optical preamplifier receivers at low data rates [11].

The photocurrent may now be expressed as a summation over l pulses $h_e(t - t_i)$:

$$i_p(t) = \sum_{i=1}^{l(t,t+T)} h_e(t - t_i) \tag{10.4}$$

where $l(t, t + T)$ denotes the number of generated electrons in the time interval from t to $t + T$. The time variables, $t_1, t_2, \ldots, t_l$, are stochastic variables, each of which expresses the generation of an electron.

In the case of a conditional Poisson probability distribution, the frequency spectrum of $i_p(t)$ may be found, according to [9], as

$$S_i(f) = |H_e(f)|^2[\langle\overline{n}\rangle + \overline{n}_d + S_n(f)] \tag{10.5}$$

Here $H_e(f)$ is the detector transfer function. Furthermore, the relationship with the electrical field launched at the detector is represented through the function $n(t) = \eta|f(t)|^2/h\nu_s$. $\langle n\rangle$ is the static mean value of $n(t)$ and $\langle\overline{n}\rangle$ is the time averaged mean value of $\langle n\rangle$. $S_n(f)$ is the intensity spectrum of $n(t)$, and this spectrum normally contains portions due to the desired intensity and portions due to background noise effects. $\overline{n}_d$ is the dark current term, which takes into account the random emission of the electrons that appear at a fixed rate when no optical signal is being detected. This dark current term is added, as seen from (10.5), directly to the detector output, and therefore also to the shot-noise level represented by $|H_e(f)|^2\,\langle\overline{n}\rangle$. It should, however, be mentioned that in typical system operation, values of $\overline{n}_d$ are much less than those for $\langle\overline{n}\rangle$, and the dark current level can therefore often be ignored [9].

This description is general in the respect that the role of the fiber amplifier has not been shown so far. However, we now consider the case in which an EDFA is placed in the transmission line, as shown in Figure 10.1. From the amplifier output end, a signal power of $G\cdot P_s$ is transmitted, together with the total ASE power P^+_{ASE}. It must be noted that for the sake of simplicity, at this stage, we only consider one amplifier with an input signal power of P_s and a gain G. In the case of cascaded amplifiers, these values must be found on the basis of ASE accumulation (see Section 10.3). The output from the amplifier is attenuated by a factor L before detection. If we now assume that P^+_{ASE} is unpolarized, it will be possible to separate it into two orthogonal polarization directions (one identical with the signal polarization), each with a power of $P^+_{\text{ASE}}/2$.

If for a moment we ignore the dark current, it can then, according to the description of [7, 10], be shown that (10.5) may be written as follows:

$$\begin{aligned} S_i(f) = {} & A_d^2(GP_s + P^+_{\text{ASE}})^2\delta(f) \\ & + qA_d(GP_s + P^+_{\text{ASE}}) \\ & + A_d^2\left\{\frac{1}{2}\,GP_s[S^+_{\text{ASE}}(\nu - \nu_s) + S^+_{\text{ASE}}(\nu + \nu_s)]\right\} \\ & + A_d^2\left[\frac{1}{2}\,S^+_{\text{ASE}}(\nu) * S^+_{\text{ASE}}(\nu)\right] \end{aligned} \tag{10.6}$$

for $|f| < \nu_s$. Here * denotes the convolution operator and A_d is the responsivity

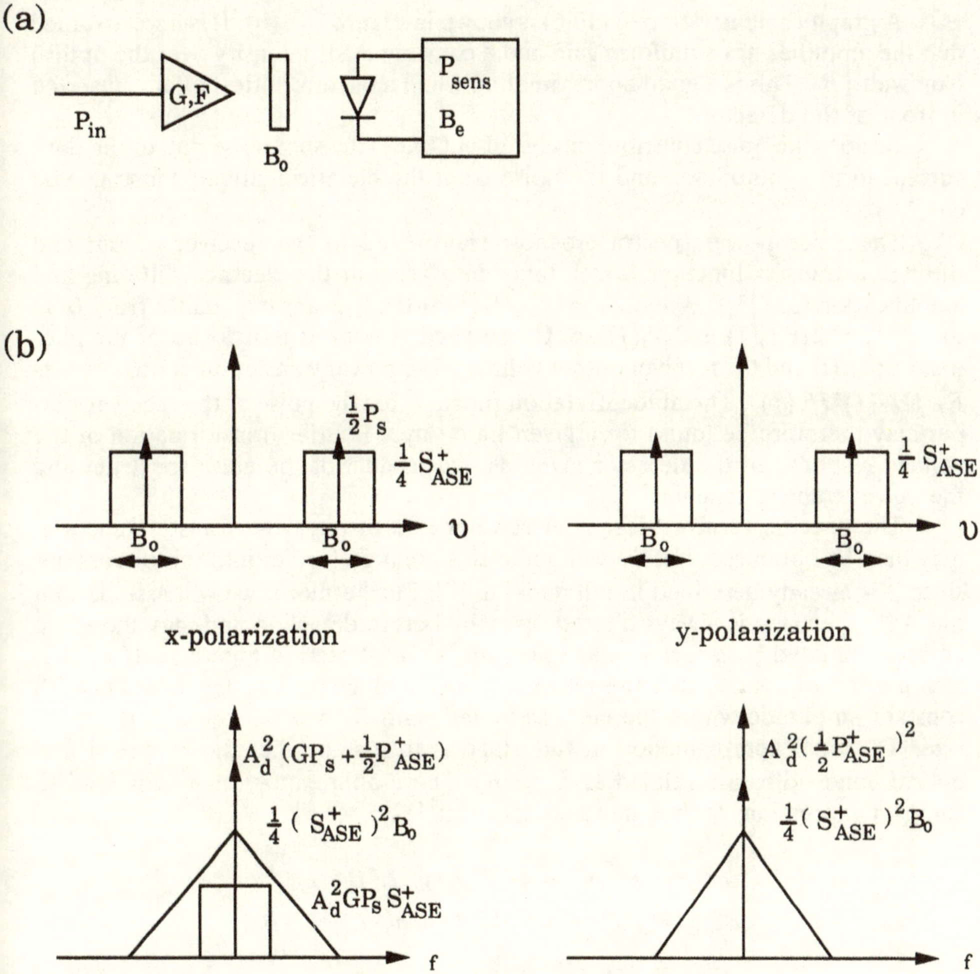

Figure 10.1 (a) Block diagram of amplifier. (b) Schematic presentation of the detection of an *x*-polarized signal and a rectangular ASE spectrum. The ASE is assumed equally distributed in the two orthogonal polarization directions *x* and *y* [10].

of the photodiode (see [9]). In (10.6), the first term on the right-hand side represents the detected signal and ASE, and the second term is the shot noise resulting from the detection of them. The third term takes into account the beating of signal and components of the ASE, and, finally, the fourth term describes the beating between the different frequency components of the ASE. Note here that (10.6) is derived under the assumption that the photon statistics of the ASE are Gaussian.

A graphical illustration of (10.6) is shown in Figure 10.1(b). It is here assumed that the amplifier has a uniform gain and a constant ASE intensity over the optical bandwidth B_0. This is a good approximation when a narrow optical filter is inserted in front of the detector.

Besides the noise contributions listed in (10.6), the shot noise due to the dark current in the photodiode and the noise from the electrical circuits must also be considered.

The noise power spectra are now transferred to the receiver output end through a transfer function, which takes into account the electrical filtering and amplification (see [8]). When $Z_T = R_T \cdot H_T(f)$ is the transfer impedance from $I_p(f)$ to $V_s(f)$, where $I_p(f)$ and $V_s(f)$ are the respective Fourier transforms of the photocurrent $i_p(t)$ and the receiver output voltage. The power transfer function becomes $R_T^2 \cdot (H_T(f)^* H_T(f))$. The autocorrelation function for the noise at the receiver output may therefore be found for a given time t by a Fourier transformation of the voltage response of the detector given as the product of the noise spectrum and the power transfer function.

The specific receiver design, which appears through the transfer function, may then be optimized, but we will not at this point go further into this discussion, since it is already described in full detail in [8]. Furthermore, we will assume that the ASE spectrum is always filtered optically before detection and may therefore be approximated by a rectangular spectrum, as illustrated in Figure 10.1(b). It is also normal to assume that the receiver transfer function is a step function with constant amplitude within the electrical bandwidth $B_e = B/2$, where B is the bit rate. The final approximation at this stage is to assume that the electrical and optical bandwidths are related as $B \ll B_0$. These approximations finally lead to compact expressions for the noise components [12]:

$$\begin{aligned}
\sigma^2_{s\text{-}sp} &= P_s \cdot (G - 1) \cdot F \cdot h \cdot \nu_s \cdot L \cdot B \\
\sigma^2_{sp\text{-}sp} &= (1/2)[(G - 1) \cdot F \cdot h \cdot \nu_s \cdot L]^2 \cdot B_0 \cdot B \\
\sigma^2_{sh,s} &= q \cdot P_s \cdot B \\
\sigma^2_{sh,sp} &= q \cdot (G - 1) \cdot F \cdot h \cdot \nu_s \cdot L \cdot B_0 \cdot B \\
\sigma^2_{crc} &= N_{th} \cdot B
\end{aligned} \tag{10.7}$$

Note that (10.7) also includes the parameter L, which is the attenuation between fiber amplifier and detector. Here the noise contributions are signal-spontaneous beat noise ($\sigma^2_{s\text{-}sp}$), spontaneous-spontaneous beat noise ($\sigma^2_{sp\text{-}sp}$), signal shot noise ($\sigma^2_{sh,s}$), amplified spontaneous emission shot noise ($\sigma^2_{sh,sp}$) and circuit noise (σ^2_{crc}) from the electrical part of the receiver. N_{th} is the electrical spectral density of the circuit noise.

In [13] it is shown that (10.7) may be used for a system operating at a bit rate of 2.5 Gbit/s as long as the optical filter has a bandwidth smaller than 5 nm. For filter bandwidths larger than this limit, the above listed approximations are not valid and (10.7) will overestimate the noise. The reason for this overestimation is that the ASE intensity distribution in practical situations may only be considered rectangular within a certain bandwidth, after which it starts to decrease. The spontaneous-spontaneous beat noise component will therefore not continue to increase linearly with the optical bandwidth such as that stated in (10.7).

For an amplitude-modulated signal of average power P_s, 50% duty cycle, and an extinction ratio of r, the photocurrent equivalents of the input powers for a mark ($I_s(1)$) and a space ($I_s(0)$) are

$$\begin{aligned} I_s(1) &= \frac{qP_s}{h\nu}\frac{2r}{(r+1)} \\ I_s(0) &= \frac{qP_s}{h\nu}\frac{2}{(r+1)} \end{aligned} \tag{10.8}$$

The BER may then be determined to be [7]

$$\text{BER} = \frac{1}{\sqrt{2}}\,\text{erfc}\left(\frac{Q}{\sqrt{2}}\right) \approx \frac{1}{\sqrt{2\pi}Q}\exp\left(-\frac{Q^2}{2}\right) \tag{10.9}$$

where

$$Q = \frac{S_1 - S_0}{\sigma_1 + \sigma_0} \tag{10.10}$$

$S_1 = GP_sI_s(1)$ and $S_0 = GP_sI_s(0)$ are the photocurrent in a mark and a space, respectively. σ_1 and σ_0 are the corresponding rms noise terms, which from the previously defined noise terms are given by

$$\sigma_i = \sqrt{\sigma^2_{s\text{-}sp} + \sigma^2_{sp\text{-}sp} + \sigma^2_{sh,s} + \sigma^2_{sh,sp} + \sigma^2_{crc}} \tag{10.11}$$

where the suffix i may equal either 0 or 1. A unity responsivity (or quantum efficiency η) of the photodetector has been assumed. The reader is also reminded that the noise figure F may be evaluated from the optical power spectral density of the forward amplified spontaneous emission $S^+_{ASE}(\nu_s)$ at the signal frequency ν_s, as shown in Section 5.1.4.

A normally applied assumption is to consider the extinction ratio r very large. This finally results in the simplification that $S_0 = 0$, which will be used in the following description.

10.1.2 Additional Properties Limiting the Performance of Optical Direct Detection Systems

We now have seen from the description in Section 10.1.1 how essential parameters such as receiver sensitivity and BER may be calculated based on the knowledge of amplifier gain and noise figure. This description is for most applications sufficient to evaluate the impact of fiber amplifiers on the performance of direct detection systems. However, since the amplifiers are used to overcome the loss limitation of the systems, very long transmission distances are realistic. In such cases, the importance of other properties may have to be reconsidered, and the dispersion is relevant here especially.

Note also that for a complete description of all the system elements, properties such as laser phase noise and chirp must be taken into account. These additional system properties are not directly related to the fiber amplifier parameters, and thorough descriptions have been given elsewhere. We will for this reason not include the calculations of the resulting system penalties here, but merely add a few comments. With regard to the consequences of laser phase noise, the reader should consult references such as [14] for an evaluation. In [14], closed-form expressions for the BER and receiver sensitivity are also derived based on the approximation that a Gaussian distribution may be used to describe the probability density function of the voltage at the decision gate.

In relation to the fiber dispersion, it is important to notice the difference between systems with either "standard" single-mode fibers (i.e., fibers with zero dispersion around 1310 nm) and systems with dispersion-shifted fibers. Standard single-mode fibers, which have typical dispersions around 17 ps/km/nm at 1550 nm, are normally found in already existing systems that are to be upgraded by optical amplifiers. In future long-distance systems, however, the choice of transmission fiber is likely to favor dispersion-shifted fibers that have intrinsic fiber losses, typically increased by 0.03 dB/km compared to standard single-mode fibers.

For a single-mode fiber, the total pulse broadening after transmission over the distance l_{tot} may be calculated as $\Delta\tau = D_s l_{tot} \delta\lambda$, where D_s is the fiber dispersion and $\delta\lambda$ is the linewidth of the laser. From [15] it is known that $\Delta\tau$ should be less than 30% of the time slot T to avoid degradation of the receiver sensitivity of more than 1 dB. Since the time slot is given by the bit rate B by the relation $T = 1/B$, the following requirement for the bit rate distance product may be established

$$l_{tot} \cdot B \leq \frac{0.3}{D_s \cdot \delta\lambda} \tag{10.12}$$

The laser linewidth may have values down to only a few megahertz (see, for example, [16]). For bit rates in the gigabit per second range, it is the modulation bandwidth ($\approx B$) that defines the efficient laser linewidth, so $\delta\lambda = \lambda^2 Bn/c$. Here c is the velocity of light in a vacuum and n is the effective refractive index of the fiber. Note that this value should only be considered as an indication of the magnitude of the effective laser linewidth, and that for detailed calculations we probably will arrive at a somewhat smaller $\delta\lambda$. However, the object of the present description is merely to indicate general limitations, and we will consequently not go into further detail at this point. If this expression for the laser linewidth is introduced in (10.12), the following relation appears:

$$l_{\text{tot}} \cdot B^2 \leq \frac{0.3c}{D_s \lambda^2 n} \tag{10.13}$$

By choosing parameters $\lambda = 1550$ nm, $n = 1.5$, and $c = 3 \cdot 10^8$ m/s, we find that the rule of thumb given in equation (10.13) expresses the following limitation: $l_{\text{tot}}B^2 \leq (2.5 \cdot 10^{19}\ \text{m}^{-1}\text{s}^{-1})/D_s$.

If we, for example, look at a system with a bit rate of 2.5 Gbit/s, this will mean that in the case of a standard singlemode fiber (with $D_s = 17$ ps/km/nm), we will estimate that $l_{\text{tot}} \leq 235$ km. If we, on the other hand, assume that the transmission fiber is dispersion-shifted so that $D_s(1550\ \text{nm}) = 0.1$ ps/km/nm, the maximum transmission distance becomes 40,000 km. We may therefore conclude that the dispersion requirement estimated by (10.13) is not a very restrictive one.

An equally severe problem is, however, represented by the chirp (or wavelength fluctuation) of the directly modulated injection laser diode, which provides the signal. This phenomenon has been investigated in [17], where the degradation in receiver sensitivity due to chirp has been studied for a 4-Gbit/s signal. The penalty (in decibels) has in [17] been found as a function of the dispersion of the total fiber length ($D_s l_{\text{tot}}$), and an approximation of this is as follows:

$$\text{Penalty(dB)} = 0.12 \cdot (D_s l_{\text{tot}}) + 0.44 \cdot (D_s l_{\text{tot}})^2 \tag{10.14}$$

where the product $D_s l_{\text{tot}}$ should be given in nanoseconds per nanometer. It must be stressed that the relation of (10.14) is found for a specific laser structure, and may therefore not be used in estimating penalties for other types of devices. However, it is included in the present form as an example of a realistic relation between chirp penalty, dispersion, and fiber length. If the penalty expressed in (10.14) should be kept below 1 dB, we then have to choose $D_s l_{\text{tot}} \leq 1.4$ ns/nm. This means that for a mean dispersion $D_s = 0.1$ ps/km/nm, the total transmission length for a 4-Gbit/s signal should be kept below 14,000 km. Compared with this, the dispersion requirement expressed in (10.13) gives the limit $l_{\text{tot}} \leq 15{,}800$ km.

We may therefore conclude that the dispersion and the chirp penalties may be comparable for realistic system parameters. In addition, we have also seen that these limits are not restrictive, as long as dispersion-shifted fibers and high quality signal lasers are used.

10.2 SYSTEMS WITH ONE ERBIUM-DOPED FIBER AMPLIFIER

Having established the basics of how to evaluate the EDFA in terms of system improvements, the different ways the EDFA can be used should now be considered. The interesting point is that depending on whether the EDFA is used as a power booster, an in-line amplifier, or a preamplifier, the gain and noise figure of the amplifier have different importance.

10.2.1 Optimum Placement of a Fiber-Optical Amplifier in High-Bit-Rate Direct Detection Systems

Basically, there are three system configurations that EDFAs can be used in: transmitter postamplifiers (normally also called power booster amplifiers), receiver preamplifiers, and in-line amplifiers. For the latter, it should further be noted that the possibility of either local or long-distance (remote) pumping exists. We will, however, at this point only consider the configurations in which erbium-doped fibers and pump laser diodes are placed together, and then return to the subject of remote pumping in Section 10.4.

The primary subject of this section will be comparisons between an EDFA placed either as a power booster, a preamplifier, or an in-line amplifier. To limit the length of the description, the pump wavelength of 1480 nm has been chosen, and as a measure of the system performance, the maximum bit rate is used. This maximum bit rate is found by increasing the bit rate for a given transmission distance and output power from the signal laser until a BER of 10^{-9} is reached. In the evaluations, we consider both co- and counter-propagating pumping, and the fiber lengths are optimized either with respect to maximum gain or by shortening the length so that maximum BER is obtained [18].

Our goal now is to describe how an optical amplifier acts as a system-improving component when it is located at different positions along the transmission line. The model used for calculating the BER of the system is the one described in Section 10.1.1, extended to include the exact spectral shape of the ASE instead of a band-limited uniform ASE spectrum [13]. These accurate ASE spectra have been calculated using the full model presented in Chapter 5 [19], and the absorption and emission spectra used in the present analysis have been determined experimentally as described in [20]. The signal wavelength is 1532 nm.

In the calculations, the average input power to the system P_{in} is 0.1 mW for a 1 and 0 mW for a 0. The total transmission distance is l_{tot} = 150 km, and both fiber couplers and filters result in signal power loss. For this description, the coupler has a coupling efficiency of 0.8. Where nothing else is mentioned, an optical filter of bandwidth 0.5 nm and coupling efficiency of 0.6 is used. The active fiber has a step-index profile with a core radius of 1.8 μm and NA of 0.21 [20]. Half duty cycle, rectangular pulses are assumed, and the receiver is assumed to convert these to raised cosine pulses as described in [8]. The base receiver sensitivity is given as $P_m = 5 \cdot \log_{10}(0.2655 \cdot 10^{-3}\ B^3)$, with the bit rate B in gigabits per second and the receiver input power P_m in decibels above 1 mW [13].

Figure 10.2 shows the maximum allowable bit rate versus the placement of the EDFA, calculated under the requirement that BER = 10^{-9}. The curves A and B are both found for an EDFA pumped co-directionally with a power of 20 mW. Curves C and D represent counter-propagating pump and signal, and the fiber lengths have been gain-optimized and BER-optimized, respectively.

The best result is found for an in-line amplifier placed in the middle of the transmission link and co-directionally pumped. This can be understood from the following considerations. The performance degrading factor in a booster (corre-

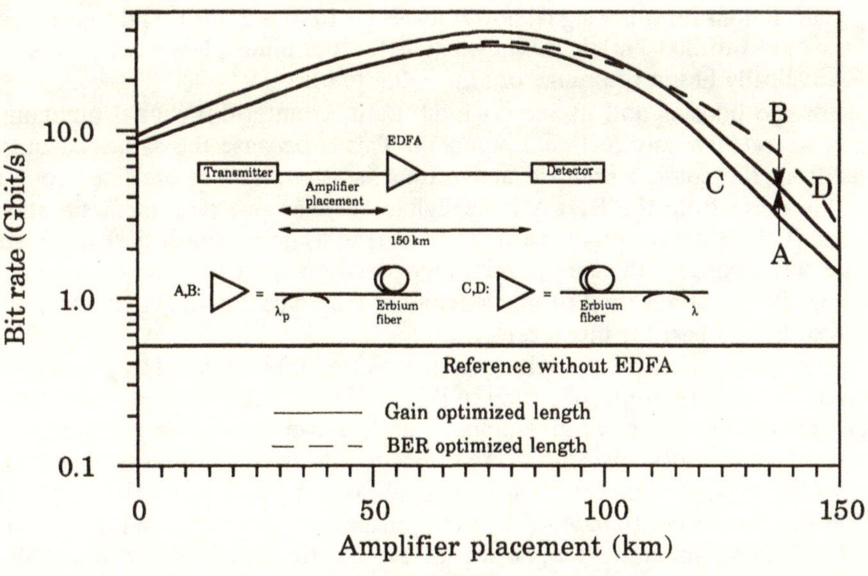

Figure 10.2 Maximum allowable bit rate versus placement of EDFA in the transmission link (P_p = 20 mW, P_{in} = −10 dBm). A,B: co-directional pumping at EDFA; C,D: counter-directional pumping at EDFA. (Data from [13].)

sponding to 0 km in the figure) is the saturation of the output power because of high input power. In the preamplifier (corresponding to 150 km), the main degrading factor is the beat noise between signal and ASE. However, for the in-line amplifier, the input power is low, so almost no saturation effect is present and the passive fiber between amplifier and receiver attenuates the noise components that originate from the amplifier. BER optimization of the active fiber length is only meaningful in the case of the amplifier noise being dominant.

From Figure 10.2 it can be seen that the co-propagating pump is better than the counter-propagating pump for gain-optimized lengths. For the preamplifier, this is due to the higher noise figure for the counter-propagating pump. For the booster, the difference is caused by the insertion loss in the coupler that combines the signal and pump power, since this attenuates the saturated output power in the counter-propagating case but not in the co-propagating case. For the co-propagating pump configuration, almost equal noise figures are obtained for the gain-optimized length and the BER-optimized length. Therefore, curve B has a course similar to that of curve A. For the counter-propagating pump, however, a low-noise figure similar to that for the co-propagating configuration may be achieved by BER-optimizing the length. Though equal noise figures are obtained in the erbium fiber, the coupler between the fiber and the receiver will lower the noise for the counter-propagating configuration, yielding a higher allowable bit rate in this case.

Calculations for other signal power levels (-10 to -2 dBm) and transmission distances (150 to 300 km) have shown that the optimum placement of a single EDFA typically lies in the range of l/l_{tot} = 0.4 to 0.6.

For the booster and in-line configurations, counter-directional pumping is slightly worse than co-directional pumping. This is because the saturated output power from the counter-pumped active fiber is attenuated in the fiber coupler. Since the noise from the EDFA is negligible in these two placement situations, gain and BER optimization of the active fiber lengths gives almost the same result. For the preamplifier with gain-optimized length co-pumping is better than counter-pumping. BER optimization of the active fiber length causes the counter-pumped EDFA to be the best for this receiver.

Additional results for how the system performance is affected by pump power and filter bandwidth are to be found in [13], and only a few central results are to be mentioned here. For the preamplifier, an increase of the pump power above 15 mW (G = 23.5 dB) gives no overall system improvement. This is because the dominant beat noise from the EDFA increases as fast as the gain. The in-line and booster performances are independent of optical filter bandwidth, since the amplifier noise is nondominant. Note, however, that the attenuation of the optical filters affects the performance of systems with boosters or in-line amplifiers, and degradations of up to 20% in maximum allowable bit rate is seen when a filter with coupling efficiency of 0.6 is used.

The filter may not, however, be omitted in the preamplifier configuration, since the performance is degraded up to 80% when the optical filter is removed. For the preamplifier, the insertion loss of the filter gives no penalty, since both the signal and the dominant noise are decreased by the same amount.

10.2.2 EDFAs for Multiwavelength Applications

The subject of multi-wavelength applications of optical amplifiers is not only one of the most diverse, but also one of the most promising ways to achieve even further extension of system capability. The reason is that multi-channel applications are relevant for both long-span digital transmission and digital and analog broadcasting in the subscriber loop. The strength of such applications is that in high-density WDM optical broadband distribution networks, multiple optical channels at different wavelengths can be used to carry a variety of broadband signals. Each of these wavelength channels may even apply different transmission bit rates or modulation formats. This is possible because of the long fluorescence lifetime in EDFAs, which leads to negligible gain-saturation-induced crosstalk [21].

The first high-density WDM system experiment using an EDFA for simultaneous amplification of multiple optical channels was reported in [22,23], where it was demonstrated that 16 optical channels could be distributed over the entire gain bandwidth of the EDFA. As pointed out in [24], this illustrates the important impact EDFAs are likely to have on future broadband distribution networks. The EDFA will, namely, allow service segregation with flexible service evolution, so new broadband services may be added without affecting the existing services. This is, however, assuming that the network is planned properly. It should also be noted that if a mixture of analog and digital channels is used, there will be optical power margin differences for the different optical channels, which need to be taken into consideration when optimizing the individual power budget and sharing factor.

The majority of applications of optical amplifiers in multi-channel WDM systems are yet to be seen, and we will therefore not go into further detail at this point. However, for further details, a very good review of the current status may be found in [24]. A number of important subjects for future research are also pointed out here. These areas follow include flattening of the gain spectrum, gain spectrum imbalance due to gain competition in gain-saturated EDFAs, spectral noise figure dependence on pump and signal powers, and possible methods for widening the gain spectrum.

10.3 SYSTEMS WITH CASCADE COUPLED AMPLIFIERS

In the past few years, the EDFAs have emerged from laboratory test models to become commercially available systems with either booster, in-line, or preamplifiers. A natural next step in the increase of the distance between transmitter and detector is to cascade coupled EDFAs to longhaul multi-repeater optical amplifier systems. However, in such systems, difficulties arise as the ASE accumulates. This will either result in amplifier saturation and high ASE power at the detector [25], or additional optical components might be necessary to limit the mutual influence of ASE generated in one amplifier and enlarged by the gain of the following.

The larger complexity of systems with more than one amplifier may also be seen from other technical aspects, one of which is the system sensitivity to polarization variations. These are not introduced by the erbium-doped fiber itself, since it is polarization-insensitive, but by polarization-sensitive couplers and other additional components. Similarly, variations in the input signal power or variations in the launched pump power, caused by temperature variations or laser degradation may have a significant influence on the detected signal after amplification in the cascade [26]. Therefore, some types of AGC loops have to be incorporated to ensure system reliability [26].

Due to all these extra complexities, it is a natural wish to attempt to limit the number of amplifiers in the cascade, which tends towards amplifiers with large gain values. However, as we shall see in Section 10.3.3, an upper practical gain limit will be set by reflections and Rayleigh backscattering. Moreover, it must be considered that amplifiers operating with some reasonable degree of saturation are much less sensitive to input power changes than high-gain amplifiers. Therefore, it may not always be optimal to reduce the number of amplifiers as much as possible.

10.3.1 Evolution of ASE Along the Transmission Link

To illustrate some of the basic properties of systems with cascaded fiber amplifiers, we will use as a point of reference the description in [27]. More specifically, the stability of an EDFA cascade with respect to signal and pump power variations are investigated at a 520-km-long transmission line. This length is chosen because it covers the longest span between two Atlantic islands (Iceland and the Faroe Islands) in a line from the U.S. to Europe. Furthermore, the example represents a distance long enough to demonstrate the fundamental properties of cascaded amplifier chains.

Let us first point out the principal differences between the approach of [27] and those presented in previous models for a system of EDFAs such as those presented in [7,25]. Compared with these, the accurate model of [27] takes into account not only the forward-traveling ASE from the neighboring amplifiers (as in [25]), but also the backward-traveling ASE at the determination of the properties of each amplifier, as well as the overall system performance. Furthermore, the self saturation of each amplifier caused by the ASE produced within the amplifier itself is also included in this model. That part, which concerns each single amplifier without the ASE influence from the neighboring EDFAs, was reported earlier and was verified [19]. In addition, the model takes into account both insertion loss and isolation for a number of isolators, as well as the insertion loss for each WDM coupler between the amplifiers.

To see how a realistic ASE buildup takes place within the transmission line, we will now choose the following system parameters. Each amplifier consists of a length of erbium doped fiber with a dopant concentration of $5 \cdot 10^{24}$ m^{-3} at a confinement of 0.8, a core diameter of 2.4 μm, and a refractive index difference of 0.03. These data are chosen as realistic values in agreement with the optimum design proposal [28]. Each erbium-doped fiber is pumped with both forward- and backward-traveling pump power at 1480 nm (realistic values of 25 mW for each in-line amplifier pump laser and 45 mW for each booster pump laser have been chosen [29]) to ensure system reliability and to allow for a local control loop between the pump sources. An isolator is placed on the output end of each EDFA to minimize the backward-traveling ASE. An insertion loss of 1 dB and an isolation of 40 dB is used for these isolators, and, in addition, a 1-dB insertion loss is assumed for each of the couplers. The input power of the booster is 0 dBm and the input power for the in-line amplifiers is higher than -16.5 dBm in order to limit the buildup of ASE [29]. The bit rate is 2.5 Gbit/s.

As can be seen in Figure 10.3(a), which illustrates the system setup, the detector was located after a transmission distance of 520 km, at which the signal level was lowered to -25 dBm. It is assumed that the average zero dispersion wavelength of the transmission fiber is identical with the signal wavelength at 1550 nm and that the average loss of the fiber, including splicing and connecting, was 0.225 dB/km. To reduce the ASE power and to separate the signal light from eventually remaining pump light, a filter was placed on the output end of the transmission line.

The gradual ASE buildup in the cascade system is illustrated in Figure 10.3(b), in which the ASE spectrum on the output end of each amplifier is shown. As can be seen from the figure, the ASE mainly builds up at the emission peak at 1530 nm. The 3-dB bandwidth narrows from 8.5 nm after the booster and 6.0 nm after the first in-line amplifier (EDFA2) to 3.5 nm after EDFA4. An increase of 11 dB is observed between the second and the fourth amplifier at 1530 nm, while only a 2-dB ASE increase is observed at 1550 nm.

10.3.2 Automatic Gain Control in Cascade EDFA Systems

To operate a practical cascaded system of optical amplifiers in practice, it is necessary to perform some type of AGC. In agreement with [27], two different control systems are therefore examined. These are, respectively, a pump-to-pump and a signal-to-pump regulation scheme. The influence of pump laser degradation on the

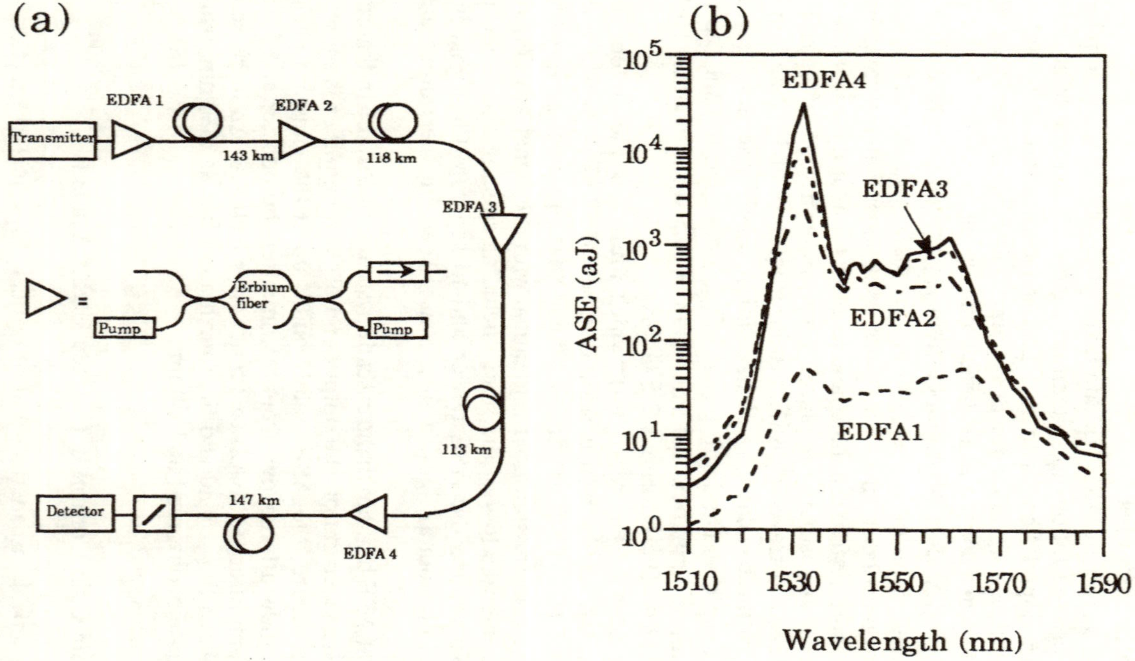

Figure 10.3 (a) Transmission line setup. (b) ASE in aJ versus the wavelength in nanometers on the output end of each amplifier. The signal wavelength is 1550 nm. (Data from [27].)

BER shows a large amount of dissimilarity for the two systems and it is clear how necessary a control system is.

To reduce the system sensitivity to pump laser degradation, an AGC loop is assumed for each of the amplifiers in the cascade. Both the pump power and the signal output power from the EDFA are measured in the loop, and the signal level will be adjusted if a change in the launched pump power is observed. As a relative measurement of the pump laser fallout, the pump laser performance is defined as the fraction between the launched pump power from a given laser and the amount of launched pump power, which was originally designed for the laser (25 mW for an in-line EDFA and 45 mW for the booster). Hence, a pump laser performance between 0 and 1 corresponds to variations in the launched pump power between 0 and 45 mW for the booster and variations between 0 and 25 mW for the in-line amplifier pumps. How critical the overall system BER is to the forward-pumping laser fallout for each EDFA is illustrated in Figure 10.4. The BER is shown versus variations in the laser performance between 0 and 1, and only one laser performance is changed at the time (i.e., only the pump powers launched into one amplifier at a time are changed whereas the other amplifiers are operated with constant pumping). The other amplifiers are therefore not used to compensate for change in one specific amplifier.

The EDFAs in the system are seen to have a much different influence on the BER, since the booster (EDFA1) and the last in-line amplifiers (EDFA4) cause only minor changes while the system acts with even a large BER increase for pump power changes to the first in-line amplifier (EDFA2). These differences are indirectly caused by the unfavorable ASE buildup in the system.

The ASE generation in the amplifier is strongly dependent on the inversion in the signal input end of the active fiber. Since the inversion in this end decreases simultaneously with the shift to backward pumping, the ASE will increase. However, since the ASE mainly builds up around the emission peak at 1530 nm, only minor changes will be observed within the optical detector bandwidth at 1550 nm. The latter may be seen from Figure 10.4, since for the last EDFA in the cascade, a constant output power corresponds to a constant signal power at the detector. The negligible changes in the BER therefore reflect the direct influence of the increased ASE. When the pump degradation occurs for one of the other in-line amplifiers, the larger amount of ASE will furthermore increase the saturation of the successive amplifiers. This will reveal gain reduction and thereby a gradual lowering of the input signal power for the following amplifiers in the cascade. However, for a booster amplifier, the highly suppressed ASE will not have an influence on the other amplifiers after the reduction in the passive fiber between the booster and EDFA2.

For the first in-line amplifier (EDFA 2), variations in the overall system BER is shown in the upper right corner of Figure 10.4 for variations in, respectively, the forward and the backward pump laser performance between 0 and 1. A reduc-

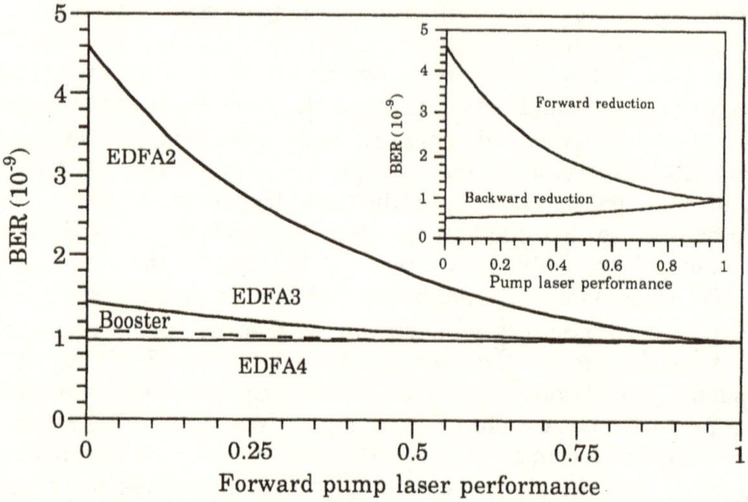

Figure 10.4 BER versus the forward pump laser performance. The performance for each laser is defined as the ratio between pump power and the original designed pump power. Inset: BER versus pump laser performance for pump lasers in EDFA2. (Data from [27].)

tion in the backward pumping, and thereby an increase in forward pumping power, produces a higher inversion in the signal input end of the fiber, yielding ASE reduction. In this case, the successive amplifiers will be less saturated and thereby produce a higher gain. In all, higher signal power and less ASE reach the detector, yielding a minor BER decrease.

Instead of a local AGC loop, gain degradation caused by pump power changes may be compensated for by simultaneous variations of the overall system signal input power. This means that the pump power degradation is compensated for by increasing the transmitted signal power. In order to illustrate this principle, we focus on EDFA2, which was identified as the most critical in the previous examples. The cascade system sensitivity to pump power changes to EDFA2 is illustrated in Figure 10.5(a). Fixed values of the BER between 10^{-10} and 10^{-2} are shown as contour lines for variations in the signal input power between -14 and $+10$ dBm, together with variations in the pump laser performance between 0 and 1.4.

The worst case is considered, which, as shown above, arises with changes in the forward-pumping laser in the first in-line amplifier (EDFA2). For a launched pump power decrease of 50%, it is possible to reestablish the BER at the detector by increasing the transmitted signal power by 5.5 dB. However, when the pump laser is almost turned off (performance ≈ 0), it will not be possible to regulate the BER by the input signal alone. For a fixed signal level at 0 dBm, the BER is 10^{-6}

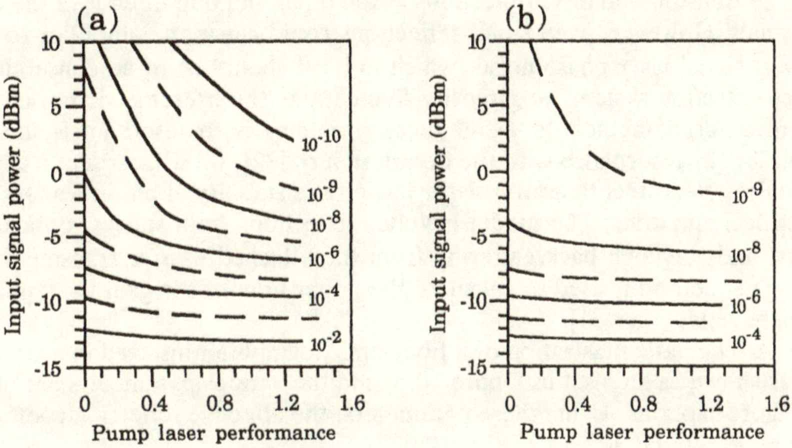

Figure 10.5 (a) Input signal power in decibels over 1 mW versus the forward pump laser performance for EDFA2 for a system with four amplifiers. The BER is represented by contour lines. (b) The only difference from (a) of the figure is that a total of five amplifiers is used in the system. (Data from [27].)

(compared to approximately $5 \cdot 10^{-9}$ by using the AGC loop) when the laser performance is zero.

The BER sensitivity to pump laser degradation may, however, be reduced remarkably by the insertion of an additional in-line amplifier in the system. Assuming unchanged overall system signal input and output powers and unchanged transmission length, the extra EDFA allows for shorter amplifier spacing. This results in a higher average signal input power from the EDFA, with saturation and a more insensitive response to pump power changes as a consequence. Figure 10.5(b) shows the input signal power versus forward pump laser performance for EDFA2 when a total of five amplifiers are used in the transmission link. A minor decrease in BER from $1 \cdot 10^{-9}$ to $0.8 \cdot 10^{-9}$ is obtained as a result of the more suppressed ASE for increased signal input power at the in-line EDFAs, and, as expected, a much lower dependence on pump laser variations is observed. A change of 50% in pump power to EDFA2 will only need a change of 2.5 dB in the signal input power.

10.3.3 Simple Stability Predictions for Systems With Optical Amplifiers

Having examined some of the problems of ASE accumulation in cascaded fiber amplifiers and the special need for AGC, we will now look at a third fundamental property of such systems. What is in question is the limitations given by backreflection of light into the amplifier module. Since the amplifier is spliced directly into the transmission link, reflections at the input and output ends of the amplifier are small. However, even small reflections could cause the amplifier to lase and convert signal laser phase noise (which in itself should be of no consequence in a direct detection system) to intensity fluctuations (interferometric noise) [30, 31]. An often used method to avoid these problems is to insert an isolator in the amplifier. In accordance with the description of [32], we will present a simple way to analyze the effect that an isolator has on the stability of an optical system with an optical amplifier. The model includes reflections from splices, optical components, and Rayleigh backscattering from the attached passive transmission fibers [33]. The method is used to calculate the upper limit of the gain for typical values of these reflections [31].

A schematic illustration of a fiber-optical amplifier inserted in a transmission link is shown as an inset in Figure 10.6, and the various parameters are defined in the figure captions. Using these parameters, the effective reflection coefficient R_{eff} may be expressed as

$$\begin{aligned} R_{\text{eff}}^2 = {} & [R_c + \eta_c^2(1 - R_c)^2 R_{\text{ray}}] \\ & \times [R_I + \eta_I^+ \eta_I^- (1 - R_I)^2 (R_s + \eta_s^2(1 - R_s)^2 R_{\text{ray}})] \end{aligned} \tag{10.15}$$

The important figure of merit for stability considerations is $x = GR_{\text{eff}}$, where G is the internal gain of the amplifier. To avoid lasing, x must be less than 1, but to ensure a low power penalty from interferometric noise, the demand on x may be far stricter (e.g., $x < 0.05$ for a power penalty of less than 1 dB [31]). The maximum allowable gain may be defined as $G_{\max} = x/R_{\text{eff}}$.

As described in [33], the Rayleigh reflection is around −30 dB, which is used in the following. The splices are assumed to have reflections of −25 dB and coupling efficiencies of 0.2 dB. Using a dichroic coupler with a reflection level of 17 dB, coupling efficiency of −1 dB, and no isolator, $G_{\max}$ may be calculated to be 7.4 dB. In Figure 10.6, $G_{\max}$ is shown as a function of the isolation of an isolator placed as shown in the inset, with the isolator reflection as a parameter. The isolator is assumed to have an insertion loss of 4 dB in the forward direction. The discrepancy between the 0-dB isolation case (9.4 dB in the figure) and the 7.4-dB previously mentioned is due to this insertion loss. Note that the specific choice of parameter values is very strict in this example, which results in relatively low maximum gain values in the case without isolation. The reason for this choice is to illustrate that even a poorly operating system may be brought to a reasonable quality level by the introduction of an isolator. Additionally, the principal idea of isolation is illustrated.

As can be seen from Figure 10.6, a maximum obtainable value of $G_{\max}$ exists for a given isolator reflection. This maximum value of $G_{\max}$ is determined by R_I and R_c as $G_{\max}$ $x/\sqrt{R_I R_c}$. For instance, if an isolator with a reflection of −40 dB

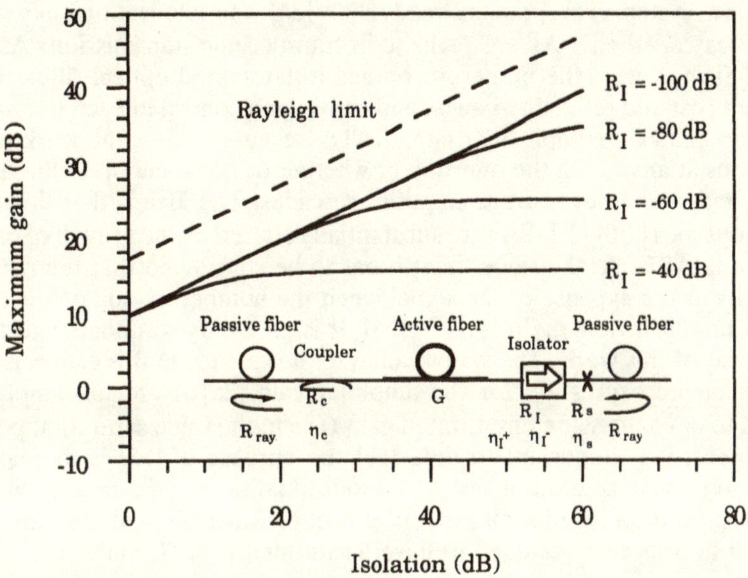

Figure 10.6 Maximum allowable gain versus isolation, calculated with $x = 0.05$. $R_{ray} = -30$ dB, $R_c = -17$ dB, $\eta_c = -1$ dB, $\eta_I^+ = -4$ dB, $R_s = -25$ dB, $\eta_s = -0.2$ dB. (Data from [32].) Inset: Fiber amplifier spliced into an optical transmission link. The significance of the parameters is as follows: R_{ray}: Rayleigh reflection coefficient, R_c: coupler reflection coefficient, η_c: coupler efficiency, G: internal amplifier gain, R_I: isolator reflection, η_I^+: isolator forward efficiency, η_I^-: isolation, R_s: splice reflection coefficient, η_s: splice coupling efficiency.

is used, a maximum gain of 15.4 dB can be obtained with an isolation of 22 dB. If a higher value of gain is desired, one has to choose an isolator with a lower reflection and a higher isolation. Choosing $R_I = -60$ dB and an isolation of 40 dB would increase the maximum gain to 25.3 dB. The dotted line in Figure 10.6 corresponds to a calculation with pure Rayleigh reflections and an ideal isolator, which represents the upper limit for G_{max}. It can be seen that the penalty from using nonideal components is 7.6 dB.

In conclusion, a simple method for calculating the upper limit of the gain, determined by the reflections at both ends of an optical amplifier, has been presented. Using this method for a highly conservative choice of system parameters, it is shown that the maximum allowable gain is around 7.4 dB when no isolators are used, and up to 40 dB, depending on the isolator reflection, when an isolator is used.

These results on the limitations of amplifier gain due to Rayleigh scattering implies that isolators are essential as long as gains above 7.4 dB are considered.

However, as system experiments already show [34], communication links with more than 250 cascaded EDFAs are realistic in transoceanic transmission. As pointed out in [35], however, the numerous optical isolators and optical filters will raise significant cost and reliability issues, and active gain control in each fiber amplifier will increase system complexity dramatically. Because of this, the work presented in [34] aims at answering the question of whether or not some or all the filters and isolators may be removed from amplifier cascades. The idea is that the self-regulating properties of the EDFAs are substantially altered by the growth of backward-propagating ASE. At the same time, it has to be considered that the noise grows necessarily at the expense of the signal when the number of isolators is reduced.

To briefly review the results of [35], it is necessary to understand the theoretical basis of this work. The way to analyze the cascade in this case was to apply an approximate expression for the amplifier gain [25]. Here, all amplifiers are assumed to have the same unsaturated gain G_0 and the same saturation power P_{sat}. Furthermore, it is important to note that the purpose of [35] is to evaluate the degradation due to saturation and ASE-induced noise, and limitations due to fiber nonlinearities, dispersion, multiple reflections, or Rayleigh scattering are omitted. Amplifier no.i. is then assumed to have a saturated gain G_i, satisfying

$$G_i = G_0 \exp\left((1 - G_i)\frac{P_{in,i}}{P_{sat}}\right) \tag{10.16}$$

where $P_{in,i}$ is the total input power to amplifier i, including contributions from both signal and ASE.

When isolators are installed in the amplifier chain, the propagating signal and ASE power satisfy the recursive equations [25]

$$P_{s,i} = LG_iP_{s,i-1} \tag{10.17}$$

$$P^+_{ASE,i} = LG_iP^+_{ASE,i-1} + 2n_{sp}(G_i - 1)h\nu_sB_o \tag{10.18}$$

where $P_{s,i}$ is the output signal power at the ith amplifier, and $P^+_{ASE,i}$ is the forward-traveling ASE at its output. B_o is the optical bandwidth of the amplifier, L is the loss between each amplifier, and n_{sp} is the spontaneous-emission factor, which is related to the noise figure as $F = 2n_{sp}$ [25]. In the present analysis, n_{sp} is assumed to have a constant value of 2 for each amplifier. To analyze the behavior of systems without isolators, a backward-traveling ASE wave is included and given by

$$P^-_{ASE,i} = LG_iP^-_{ASE,i-1} + 2n_{sp}(G_i - 1)h\nu_sB_o \tag{10.19}$$

The inclusion of (10.19) physically implies that the gain of each amplifier depends not only on all previous, but also all successive stages.

Based on the simplified model described in (10.16) to (10.19), it was found in [35] that in systems containing isolators, the signal power steadily declines as the forward-propagating ASE grows. On the other hand, when the isolators are removed, this no longer occurs, since the total ASE power incident on a given amplifier is nearly constant, independent of its position on the chain. Consequently, the saturated gains of all amplifiers become nearly the same, and the signal will propagate with nearly constant power. For the specific example of an unsaturated gain $G_0 = 13$ dB, it was found that the signal-to-noise ratio at the receiver was degraded by approximately 6 dB due to the removal of isolators. If such a degradation of the signal-to-noise ratio is acceptable in the actual case, one obtains the advantages of cost and complexity improvement in the system.

In conclusion, we may therefore say that for cascades of low-gain ($\approx$10 dB) amplifiers, the need for isolators will be governed by reflection effects and not by saturation and ASE-induced degradations. This is even more noteworthy because the backward-propagating ASE in a system without isolators results in a self-regulating system whose signal power is held more nearly constant than in systems with isolators.

10.4 REMOTE PUMPING OF FIBER AMPLIFIERS

As shown in Section 10.2, a promising application of EDFAs is as in-line amplifiers. However, for submarine cable systems [36], the power supply may be a problem, and reliability demands to the pump sources are very strict. To overcome this problem, feeding 1480-nm pump light to an in-line fiber amplifier through the transmission line [37] has been proposed. The purpose of this section is to investigate this remote pumping approach in detail and demonstrate that we not only solve the listed problems, but actually obtain additional system advantages. Compared to the more obvious preamplifier application, we find that the beat-noise components, which are dominant [7] when the amplifier is placed close to the detector, will decrease proportionally to the square of the loss between the in-line amplifier and the detector. For a fixed total transmission length and output power from the signal laser, this means that the signal-to-noise ratio will increase with increasing distance to the amplifier, provided that enough pump power reaches the amplifier.

10.4.1 Transmission Length Improvement by Remote Pumping

We will first start by giving a more quantitative description of the advantages of remote pumping, and for this purpose we refer to [37]. The inset of Figure 10.7

illustrates this idea, where for a fixed pump power we add a fiber with length L between the detector and active fiber. To maintain a BER of 10^{-9}, the signal input power to the amplifier must be increased, which corresponds to reducing the original transmission distance by L_s. A measure of the system improvement is therefore the difference $(L - L_s)$. Note here that the counter-propagating pump configuration normally gives the best performance. The reason is that the noise figure improvement for co-propagating pumping is not sufficient to outbalance the power loss in the coupler when this is placed in front of the amplifier.

To quantify the advantages of remote pumping, the performance of a 5-Gbit/s direct detection system is evaluated using an accurate fiber amplifier model [20] and the receiver model described in Section 10.1. The fiber loss is 0.25 dB/km at 1480 nm, and at the signal wavelength, 1535 nm, the attenuation is 0.20 dB/km. Active fibers with normalized frequency $V = 1.6$ are considered [38].

The gain curve shape indicates if system improvements may be achieved by moving the active fiber away from the detector. To demonstrate this, typical gain

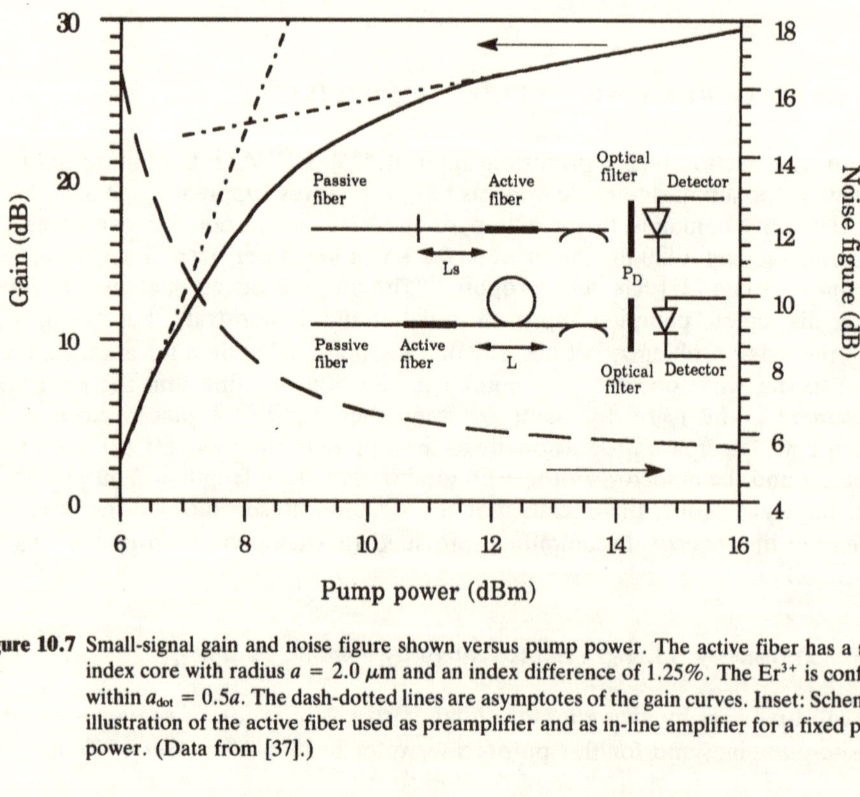

Figure 10.7 Small-signal gain and noise figure shown versus pump power. The active fiber has a step-index core with radius $a = 2.0\ \mu m$ and an index difference of 1.25%. The Er^{3+} is confined within $a_{dot} = 0.5a$. The dash-dotted lines are asymptotes of the gain curves. Inset: Schematic illustration of the active fiber used as preamplifier and as in-line amplifier for a fixed pump power. (Data from [37].)

and noise figure curves are shown in Figure 10.7 versus decreasing pump power. For pump power levels above 12 dBm, the small-signal gain decreases slowly with decreasing pump power, and since the noise figure also varies slowly, system improvements may therefore be expected for in-line amplifiers placed within 20 km of the detector. As the pump power is further decreased, the gain starts to drop off much faster. This defines a lower limit for the pump power, under which no system improvement can be found.

The transmission distance improvement, $(L - L_s)$, is shown in Figure 10.8(a). For the example, we find that the transmission distance may be increased by up to 16 km only by moving the active fiber 25 km away from the detector. If the distance L is further increased, the system improvement decreases and will eventually disappear, corresponding to section two on the gain curve (Figure 10.7).

An even clearer picture may appear by looking at the noise contributions for the remote pumped system. These noise contributions are shown in Figure 10.8(b) versus the pumping distance from the receiver to the remote pumped in-line amplifier. The input power to the detector is adjusted to ensure a BER of 10^{-9} for all

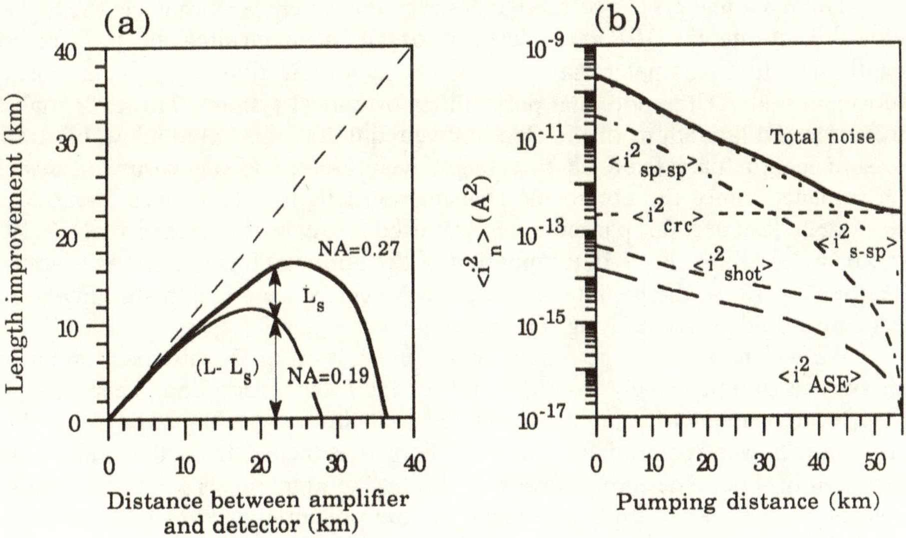

Figure 10.8 (a) Transmission length improvement shown as a function of the distance L between amplifier and detector. Both active fibers have a confined core $a_{dot} = 0.5a$. (Data from [37].) (b) The noise components as a function of the pumping distance in kilometers to the in-line remote pumped amplifier. The components are signal-spontaneous beat noise, spontaneous-spontaneous beat noise, signal shot noise, amplified spontaneous emission shot noise, and circuit noise. The solid line illustrates the total noise. (Data from [39].)

pumping distances. For small distances, the signal-spontaneous beat noise dominates the total noise, while for larger distances the circuit noise becomes dominant.

10.4.2 Transmission Lines Pumped Solely from the Ends

The approach of remote pumping fiber amplifiers as explained in the previous section will be especially attractive for submarine cable systems, where the power supply may be a problem and reliability demands to the pump sources must be very strict. As the figures have already indicated, remote pumping may not give rise to very long length improvements, but the approach is still of relevance since length increases of 20 to 40 km are also highly relevant for submarine systems. To give an idea of the limitations of this kind of systems, we will now focus on transmission links that are pumped solely from the ends. However, in contrast to the discussion of Section 10.4.1, we will consider totally land-pumped transmission systems, using a power booster, a remote pumped in-line amplifier, and a preamplifier. Another way of avoiding the in-line placement of the pump source is by using a distributed fiber, and we will in the following also compare such an amplifier with the other land-pumped configurations.

Three parameters can be used to describe the system performance: the highest allowable bit rate, the BER at the detector, or the maximum allowable transmission length. The first parameter may, as we have seen in Section 10.2, give important information about the potential possibilities of optical systems. However, due to present speed limitations of the electronic circuits, the maximum allowable transmission length for a fixed bit rate might be preferred as a measure of system performance. Since the obtainable transmission length gives a direct measure to the system designer, this parameter will be used for realistic values of BER $= 10^{-9}$ and $B = 2.5$ Gbit/s. For a determination of the optimum location for an amplifier (see Section 10.2), the bit rate was used, however, since a length specification in this case would be a confusing relative value.

We will now aim at getting a clearer impression of the distance limitations for systems pumped solely from the ends of the transmission line. Therefore, we have, in agreement with [39], investigated four different configurations in which the pump power required for the amplifiers is launched from the ends of the transmission line. The pump power to the in-line amplifier is kept at a realistic level of 50 mW, while the pump power to the preamplifier is 25 mW, since calculations with higher pump powers show no further sensitivity improvement. Finally, the pump power for the booster is 20 mW, yielding a gain of 12 dB for a realistic input signal power of 0 dBm. The electrical circuit is specified by a base receiver noise of 10 pA/$\sqrt{\text{Hz}}$.

The four different configurations are shown in Figure 10.9. In configuration A, a single in-line amplifier is assumed, but the pump power is launched into the

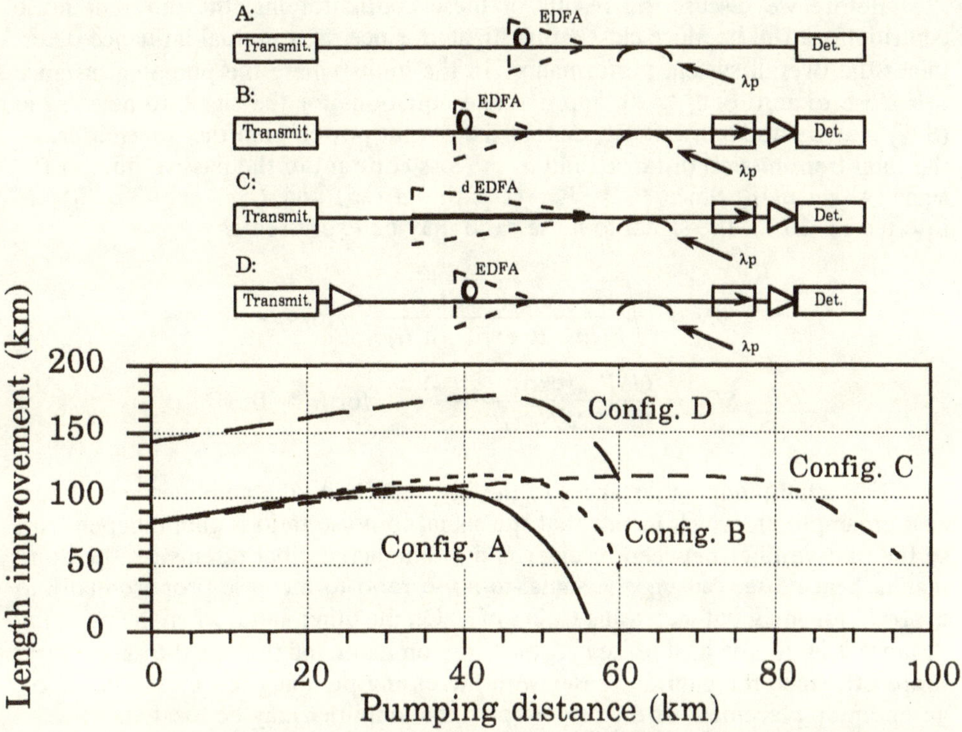

Figure 10.9 Upper part: different amplifier configurations. A: in-line remote pumped EDFA; B: in-line remote pumped EDFA and a preamplifier; C: distributed EDFA and a preamplifier; D: booster, in-line remote pumped EDFA, and a preamplifier. Lower part: transmission length improvement shown as a function of the pumping distance between amplifier and detector for the configurations. (Data from [39].)

end of the transmission line and guided through a distance of passive fiber before it reaches the doped area. Configuration B illustrates a corresponding configuration, where a preamplifier is also inserted immediately after the coupler for the in-line pump power supply. The isolator is included to block the backward-traveling ASE from the preamplifier, which otherwise would cause saturation of the amplifier, resulting in gain degradation. Omitting the isolator would yield a 5-km reduction of the above-mentioned length improvement. In configuration C, the in-line remote pumped amplifier has been replaced by a distributed fiber. To allow for a proper comparison with configuration B, the same amount of erbium ions has been used for the distributed amplifier as used for the in-line amplifier, and they are compared for equal pump power conditions. Finally, configuration D shows the combination of a booster, an in-line remote pumped amplifier, and a preamplifier.

Before we discuss the results of these configurations, the different noise contributions will be more closely investigated, since their mutual influence determines the overall system performance. In the limits where the pumping distance l_1 is close to zero or $l_1 \gg 0$, approximate equations for the signal-to-noise ratio (S/N) may be derived. Let P_{out} denote the output power from the transmitter, l_{tot} the total transmission distance, and α_s the loss constant of the passive fiber at the signal wavelength. Since $P_s = P_{out} \cdot G \cdot \exp(-\alpha_s \cdot l_{tot})$ and $L = \exp(-\alpha_s \cdot l_1)$ are inserted in (10.7), the signal-to-noise ratio may be expressed as

$$S/N = \sqrt{\frac{(P_{out} \cdot \exp(-\alpha_s l_{tot}))}{F \cdot h\nu_s \cdot B \cdot \exp(-\alpha_s l_1)}} \qquad \text{for } l_1 \approx 0$$
$$S/N = \frac{G \cdot P_{out} \cdot \exp(-\alpha_s l_{tot})}{\sqrt{N_{th} \cdot B}} \qquad \text{for } l_1 \gg 0 \tag{10.20}$$

For small pumping distances to the in-line EDFA, the amplifier acts almost as a preamplifier, which means that the signal-to-noise ratio is gain-independent, so the passive fiber between receiver and in-line active fiber attenuates the dominating beat noise, causing the signal-to-noise ratio to increase proportionally to $\exp(\alpha_s l_1)$ for small but increasing values of l_1. On the other hand, when the pumping distance is large, the beat noise becomes nondominant and the signal-to-noise ratio drops off, since the gain decreases with increasing pumping distance. Therefore, an optimum placement of the remote pumped amplifier may be located.

Assuming that the noise figures for the preamplifier and the in-line amplifier are constant, that the gain of the in-line EDFA decreases as $G_1(l_1) = G_{10} \cdot (-p \cdot l_1)$ and that $G_1 \gg 1$ at the optimum pumping distance, then one may derive the following approximate expression for the optimum location of the remote pumped active fiber:

$$l_1 = \frac{l}{\alpha_s + p} \cdot \ln\left(\frac{\alpha_s \cdot G_{10} \cdot F_1}{p \cdot F_2}\right) \tag{10.21}$$

p can be derived from the small-signal gain G_{10} and the pumping threshold $P_{p1,th}$ for the in-line EDFA as $p = \alpha_p \cdot \log(G_{10})/(\log(P_{p1}) - \log(P_{p1,th}))$. α_p is the passive fiber attenuation constant at the pumping wavelength. F_1 is the noise figure of the remote pumped in-line amplifier and F_2 is the noise figure of the preamplifier.

For each of the four system configurations shown in Figure 10.9, the length improvement (that is, the extra transmission distance), which can be added to the unamplified link for a BER of 10^{-9} and a bit rate of 2.5 Gbit/s, is shown in the lower part of Figure 10.9 as a function of the pumping distance for the in-line amplifier. In this case, the erbium doped fiber has a core diameter of 2.8 μm, a refractive index difference of 0.03, and the same dopant concentration and confinement as the fiber used for calculations in Section 10.2.1. For the distributed amplifier, the pumping distance corresponds to the total length of the amplifier.

As expected, an optimum amplifier placement is found. For a location of the in-line amplifier just before the receiver, a total system length improvement of 82 km may be obtained. But by moving the amplifier into the transmission link, for an unchanged pump location just before the receiver, a further improvement of 25 km may be achieved. Insertion of a preamplifier in this system gives another 11 km, and the insertion of a booster another 62 km, yielding a total of 180 km of length improvement for the whole transmission system. Note that the saturation of the booster amplifier, caused by the high input signal power, yields a lower gain and thereby a shorter length improvement than the small-signal preamplifier.

Replacing the in-line amplifier with a distributed fiber, no further length improvement is obtained. To understand this, we have to look at the advantages and disadvantages of the distributed amplifier. One of the most obvious advantages is that the ASE buildup is lowered by intrinsic light scattering in the amplifier itself. However, the ASE power of the distributed amplifier in the present configuration has to be compared to the attenuated ASE power ($P_{ASE} \cdot L$) of the in-line amplifier. One of the disadvantages is that the pump in the signal input end of the active fibers will be more attenuated in the case of a distributed amplifier, since it has been traveling a longer distance. We therefore conclude that the effect of the reduced ASE power out of the distributed active fiber itself is too weak to overthrow the advantages of optimum placement of the amplifying medium as argued above. We will, however, return to a thorough discussion of the distributed amplifier in Chapter 11, where its superiority as a medium for nonlinear transmission over ultralong distance is argued.

The changes in length improvement and in the optimum pumping distance are shown in Figure 10.10 versus the pump power for the remote pumped in-line amplifier placed in a system also using a booster and a preamplifier (configuration D). Also shown are optimum pumping distances calculated by (10.21) and the corresponding length improvement, showing less than 2% deviation of the total length improvement. The increase in pump power both yields an increase in the length improvement and causes the optimum placement of the active fiber to move towards the transmitter. A length improvement of 13 km may be obtained by moving the remote pumped EDFA 14 km further and by increasing the pump power from 50 to 100 mW.

10.5 RESULTS OF SYSTEM PERFORMANCE TODAY

When the aim is to give a short overview of the performance improvement already obtained by fiber amplifiers, the definition of a single determinant is not evident.

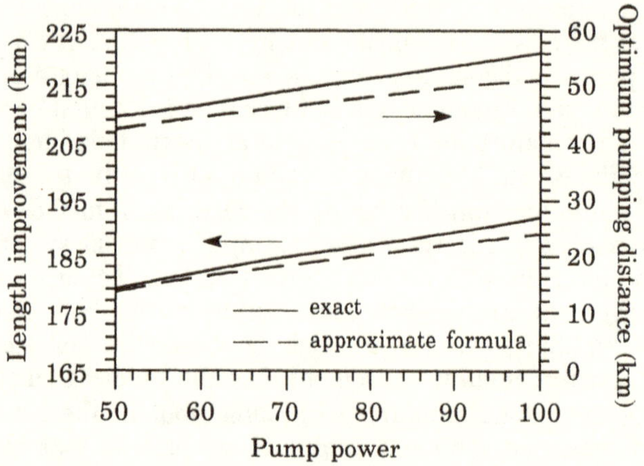

Figure 10.10 Transmission length improvement and optimum pumping distance versus the launched pump power to the in-line remote pumped EDFA. The solid lines represent exact calculation of the optimum pumping distances, while the dashed lines represent optimum pumping distances when calculated from (10.21). (Data from [39].)

However, as a measure of the performance, it is usual to choose the bit rate distance product achieved without electronic signal generation at a given time. This measure of system capability relates to the laboratory system trial phase, and according to [3] there is usually a time lag of three to four years before the capability is put to good use in the real network.

Some of the most significant system capability results achieved over the last few years are plotted in Figure 10.11(a) as a function of the publication date. For comparison, the highest performance results of direct detection systems without optical amplifiers are also included. Compared to these, it is obvious that only by including EDFAs at the ends of the transmission link (either as power boosters or preamplifiers) is a significant bit rate distance improvement obtained. One of the highest capacities achieved with only one EDFA was reported in [40], where 2000 Gbit/s·km was demonstrated by transmitting 20 Gbit/s over a distance of 100 km. This is equivalent to the transmission of 300,000 simultaneous telephone channels over 100 km [3].

Not surprisingly, most of the later hero system experiments include one or a multiple of in-line EDFAs, and transmission over transoceanic distances is obtained in this way. The breakthrough for capacity increase that happened with the availability of optical amplifiers and especially EDFAs in 1989 is therefore evident. It should also be mentioned that all reported hero system experiments today use optical amplifiers. The present record is as much as 45 Tbit/s·km [41], with which,

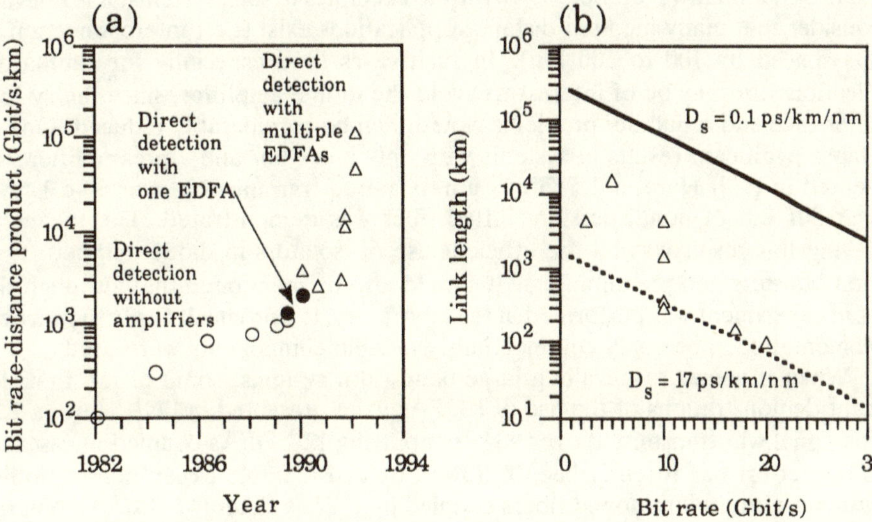

Figure 10.11 (a) Capability of optical direct detection systems measured as reported bit rate distance product against data of report. (b) Relationship between transmission distance and bit rate for the same system reports. (Data references in [3,39,40,42–45].)

according to Figure 10.11(a), we have obtained a bit rate distance product improvement of a factor of 100 compared to systems without EDFAs.

However, limits also exist for direct detection systems with EDFAs. In contrast to electronic repeaters, the EDFA cannot regenerate the shape of the optical pulses. Thus, in dispersive fiber links, the transmission length will be dispersion-limited, especially at high bit rates. In Figure 10.11(b), we have shown the dispersion limit for information bandwidth-limited signals obtained by chirp-free external modulation. The dispersion limit is given in (10.13) and calculated for dispersions of 17 ps/km/nm and 0.1 ps/km/nm.

As we can see in Figure 10.11(b), some of the obtained results are, however, beyond the limits given by the dispersion. The reason is that optical amplifiers can also help to overcome the dispersion limitation by exploiting the nonlinear effects. As we shall see in Chapter 11, it is especially the Kerr nonlinearity that may be used in optical fibers.

From the results in Figure 10.11, it is evident that optical amplifiers have demonstrated their relevance in high speed, long-distance repeaterless communication systems. The applications here are many, such as undersea and off-shore links, island jumping, and trunk lines between main exchanges. The very dramatic results of bit rate distance products are normally obtained by the use of EDFAs in a series chain. However, before we turn to examples of such systems, it is relevant to consider that many medium-distance applications exist (i.e., interconnection of points spaced by 100 to 300 km). In such cases (and especially for submarine applications), it may be of interest to avoid the in-line amplifier, since the system maintenance and reliability problems would then be considerably reduced. One of the most significant results in systems with only a booster and a preamplifier was presented in [46]. Here, a 2.5-Gbit/s unrepeatered transmission over 59-dB fiber loss or 301 km of nondispersion-shifted fiber was demonstrated. The means of achieving this result was by the efficient use of isolators in diode-pumped, two-section boosters and preamplifiers. It should also be mentioned that although the present experiment was performed in the laboratory, it lends itself easily to practical implementation, since only commercially available components were used.

When it comes to ultralong large-bandwidth systems, some of the first significant demonstrations of the use of EDFAs were presented in [29]. Here, a 1.2-Gbit/s signal was transmitted over 904 km applying 12 EDFAs coupled in cascade. This first effort has recently been followed by considerable experimental studies of systems with erbium-doped fibers coupled in a series chain [47–49]. Among the significant results of these demonstrations, it can be mentioned that [47] reported a transmission experiment using a 2.488-Gbit/s coherent detection system, 25 EDFAs placed at approximately 80 km intervals, and dispersion-shifted fiber as a transmission medium. A transmission distance of more than 2223 km was achieved with a receiver sensitivity of -42.0 dBm and a power penalty of only 4.2 dB from the initial sensitivity achieved by the back-to-back configuration.

The aim of [48] was to demonstrate an optical bit rate flexible system using multiple in-line EDFAs, and the transmission capacity was theoretically and experimentally verified to be more than 5 Tbit/s · km. The fact that the examined system operated both at 5 Gbit/s (over 523 km) and at 10 Gbit/s (over 505 km) points out the important feature of the bit rate insensitivity of the EDFAs, whereby their ability of efficiently enhancing system transmission capacity becomes even more evident. Finally, it is here drawn forward that [49] demonstrates the aspect of wavelength-multiplexed signal transmission. More specifically, four wavelengths were used in [49] and the 2.4-Gbit/s signals were transmitted over 459 km using six EDFAs operating in the 1550-nm region with a spectral spacing of $\approx$2 nm between the channels.

These results have also been followed by circulating loop experiments, which have confirmed the feasibility of a 9000 km nonregenerative system at 2.5 and 5 Gbit/s [50, 51]. Also, system experiments at even larger bit rates have been performed and an interesting example is given in [45], where a 10-Gbit/s signal is transmitted nonregeneratively over 1520 km, having no indication of an error floor down to a BER of 10^{-11}. These results show that 10-Gbit/s nonregenerative systems using conventional intensity modulation and direct detection are feasible over distances larger than 1500 km.

As pointed out in [3], we have therefore arrived at the point where properly designed EDFAs in the near future may provide gain to completely overcome intrinsic system losses over any distance on earth. In combination with soliton transmission techniques used to overcome the limitations of pulse dispersion (see Chapter 11), there is in principle and in practice no limitation on the ability of optics to provide practically unlimited guided-wave span and information capacity. This is supported by recently reported loop experiments, where 10-Gbit/s soliton data transmission is achieved over 1 million kilometers [52]. The loop, which was 510 km long, included five EDFAs, and an important additional optical retiming component was used to remove the timing jitter limitation (see Section 11.1) of the system.

10.6 ADDITIONAL SYSTEM ASPECTS OF USING ACTIVE FIBERS

There is no doubt that the dominating applications of optical-fiber amplifiers in the near future are going to be their use power boosters, in-line amplifiers, or preamplifiers in optical communication systems. In a longer term perspective, the active fibers are, however, expected to have use far beyond these somewhat limited applications, and already today alternative suggestions have been made.

The aim here is to give an idea of the possible types of new applications and problems that may appear with the use of optical amplifiers. Therefore, we will review two alternative aspects of the use of EDFAs, and even though this in no way covers the full range of applications, the present choice has been made to limit the description within the field of optical communications. An example of a new concept is the all-fiber-optical routing switch presented in [53], whereas Section 10.6.2 discusses the new problems of error detection along ultralong transmission links.

10.6.1 Active Fibers as Routing Switches

As the optical communication systems in the future develop from point-to-point transmission systems to more advanced fiber-optic networks, the optical switching and routing technology becomes essential. A key component here would be a lossless optical cross-point switch [54]. Such a component has been realized using Er^{3+} fiber amplifiers to obtain a 2×2 optical routing switch that is polarization independent and nonblocking [53].

The implementation of the Er^{3+} fiber switch is similar to that proposed for systems with semiconductor amplifiers [55] and is based on high loss in the amplifier when unpumped and gain from the amplifier when pumped. This gain may be used to overcome splitting and coupling losses of the switch.

Figure 10.12 shows the configuration as proposed in [53]. Here, the two input signals are equally split using biconic fiber couplers, and then the fractions are passed individually through separate fiber amplifiers. Each fiber amplifier module consists of a dichroic coupler used for combining signal and pump light. The outputs from the fiber amplifiers are then cross-coupled, creating a 2×2 switch with redundant output ports. From Figure 10.12 it is also clear that the signal has passed two 3-dB couplers, and the total insertion loss is consequently 6 dB plus the sum of the loss/gain through the amplifier fiber, the loss through the WDM coupler, and the splice losses. In [53] it is explained that the length of the Er^{3+}-doped fiber is determined so that, with the amplifier unpumped, the total loss through the switch is larger than 20 dB. When the amplifier is pumped, the goal is to achieve 6 to 7 dB gain so that the switch becomes lossless. By the chosen length the necessary pump power remains small. In Figure 10.12, appropriate choices of signal and pump wavelengths are shown.

It is reported that the specified switch is polarization-independent, has zero insertion loss, and has a contrast ratio ≥20 dB over a 13-nm optical bandwidth. Furthermore, pump power requirements of around 12 mW for 980 nm pumping and 20 mW for 1480-nm pumping are found.

10.6.2 OTDR Measurements

Another important issue for the systems applying EDFAs is the amplifier impact on the possibility of locating transmission link damage by means of *optical time domain reflectometry* (OTDR). The central problem is that the amplifiers, as we have seen in this and previous chapters, normally includes one or more isolators. This prevents the transmission of backreflected signals over distances beyond the amplifier spacing.

Even though we will not attempt at this point to outline the general solution of this problem, it should be mentioned that one solution is to include an optical circulator in each amplifier module [56]. The fault location is then made by using a return optical transmission line as a propagation medium for the backreflected signal. By locating such extra components at the amplifier input and output ends, backward reflected light may therefore be routed back to the OTDR equipment. It should, however, be noted that unless the amplifier module also amplifies the reflected probe signal, detection will be ruled out anyway due to the loss over a long transmission distance.

Also, factors such as system architecture, control signaling, and amplifier accessibility become important. At this point, therefore, it may only be concluded that if the OTDR test facility is to be maintained, the complexities of the amplifiers are increased severely.

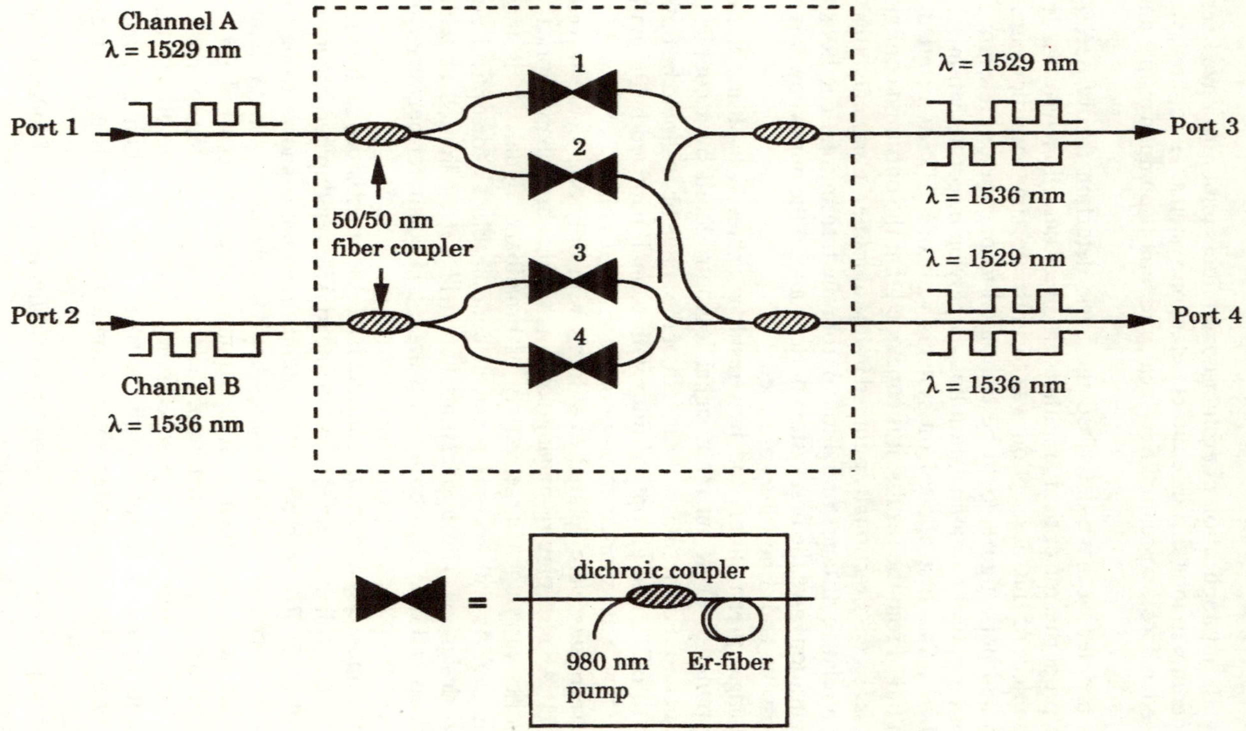

Figure 10.12 Experimental configuration of a 2 × 2 lossless all-fiber routing switch as described in [53].

10.7 SUMMARY AND PERSPECTIVES

The most significant application of erbium-doped fibers until now have been as an amplifying medium in long-distance direct detection systems. Here, however, they have been used in booster, preamplifier, local, and remote pumped in-line amplifier configurations.

In this chapter we have described the basic detection theory necessary to describe the properties of optical amplifiers in direct detection systems. In other words, the subject of this description was to illustrate how amplifier parameters such as gain and noise figure could be translated to parameters relevant for the evaluation of an optical communication link employing direct detection.

This theory was first applied on systems with one EDFA included in the transmission link. From the results, it is concluded that the optimum configuration of an EDFA is as a co-directionally pumped in-line amplifier. Optimally, this should be placed at a relative distance of 0.4 to 0.6 from the transmitter. The booster and in-line configurations give the best system performance if the pump power is chosen as high as possible and if no filters are used.

For a single preamplifier, the best result is achieved by using a counter-propagating pump to avoid an input signal loss introduced by a coupler in the amplifier input end. Furthermore, the length of the active fiber should be optimized with respect to BER, and an optical filter with a small bandwidth (0.5 nm) must be applied to lower the spontaneous-spontaneous beat noise.

All these results were related to systems with only one signal wavelength. However, EDFAs also play promising roles in various broadband optical distribution networks carrying a large variety of broadband signals. It is therefore believed that EDFAs will become practical building blocks for optical network planners and designers. The applications here will not be limited to wide-area subscriber video and data distribution networks, and computer networks are also likely to benefit from the fiber amplifier technology.

When it comes to systems with multiple EDFAs, other issues such as stability, automatic gain control, and gain saturation due to ASE buildup become important. The most critical EDFA with respect to pump reduction is shown to be the first in-line amplifier. When a pump in this amplifier is totally turned off, it will influence the BER seven times as much as if the pump reduction occurred for another amplifier. Furthermore, the reduction of the forward pump is much more critical than that of the backward. For a launched pump power decrease of 50% from one of the lasers, it is possible to reestablish the BER with a maximal increase of the signal input of 5.5 dB. By inserting an additional amplifier, this requirement may be reduced to 2.5 dB.

Also, the question of limitations due to reflections and Rayleigh backscattering has been raised, and a simple method for calculating the upper limit of the gain, determined by the reflections at both ends of an optical amplifier, has been presented. Using this method for a representative case, it is shown that the maximum allowable gain is around 7.4 dB when no isolators are used, and up to 40 dB, depending on the isolator reflection, when an isolator is used.

The subject of remote pumping has furthermore been examined in this chapter. It has been demonstrated that a transmission distance for a 2.5-Gbit/s direct detection system may be improved by more than 25 km simply by moving the active fiber away from the detector while pumping through the transmission line. It is thereby demonstrated that long-distance transmission of 1480-nm pump power not only can be done, but it actually may improve the overall system properties. Further length improvement can be obtained when a booster and a preamplifier are also added to the system. It is shown that there is no obtainable improvement by using a distributed preamplifier compared to a remote pumped in-line amplifier with the same amount of erbium. A simple analytical expression for the optimum placement of the remote pumped in-line amplifier is tested and is shown to give results in good agreement with results calculated by an accurate method.

In the last two sections of this chapter, the actual impact of EDFAs on direct detection systems was first discussed, showing that the obtainable bit rate distance product has been increased by two decades since 1989. Finally, a few alternative properties of fiber amplifiers were discussed. Here, especially, the use of erbium-doped fibers as an amplifying as well as attenuation medium in a routing switch should be mentioned.

REFERENCES

[1] Giles, C. R., E. Desurvire, J. R. Talman, J. R. Simpson, and P. C. Becker, "2 Gbit/s Signal Amplification at 1.53 μm in an Erbium-Doped Single-Mode Fiber Amplifier," *IEEE Journal of Lightwave Technology*, Vol. 7, No. 4, 1989, pp. 651–656.

[2] Edagawa, N., K. Mochizuki, and M. Wakabayashi, "1.2 Gbit/s, 218 km Transmission Experiment Using Inline Er-Doped Optical Fibre Amplifier," *IEE Electronics Letters*, Vol. 25, No. 5, 1989, pp. 363–365.

[3] Millar, C. A., "Fibre Amplifiers in Optical Telecommunications—A Perspective," *BT Technology Journal*, Vol. 9, No. 4, 1991, pp. 5–11.

[4] Saito, S., T. Imai, and T. Ito, "An Over 2200-km Coherent Transmission Experiment at 2.5 Gb/s Using Erbium-Doped-Fiber In-Line Amplifiers," *IEEE Journal of Lightwave Technology*, Vol. 9, No. 2, 1991, pp. 161–169.

[5] Walker, G. R., N. G. Walker, R. C. Steele, M. J. Creaner, and M. C. Brain, "Erbium-Doped Fiber Amplifier Cascade for Multichannel Coherent Optical Transmission," *IEEE Journal of Lightwave Technology*, Vol. 9, No. 2, 1991, pp. 182–193.

[6] Ryu, S., S. Yamamoto, H. Taga, N. Edagawa, Y. Yoshida, and H. Wakabayashi, "Long-Haul Coherent Optical Fiber Communication Systems Using Optical Amplifiers," *IEEE Journal of Lightwave Technology*, Vol. 9, No. 2, 1991, pp. 251–260.

[7] Olsson, N. A., "Lightwave Systems With Optical Amplifiers," *IEEE Journal of Lightwave Technology*, Vol. 7, No. 7, 1989, pp. 1071–1082.

[8] Smith, R. G., and S. D. Personick, "Receiver Design for Optical Fiber Communication System, I," *Bell System Technical Journal*, Vol. 52, 1973, pp. 843–874.

[9] Gagliardi, R. M., and S. Karp, *Optical Communications*, New York: John Wiley & Sons, 1976.
[10] Rasmussen, T. P., "Fiber Amplifiers in Optical Communication Systems," Master's Thesis E433 (in Danish), Electromagnetics Institute, Technical University of Denmark, 1991.
[11] Fye, D. M., Practical Limitations on Optical Amplifier Performance," *IEEE Journal of Lightwave Technology*, Vol. 2, No. 4, 1984, pp. 403–406.
[12] Yamamoto, Y., "Noise and Error Rate Performance of Semiconductor Laser Amplifiers in PCM-IM Optical Transmission Systems," *IEEE Journal of Quantum Electronics*, Vol. 16, No. 10, 1980, pp. 1073–1081.
[13] Rasmussen, T., A. Bjarklev, B. Pedersen, and J. H. Povlsen, "Optimum Placement of a Fiber Optical Amplifier in High Bit Rate Direct Detection Systems," *Applied Optics*, Vol. 30, No. 30, 1991, pp. 4376–4378.
[14] Tonguz, O. K., and L. G. Kazovsky, "Theory of Direct-Detection Lightwave Receivers Using Optical Amplifiers," *IEEE Journal of Lightwave Technology*, Vol. 9, No. 2, 1991, pp. 174–181.
[15] Viereck, P., "Optical Teletransmission Systems," *FN75 (in Danish)*, Electromagnetics Institute, Technical University of Denmark, 1986.
[16] Saiti, S., T. Imai, T. Sugie, N. Ohkawa, Y. Ichihashi, and T. Ito, "Coherent Transmission Experiment Over 2, 223 km at 2.5 Gbit/s Using Erbium Doped Fibre Amplifiers," *Electronics Letters*, Vol. 26, No. 10, 1990, pp. 669–671.
[17] O'Reilly, J. J., and H. J. Silva, "Chirp Induced Penalty in Optical Fibre Systems," *Electronics Letters*, Vol. 23, No. 19, 1987, pp. 992–993.
[18] Vendeltorp-Pommer, H., B. Pedersen, A. Bjarklev, and J. H. Povlsen, "Noise and Gain Performance for an Er^{3+} Doped Fibre Amplifier Pumped at 980 nm or 1480 nm," *Proceedings of Society of Photo-Optical Instrumentation Engineers Conference, OE/Fibers '90*, part: Fiber Laser Sources and Amplifiers II, San Jose, September 1990.
[19] Pedersen, B., K. Dybdal, C. D. Hansen, A. Bjarklev, J. H. Povlsen, H. Vendeltorp-Pommer, and C. C. Larsen, "Detailed Theoretical and Experimental Investigation of High-Gain Erbium Doped Fiber Amplifier," *IEEE Photonics Technology Letters*, Vol. 2, No. 12, 1990, pp. 863–865.
[20] Dybdal, K., A. Bjarklev, C. Dam-Hansen, B. Pedersen, J. H. Povlsen, H. Vendeltorp-Pommer, and C. C. Larsen, "Detailed Experimental and Theoretical Investigation of High-Gain Er-Doped Fibre Amplifier," *European Conference on Optical Communication ECOC '90*, Amsterdam 1990, Proc. Vol. I, pp. 443–446.
[21] Laming, R. I., L. Reekie, P. R. Morkel, and D. N. Payne, "Multichannel Crosstalk and Pump Noise Characterization of Er^{3+}-Doped Fibre Amplifier Pumped at 980 nm," *Electronics Letters*, Vol. 25, No. 7, 1989, pp. 455–456.
[22] Way, W. I., S. S. Wagner, M. M. Choy, C. Lin, R. C. Menendez, H. Tohme, A. Yi-Yan, A. C. Von Lehman, R. E. Spicer, M. Andrejco, M. A. Saifi, and H. L. Lemberg, "Simultaneous Distribution of Multichannel Analog and Digital Video Channels of Multiple Terminals Using High-Density WDM and Broad-Band In-Line Erbium-Doped Fiber Amplifier," *IEEE Photonics Technology Letters*, Vol. 2, No. 9, 1990, pp. 665–668.
[23] Chen, C. Y., M. M. Choy, M. J. Andrejco, M. A. Saifi, and C. Lin, "A Widely Tunable Erbium-Doped Fiber Laser Pumped at 532 nm," *IEEE Photonics Technology Letters*, Vol. 2, No. 1, 1990, pp. 18–20.
[24] Lin, C., and W. Way, "Optical Amplifiers for Multi-Wavelength Broadband Distribution Networks," *European Conference on Optical Communication ECOC '91*, Paris, 1991, Proc. Vol. II, invited paper, pp. 125–133.
[25] Giles, C. R., and E. Desurvire, "Propagation of Signal and Noise in Concatenated Erbium-Doped Fiber Optical Amplifiers," *IEEE Journal of Lightwave Technology*, Vol. 9, No. 2, 1991, pp. 147–154.
[26] Ellis, A. D., R. M. Percival, A. Lord, and W. A. Stallard, "Automatic Gain Control in Cascaded Erbium Doped Fibre Amplifier Systems," *Electronics Letters*, Vol. 27, No. 3, 1991, pp. 193–195.

[27] Lumholt, O., A. Bjarklev, J. H. Povlsen, T. Rasmussen, and K. Rottwitt, "Stability of a 500 km Erbium Doped Fiber Amplifier Cascade," *IEEE Photonics Technology Letters*, Vol. 4, No. 2, 1992, pp. 183–186.

[28] Pedersen, B., A. Bjarklev, O. Lumholt, and J. H. Povlsen, "Detailed Design Analysis of Erbium Doped Fiber Amplifiers," *IEEE Photonics Technology Letters*, Vol. 3, No. 6, 1991, pp. 548–550.

[29] Edagawa, N., Y. Yoshida, H. Taga, S. Yamamoto, K. Mochizuki, and H. Wakabayashi, "904 km, 1.2Gbit/s Non-Regenerative Optical Fibre Transmission Experiment Using 12 Er-Doped Fibre Amplifiers," *Electronics Letters*, Vol. 26, No. 1, 1990, pp. 66–67.

[30] Gimlet, J. L., and N. K. Cheung, "Effects of Phase-to-Intensity Noise Conversion by Multiple Reflections on Gigabit-per-Second DFB Laser Transmission Systems," *IEEE Journal of Lightwave Technology*, Vol. 7, No. 6, 1989, pp. 888–895.

[31] Gimlet, J. L., N. Z. Iqbal, L. Curtis, N. K. Cheung, A. Righetti, F. Fontana, and G. Grasso, "Impact of Multiple Reflection Noise in Gbit/s Lightwave Systems With Optical Fibre Amplifiers," *Electronics Letters*, Vol. 25, No. 20, 1989, pp. 1393–1394.

[32] Rasmussen, T., A. Bjarklev, J. H. Povlsen, O. Lumholt, and K. Rottwitt, "Maximum Gain in Optical Amplifier-Based Systems as Determined From Reflections and Backscattering Effects," *Applied Optics*, Vol. 31, No. 8, 1992, pp. 3386–3387.

[33] Henmi, N., Y. Aoki, S. Fujita, Y. Sunohara, and M. Shikada, "Rayleigh Scattering Influence on Performance of 10 Gb/s Optical Receiver With Er-Doped Optical Fiber Preamplifier," *IEEE Photonics Technology Letters*, Vol. 2, No. 4, 1990, pp. 277–278.

[34] Bergano, N. S., J. Aspell, C. R. Davidson, P. R. Trischitta, B. M. Nyman, and F. W Kerfoot, "A 9000 km 5 Gb/s and 21,000 km 2.4 Gb/s Feasibility Demonstration of Transoceanic EDFA Systems Using Circulating Loop," *Optical Fiber Communications OFC '91*, San Diego, 1991, postdeadline paper PDP13.

[35] Eskildsen, L., and E. Goldstein, "High-Performance Amplified Optical Links Without Isolators or Bandpass Filters," *IEEE Photonics Technology Letters*, Vol. 4, No. 1, 1992, pp. 55–58.

[36] Edagawa, N., S. Yamamoto, Y. Namihira, K. Mochizuki, and H. Wakabayashi, "First Field Demonstration of Optical Submarine Cable System Using LD-Pumped Er-Doped Optical Fibre Amplifier," *Electronics Letters*, Vol. 25, No. 19, 1989, pp. 1278–1280.

[37] Bjarklev, A., B. Pedersen, J. H. Povlsen, and T. Rasmussen, "System Performance Improvement by Long-Distance Pumping of Inline Fiber Amplifiers," *Conference on Lasers and Electro-Optics CLEO '91*, Baltimore, May 1991, paper CThJ, pp. 396–399.

[38] J. H., Povlsen, A. Bjarklev, B. Pedersen, H. Vendeltorp-Pommer, and K. Rottwitt, "The Optimum Design of Erbium Fibre Amplifiers Pumped With Sources Emitting at 1480nm," *Electronics Letters*, Vol. 26, No. 17, 1990, pp. 1419–1421.

[39] Bjarklev, A., J. H. Povlsen, K. Rottwitt, O. Lumholt, and T. Rasmussen, "Performance Improvement of Direct Detection Systems Using Local and/or Long Distance Pumped Fiber Amplifiers," *Proceedings of Society of Photo-Optical Instrumentation Engineers Conference, OE/Fibers '91*, part: Fiber Laser Sources and Amplifiers III, Boston, September 1991.

[40] Marshall, I. W., D. M. Spirit, G. N. Brown, and L. C. Blank, "20 Gbit/s 100 km Nonlinear Transmission With Semiconductor Source," *Optical Fiber Communications OFC '90*, San Francisco, 1990, postdeadline paper PD6.

[41] Taga, H., N. Edagawa, Y. Yoshida, S. Yamamoto, M. Suzuki, and H. Wakabayashi, "10 Gbit/s, 4500 km Transmission Experiment Using 138 Cascaded Er-Doped Fiber Amplifiers," *Optical Fiber Communications OFC '92*, San Jose, 1992, postdeadline paper PD12.

[42] Wedding, B., "Breakthrough in Multigigabit Transmission Systems Technology Due to Optical Amplifiers," *Topical Meeting on Optical Amplifiers and Their Applications OAA '92*, Proceedings Vol. 1, June 1992, Santa Fe, paper FC4, pp. 196–199.

[43] Mollenauer, L. F., E. Lichtman, G. T. Harvey, M. J. Neubelt, and B. M. Nyman, "Demonstration of Error-Free Soliton Transmission Over More Than 15,000 km at 5 Gbit/s, Single-Channel, and

Over 11,000 km at 10 Gbit/s in a TwoChannel WDM," *Optical Fiber Communications OFC '92*, San Jose, 1992, postdeadline paper PD10.

[44] Saito, S., M. Murakami, A. Naka, Y. Fukuda, T. Imai, M. Aiki, and T. Ito, "2.5 Gbit/s, 80–100 km Spaced In-Line Amplifier Transmission Experiments Over 2,500–4,500 km," *European Conference on Optical Communication ECOC '91*, Paris, 1991, Proc. Vol. III, postdeadline paper, pp. 68–71.

[45] Edagawa, N., H. Taga, Y. Yoshida, M. Suzuki, S. Yamamoto, and H. Wakabayashi, "10 Gbit/s, 1500 km Transmission Experiment Using 22 Er-Doped Fibre Amplifier Repeaters," *European Conference on Optical Communication ECOC '91*, Paris, 1991, Proc. Vol. III, postdeadline paper, pp. 76–79.

[46] Joergensen, C. G., B. Mikkelsen, C. F. Pedersen, K. Schüsler, O. Froelich, C. C. Larsen, J. A. Berk, T. Durhuus, and K. E. Stubkjaer, "Refined Amplifier Design for Unrepeatered 2.5 Gbit/s Transmission Over 59 dB of Fibre Loss," *Topical Meeting on Optical Amplifiers and Their Applications OAA '92*, Santa Fe, 24–26 June 1992, paper FC3, pp. 192–195.

[47] Saito, S., T. Imai, and T. Ito, "An Over 2200-km Coherent Transmission Experiment at 2.5 Gb/s Using Erbium-Doped-Fiber In-Line Amplifiers," *IEEE Journal of Lightwave Technology*, Vol. 9, No. 2, 1991, pp. 161–169.

[48] Hagimoto, K., S. Nishi, and K. Nakagawa, "An Optical Bit-Rate Flexible Transmission System With 5-Tb/s·km Capacity Employing Multiple In-Line Erbium-Doped Fiber Amplifiers," *IEEE Journal of Lightwave Technology*, Vol. 8, No. 9, 1990, pp. 1387–1395.

[49] Taga, H., Y. Yoshida, N. Edagawa, S. Yamamoto, and H. Wakabayashi, "459 km, 2.4 Gbit/s Four Wavelength Multiplexing Optical Fibre Transmission Experiment Using Six Er-Doped Fibre Amplifiers," *IEE Electronics Letters*, Vol. 26, No. 8, 1990, pp. 500–501.

[50] Bergano, N., J. Aspell, C. R. Davidson, P. R. Trischitta, B. M. Nyman, and F. W. Kerfoot, "A 9000 km 5 Gb/s and 21,000 km 2.4 Gb/s Feasibility Demonstration of Transoceanic EDFA Systems Using a Circulating Loop," *Optical Fiber Communications OFC '91*, Anaheim, 1991, postdeadline paper PD13, pp. 61–64.

[51] Taga, H., M. Suzuki, Y. Yoshida, H. Tanaka, S. Yamamoto, and H. Wakabayashi, "2.5 Gbit/s Optical Transmission Using Electroabsorption Modulator Over 11,000 km EDFA System," *Conference on Lasers and Electro-Optics CLEO '91*, Baltimore, May 1991, paper CPDP38, pp. 650–651.

[52] Nakazawa, M., et al., "10 Gbit/s Soliton Data Transmission Over One Million Kilometers," *IEE Electronics Letters*, Vol. 27, No. 14, 1991, pp. 1270–1272.

[53] Eichen, E., W. J. Miniscalco, J. McCabe, and T. Wei, "Lossless, 2x2, All-Fiber Optical Routing Switch," *Optical Fiber Communications OFC '90*, San Francisco, 1990, postdeadline paper PD20-2.

[54] Midwinter, J. E., "Photonics Switching Technology: Component Characteristics vs. Network Requirements," *IEEE Journal of Lightwave Technology*, Vol. 6, No. 10, 1988, pp. 1512–1519.

[55] O'Mahony, M. J., "Semiconductor Laser Optical Amplifiers for Use in Future Fiber Systems," *IEEE Journal of Lightwave Technology*, Vol. 6, No. 4, 1988, pp. 531–544.

[56] Sato, Y., S. Nishi, and K.-I. Aoyama, "OTDR in Optical Amplifier Transmission Systems Using EDFAs Containing Optical Circulators," *Topical Meeting on Optical Amplifiers and Their Applications OAA '91*, Proceedings Vol. 1, July 1991, Snowmass CO, paper FE1, pp. 268–271.

Chapter 11
Distributed Fiber Amplifiers

The erbium-doped fibers discussed in this book have so far been of the general type, with erbium ion concentrations of 100 to 1000 ppm by weight and lengths on the order of 10 m to 100 m. The gain is therefore obtained over fiber lengths, which are considerably shorter than the transmission spans, and under these circumstances the amplification is normally referred to as being discrete, and the communication system is said to apply "lumped" gain. Such systems, therefore, subject the signal to large excursions in amplitude, which can cause significant interchannel interference in a WDM system [1].

There is also a limit to which this high gain, step, and repeat concept may be employed before the ASE will dominate [2] and limit the overall system performance. An attractive way to overcome these problems is to apply distributed amplification; a larger signal to noise ratio may be obtained when the signal is sustained at a high level throughout the system [3].

Distributed amplification therefore provides one of the most attractive methods for ultralong-distance signal transmission (over several thousands of kilometers). However, in order to be able to optimally design such systems, it is essential also to take into account the nonlinear properties of the transmission. This may be understood from the fact that, in optically amplified, unregenerated long-span optical transmission, we will find that the dispersion of the signal pulses forms one of the major limits to system lengths. For sufficiently intense pulses of the appropriate characteristics, it is, however, possible to apply pulse compression caused by the optical nonlinear Kerr effect [4]. With proper system design, the pulse compression may exactly match the pulse broadening due to normal linear dispersion. Transmission beyond the usual dispersion limit is thereby obtainable. Such optical communication systems, which make constructive use of the fiber nonlinearity, are often called *soliton systems*. As we shall see in the following discussion, this also indicates the special requirements for amplitude and shape of the signal pulses (solitons) [5].

In this chapter, we will describe the fundamental limits for ultralong-distance systems. These systems are realized by means of utilizing fiber nonlinearities together with the advantageous properties of distributed fiber amplifiers. In order to design such optical transmission links, however, it is necessary to get acquainted with a number of new concepts. Therefore, we will first sketch some basic properties of nonlinear signal transmission. The purpose is to establish a reference for the later results and apply these for system evaluation in the case of discrete amplification (see Section 11.1.1).

Thereafter, Section 11.1.2 is a clarification of some of the fundamental problems that appear in connection with soliton pulse transmission through a communication line applying lumped gain.

These fundamental problems are the reason for the large amount of attention that *distributed erbium-doped fiber amplifiers* (dEDFAs) have recently attracted by offering to be a promising transmission medium for high-bit-rate optical pulses over ultralong distances [6]. It is its low generation of ASE that is the attractive property. Furthermore, the ability of limiting the energy fluctuation along the transmission line is advantageous through the use of distributed amplifiers compared to lumped-gain systems.

We will quantify these properties in Section 11.1.3, and the noise properties of dEDFAs will be discussed in Section 11.1.4.

After we have established the necessary descriptive tools for the analysis of distributed amplifiers in Section 11.1, the focus in Section 11.2 is the design of distributed EDFAs. It is already important here to notice that the goal is different from achieving gain coefficient effective active fibers for lumped-gain amplifiers, and thus we will show the special aspects of optimization of distributed EDFAs. The optimum fiber design parameters will be demonstrated, taking into account that the fiber dispersion varies with the NA of the optical fiber. In Section 11.2.1, we look at the question of how to choose the best ratio between co- and counter-propagating pump power. Furthermore, Section 11.2.2 will deal with the question of the optimal choice of signal wavelength for dEDFAs, and Section 11.2.3 describes the impact of nonlinear transmission on a 9000-km-long communication system. Finally, Section 11.2.4 presents additional means of achieving optimal long-distance transmission links. This is based on the idea of [7], which proposes that the erbium ion concentration should preferably be distributed with a lengthwise variation in order to obtain optimal attenuation-free transmission.

In Section 11.3, a short presentation of the second main application suggested for dEDFAs will finally be given. This other application is to use the dEDFA to compensate for tapping losses as signals are removed from a bus backbone consisting of a length of erbium-doped fiber [8]. This requires the extraction of the signal at strategic points along the bus, while the pump should remain in the fiber to pump erbium ions further along the fiber.

11.1 OPTICAL COMMUNICATION SYSTEMS APPLYING NONLINEAR EFFECTS

The interest in using solitons for future high-speed communication systems has become realistic through the development of high-performance optical amplifiers, and system experiments [9] show signal transmission over long distances using in-line fiber amplifiers. A primary object in designing high-bit-rate communication systems optimally for use over medium or long distances is to increase the distance between the amplifiers (or minimize the number of these) and at the same time keep the shape of the solitons unchanged. This task is a common one and is independent of whether we deal with communication systems applying lumped or with distributed amplification.

In order to establish a clear understanding of the advantages of distributed amplification for soliton systems, we will first look at reference systems applying lumped amplification. After a short presentation of some basic concepts and parameters, we will with reference to [10] address the problem of soliton transmission with maximal pump spacing. This will include a presentation of optimal ways of cascading sections of passive single-mode fibers followed by in-line optical amplifiers. The presentation is based on the preemphasis method [6], in which there exists a distance at which the pulse width is the same as that of the input soliton, and the decrease in power can be compensated for by the lumped gain of the erbium-doped fiber.

Even with the advantages of the preemphasis technique, the inevitable signal excursions along the transmission line will still set serious limitations to the maximal spacing between the amplifiers. This will be quantified in the following text, and the properties of the more advantageous distributed EDFAs will be explained.

11.1.1 Basic Theory for Nonlinear Transmission of Optical Signals

The wish to use the longest possible distance between pump stations in ultralong-distance communication systems is the motivation for increasing the power level of the signal launched into the fiber. Due to the waveguide properties and the resulting very high electric field in the fiber, a number of nonlinear effects must be considered. Among these are stimulated scattering processes such as Brillouin or Raman scattering [11,12], which, in the case of optical communication systems, may often be ignored. This is possible because the considered pulse widths are a few picoseconds transmitted at rates of a few gigahertz, thus keeping the average power sufficiently low. As described in [13], the nonlinear property of principal concern is the field dependence of the refractive index. In a glass on a length scale at which inversion symmetry exists, the refractive index takes the form

$$n(\lambda, |E|^2) = n_0(\lambda) + n_{ke}|E|^2 \tag{11.1}$$

where n_0 is the linear part of the refractive index, which may be well approximated by the Sellmeier equation (see [14] for further details). $n_{ke} \approx 2.3 \cdot 10^{-22}$ m^2/V^2 (in mks units) for SiO_2 [15] is the nonlinear index of refraction and λ is the vacuum wavelength. The parameter E denotes the electric field in the fiber.

As described in [13], propagation of a guided mode in a perfect single-mode fiber with such nonlinearity was first considered in [16]. Here, the transverse field dependence was factored out and transverse effects ignored. An equation describing the slowly varying pulse envelope, $\Phi(t, z)$, may be derived. After a number of transformations (we will shortly return to the most central one) and the inclusion of a fiber loss term, the following dimensionless equation appears [13]:

$$i\frac{\partial \Phi}{\partial z} + i\Gamma\Phi + \frac{1}{2}\frac{\partial^2 \Phi}{\partial t^2} + |\Phi|^2\,\Phi = 0 \tag{11.2}$$

This equation is normally referred to as the *perturbated nonlinear Schrödinger* (NLS) equation, where the perturbation refers to the inclusion of the loss term Γ. For a detailed derivation of NLS, see [17].

We shall now define the different parameters of (11.2); central are the transformations in length and time leading to the dimensionless NLS. From the physical place coordinate z_p and the physical time t_p, the parameters z and t of (11.2) are defined as follows [4]: $z = \eta^2 z_p$ and $t = \eta(t_p - z_p/v_g)\sqrt{-\beta_0''} \cdot v_g$ is the group velocity, and β_0'' is the second derivative of the propagation constant with regard to the angular frequency. The coordinate system (t, z) is chosen so that it follows the pulse.

The basis of the following description is the characteristic properties of solitary wave [13] solutions that appear from (11.2). In order to achieve the most convenient description, before the definition of the last parameters of (11.2), we will look at these solutions, which are given by the initial pulse shape

$$\Phi(t) = A\,\text{sech}(t) \tag{11.3}$$

where an amplitude $A > 1$ is required. For an initial pulse shape as given in (11.3), η may be expressed as

$$\eta = \frac{\lambda}{t_{\text{FWHM}}}\sqrt{\frac{D}{2\pi c}}\ln\frac{\sqrt{2}+1}{\sqrt{2}-1} \tag{11.4}$$

where t_{FWHM} is the full pulse width at half maximum of the optical pulse intensity, D is the fiber dispersion, and c is the vacuum velocity of light. Finally, we need to define the factor related to the fiber loss:

$$\Gamma = \eta^{-2}\gamma \tag{11.5}$$

where γ is the attenuation constant of the fiber.

In order to get a better understanding of the special properties of the solitary pulse solutions to (11.2), we will in the following part of this section focus on the properties of the so-called *preemphasis* method.

It can be shown that for propagation in a medium with loss, such as a single-mode fiber with an attenuation of 0.2 dB/km at the wavelength λ = 1535 nm, the relative pulse width varies as illustrated in Figure 11.1(a). If we now look at the possibilities of changing the shape of these curves, we must note that the curves in Figure 11.1(a) are all characterized by the parameter Γ.

For a given η, the parameter Γ is determined only by the attenuation in the fiber. From [13] it is known that a significant improvement of the transmission line may be obtained by reducing Γ. Since the attenuation for typical standard fibers is 0.2 dB/km, the broadening of the pulse may only be reduced by using a lower value of η^{-2}. As also appears in Figure 11.1(a), solitons with an initial amplitude

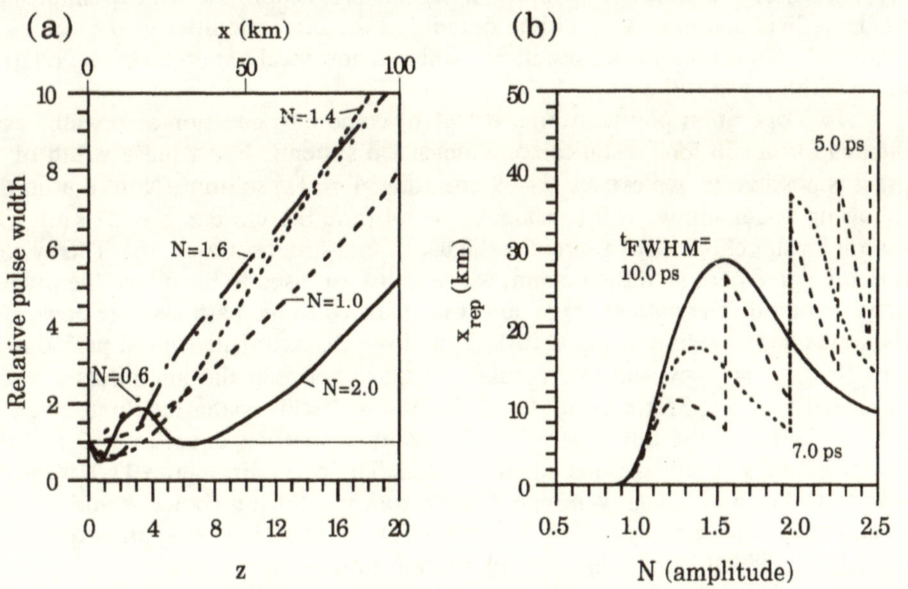

Figure 11.1 (a) The relative FWHM amplitude value of the optical pulse (normalized with respect to the input pulse width) shown versus the propagation distance along the fiber. The curves are calculated using a dispersion of 2.5 ps/km/nm and t_{FWHM} = 7 ps. The different curves refer to the soliton maximum amplitude values. (b) Repeater distance versus initial amplitude for ordinary soliton input signals with Γ as a parameter represented through variation in t_{FWHM}. (Data for both figures from [10].)

$A > 1$, after propagating a given distance, will return to a relative pulse width, which equals that of the initial pulse. For example, this is shown to happen at three different distances for $A = 2.0$. This is due to the dynamic balance between the dispersion and the nonlinearity. The idea now is to consider the possible use of these points, taking the parameter η into account, and at the same time ensure that the path average pulse power is equal to the fundamental soliton power.

When we consider the use of higher order solitons, it is important to start by identifying the soliton amplitudes A relevant for use in communication systems. Therefore, we either have to choose the value of A, which gives a first order soliton state and maximum distance between two amplifiers, or a value of A, for which we still have maximum distance between the amplifiers but at the same time use a higher order soliton. This means, for instance, that for $A = 2.0$ in Figure 11.1(a) we may choose the rightmost point, where the relative pulse width is equal to 1.

In Figure 11.1(b), the distance between the amplifiers is shown as a function of the soliton amplitude A. The curves are calculated for a dispersion-shifted single-mode fiber with an attenuation of 0.2 dB/km and a total dispersion at a constant value of 2.5 ps/km/nm. Presented are three curves for t_{FWHM} equal to 5 ps, 7 ps, and 10 ps, respectively (equivalent to three different values of the parameter η in (11.4)). These are all realistic optimum parameters, achievable with commercially available light sources. It should be noted that the discontinuities of the curves in Figure 11.1(b) represent the situations, where a new local minimum of the relative pulse width reaches the value 1.

Two operation points in Figure 11.1(b) come into question as possible candidates for use in long-distance communication systems. For a pulse width of 10 ps, it is possible to achieve $x_{rep} = 29$ km using $A = 1.5$ solitons. Note that in the following discussion we will consider $A = 1.4$ (and therefore $x_{rep} = 27$ km). The reason for this choice is to guarantee the use of first-order solitons [18]. This system is in full agreement with the system, which in [6] was used to introduce the preemphasis method. The path average power normalized to the path average power for the fundamental soliton is equal to 1.12, and the characteristic soliton period is 16 km. If $t_{FWHM} = 7$ ps, the intermediate distance between the amplifiers may be increased to $x_{rep} = 37$ km using $A = 2.0$ solitons. In this system, the path average power relative to the fundamental $A = 1$ soliton power is equal to 1.92 and the characteristic soliton period is equal to 8 km. The curves in Figure 11.1(b) might lead to the idea of using even higher order solitons. Such a choice would, on the other hand, require an optical peak power of more than 250 mW as an output from the EDFA, which is unrealistic at the present time.

For typical fiber parameters (with dispersion of 2.5 ps/km/nm and loss of 0.2 dB/km), optimal distances between the in-line amplifiers are therefore shown to be 27 and 37 km for $A = 1.4$ and $A = 2.0$ solitons, respectively [10]. Using $A =$ 1.4 solitons, it was furthermore found in [10] that the noise power level after transmission of a 10-Gbit/s signal over 486 km was 18 dB under the pulse power maximum. The necessary condition for achieving this is the use of amplifiers with gain limited to 5.4 dB, whereas higher gain would result in excitation of even higher order solitons with more restricted applications as illustrated by the use of $A = 2.0$ solitons. However, the $A = 2.0$ solitons make it possible to transmit a 10-Gbit/s signal over 111 km by using only three amplifiers. In this situation, the gain of the EDFA must be restricted to 8.1 dB and a maximum bit rate length product of 1.11 THz·km was found.

11.1.2 Problems of Discrete Amplification in Very-Long-Distance Transmission Links

As demonstrated in the previous section, the spacing between in-line amplifiers in a transmission link for solitons may be improved by the use of higher order solitons. In this connection, we should however point out one phenomenon which can prevent a successful use of higher order solitons in communication systems. This is the existence of side loops on the solitons (see, for instance, [13]), which might grow considerably if the pulse is amplified where the side loops have a large amplitude relative to the center value. The appearance of this problem is represented through the criterion $x_{rep} \gg L$ [19], where x_{rep} is the distance between electrical repeaters, and L is the distance between two amplifiers.

The illustration of this problem of side loops has been discussed in more detail in [10,20], but it should be stressed here that long-distance soliton transmission with lumped gain requires a typical amplifier gain of only 5 to 8 dB, depending on the actual situation. Furthermore, gain control is very important, since a larger amplification than required for the information carrying solitons would actually lead to the excitation of higher order solitons. We therefore may conclude that even though larger distances between amplifiers may be obtained by the application of higher order solitons, this will require very accurate power level control. This means that the method due to the existence of side loops will only be realistic for limited distances on the order of 100 km totally.

Besides the need for pulse power control, the noise properties are central when we want to evaluate a soliton communication system employing lumped gain and compare this to systems with distributed amplification. In order to be able to make a fair comparison in the following sections, we will briefly follow the derivation of [21], in which lower noise limits for systems with lumped gain are presented.

When we want to find the lower noise limit for a transmission link based on lumped gain, it is, according to Chapter 7, necessary to consider amplification achieved with the use of EDFAs pumped at 980 nm. The noise figure of each EDFA is furthermore assumed to be ideal. This means that complete inversion is assumed, and thus the noise figure $F = 2 - 1/G$ [21] (see also Chapter 5), where

G denotes the amplifier gain. This approximation is applicable, since we wish to find the lower noise limit for lumped amplification. To ensure transparency of a lumped-gain transmission link, the gain G of each amplifier has to compensate for the attenuation in the passive optical-fiber section between two amplifiers. Thus, $G = \exp(\alpha L)$, where α is the attenuation constant and L the length of the passive optical-fiber section. The path average signal power along the transmission link based on lumped gain is

$$\langle P \rangle = P_0 \cdot R_g \tag{11.6}$$

where P_0 is the launched signal power and the parameter R_g, which is a measure of the pulse energy fluctuation along the transmission link, is defined from the amplifier gain as

$$R_g = (G - 1)/(G \cdot \ln(G)) \tag{11.7}$$

In general, two configurations of blocks in a transparent lumped-gain link exist [22]. In one, the *power amplifier configuration* (conf I), the signal from its source is first amplified and then transmitted through a passive fiber section. In the other, the *preamplifier configuration* (conf II), the signal is directly launched to a passive fiber and then subsequently amplified. Each link consists of N identical blocks containing an amplifier and a passive fiber per section. In conf I, the mean signal power along the transmission link is $\langle P \rangle_I = P_I \cdot G \cdot R_g$, where P_I is the launched signal power. The number of ASE photons after one block is $n_{\text{ASE},b} = (F \cdot G - 1)/G$. After N blocks, the number $n_{\text{ASE},t} = N \cdot (F \cdot G - 1)/G$, which is equal to a total noise figure of

$$F_I = N \cdot \left(F - \frac{1}{G}\right) + 1 \tag{11.8}$$

for a link constructed on the basis of the power amplifier configuration. In conf II, the mean signal power is $\langle P \rangle_{II} = P_{II} \cdot R_g$, where P_{II} is the launched signal power and the number of ASE photons after one block is $n_{\text{ASE},b} = (F \cdot G - 1)$, resulting in $n_{\text{ASE},t} = N \cdot (F \cdot G - 1)$ ASE photons after N blocks. This is equivalent to a total noise figure for the preamplifier configuration of

$$F_{II} = N \cdot (F \cdot G - 1) + 1 \tag{11.9}$$

As mentioned, a long-distance transmission system is considered nonlinear if the path average power is high enough to support solitary pulses. It is normally considered a linear system if the path average signal power is sufficiently below the soliton power [23]. Thus, a comparison between various systems may be performed by examining signal-to-noise ratios rather than noise figures. Furthermore, a fair comparison is performed on the basis of an equal signal path average power. In conf. I, the signal-to-noise ratio at the signal output end is $P_I/F_I = \langle P \rangle_I/(F_I \cdot R_g G)$, whereas in conf. II it is $P_{II}/F_{II} = \langle P \rangle_{II}/(F_{II} \cdot R_g)$. For a mean signal power level, which is fixed at equal values in the two cases, it is suggested in [24] that the so-called *effective noise figures* be defined for easy comparison. According to [24], the two effective noise figures are defined as $NF_I = F_I \cdot G \cdot R_g$ and $NF_{II} = F_{II} \cdot R_g$, when lumped gain is considered.

For comparison with distributed amplification, a long-haul transmission line based on lumped gain is subdivided into sections, each 100 km long. One section is then further divided into N parts of length $100/N$ km. In Figure 11.2, the effective noise figure of the 100-km fiber span is shown as a function of the number of parts N equal to the number of amplifiers within the span. From the figure, it can be seen that two amplifiers give the effective noise figures NF_I = 14.4 dB and NF_{II} = 13.5 dB. If A is the total attenuation over 100 km fiber without any amplification, the asymptotic value of the effective noise figure, equivalent to an infinite number of amplifiers, is

$$F_{\text{lim}} = 2\ln(A) + 1 \tag{11.10}$$

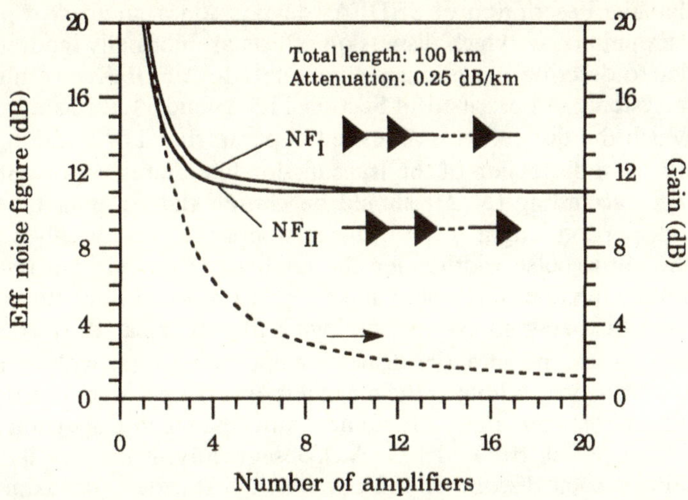

Figure 11.2 Effective noise figure NF_I and NF_{II} for the power and preamplifier configuration, respectively, shown versus the number of amplifiers within a 100-km fiber span. The dashed curve displays the gain G of each amplifier.

This value is the noise figure of an ideal transmission line, where the loss is compensated for at every point along the fiber. The noise figure of a dEDFA has this value as a lower limit.

It should finally be noted that for an amplifier number below 10, the preamplifier configuration results in a slightly lower noise figure than the power amplifier configuration.

11.1.3 Figures of Merit for the Distributed Fiber Amplifier

Section 11.1.2 discussed how soliton communication links using lumped gain have problems in respect to signal power level control. We have furthermore seen that the optimum distance between pump stations is on the order of no more than 40 km. Therefore, we will direct our attention towards the alternative way of realizing long-distance transmission links, namely, by the use of distributed EDFAs. For the completeness of the description, it should be noted, however, that the concept of distributing gain along an optical-fiber span was first considered for a different type of amplifier. In this, the mechanism of Raman gain for soliton propagation was applied [25]. However, lately the major interest in distributed amplification has been directed towards dEDFAs, which will also be the focus of the remaining part of this chapter.

Distributed amplifiers become relevant primarily in systems where the dispersion and Kerr effect are used to balance each other out. Therefore, it is not surprising that the description of dEDFAs deviates on a number of points from the discrete amplifiers, in which dispersion effects are normally ignored.

In order to describe the primary parameters for dEDFAs, we must use the lumped-gain systems as described in Section 11.1.1. and 11.1.2 as a point of reference, in which the dispersion is an essential parameter. For ultralong transmission, it is the mean dispersion of the transmission line that is important, since the repeater length according to [23] should be chosen shorter than the dispersion length. The dispersion length is the distance where the pulse width is two-and-a-half times the initial pulse width when the nonlinear Kerr effect is ignored. This length is given by the fiber dispersion and the pulse width of the optical pulse [23]. Therefore, in a lumped-gain system employing the nonlinear Kerr effect to form optical solitary pulses, the amplifier dispersion may be negative without influencing the system performance, as long as the mean dispersion is positive. In a transmission system based on homogeneously distributed gain, the mean dispersion is equal to the dispersion of one distributed EDFA. Consequently, it is essential to analyze the dispersion of each distributed EDFA. As a first-order approximation, it is assumed that the dispersion of the fiber may be calculated using the usual theory for passive optical fibers and ignoring the influence on the refractive index profile from the nonlinear refractive index. The total fiber dispersion may consequently be found using more conventional models for propagation of electrical fields in optical fibers.

The effective core area, which is essential for the pulse power and thus the timing jitter [26] (see later in this description), may also be evaluated from similar models for passive fibers by use of the relation

$$A_{\text{eff}} = \frac{\left(\int_A \psi^2(r, \theta)\, dA\right)^2}{\int_A \psi^4(r, \theta)\, dA} \tag{11.11}$$

where $\psi(r, \theta)$ is the modal distribution of the fundamental fiber mode field and A is the cross section of the fiber, and (r, θ) are polar coordinates [27]. The parameter A_{eff} is only dependent on the index profile of the fiber, since this defines the modal field distribution. Thus, A_{eff} is calculated without the influence of any of the design parameters concerning the erbium dopant profile. For an ideal transmission line, the gain is at a constant level, which is equal to 1. In a real transmission line, based on either lumped gain or distributed gain, the gain varies along the transmission line, and ASE power is accumulated. The gain variation along the transmission line is described by the relative gain $g(x)$ and results in pulse power fluctuations as a function of the length coordinate x. The fidelity of a distributed fiber is evaluated on the basis of the main errors, which affect the transmission system. These are fluctuations of the detected pulse power and uncertainty in the pulse arrival times [28].

Following the description of [29], for ultralong pulse transmission over several distributed amplifiers, we will define the distance between two pumping stations as the amplification period. When the total transmission length is much larger than the distance of one amplification period, the signal-to-noise ratio after each distributed EDFA is calculated according to values for the signal and noise power levels after the amplifier. For distributed amplification, two boundary cases exist, co-propagating pump and signal and counter-propagating pump and signal. Between these cases, the fraction of the two pump powers may be varied continuously (see Section 11.2.1 for a detailed description). Considering the case of only the co-propagating pump and signal, the power level of both signal and noise is increased in the beginning of the amplification period and attenuated at the end of the amplification period. Therefore, the mean power level along the transmission line is higher than the initial power level for both the signal and noise power. For the counter-propagating pump and signal, the situation is reversed. In this situation, the power level of both signal and noise is attenuated at the beginning of the amplification period and increased at the end of the amplification period. Therefore, in this situation the power level is lower than the initial power level for both the signal and noise power. In order to take advantage of the nonlinear Kerr effect to form solitary pulses, the mean energy along the transmission line in a signal given by a first-order soliton is [23]

$$\langle E\rangle = \ln(\sqrt{2} + 1)\cdot\frac{D}{\tau}\cdot\frac{\lambda^3\cdot A_{\text{eff}}}{\pi^2\cdot c\cdot n_{\text{ke}}^l} \tag{11.12}$$

where D is the fiber dispersion, τ is the pulse width at half maximum intensity, λ is the signal wavelength, A_{eff} is the effective core area defined according to (11.11), c is the vacuum velocity of light, and n_{ke}^l is the nonlinear Kerr coefficient expressed in terms of intensity ($= 3.2 \times 10^{-20\ \text{m}^2\text{W}}$). Assuming that the dispersion length is much larger than the amplification period, the power level for both signal and noise at point z along the transmission line is expressed as in [23]:

$$P(z) = P_0\cdot\exp\left(\int_0^z g(x)\,dx\right) \tag{11.13}$$

For distributed EDFAs, it must be noted that

$$\exp\left(\int_0^L g(x)\,dx\right) = 1 \tag{11.14}$$

where L is the amplification period. From (11.13) we see that the path average power is given by

$$\langle P\rangle = P_0\cdot\frac{1}{L}\cdot\int_0^L \exp\left(\int_0^z g(x)\,dx\right)dz = P_0\cdot R_g \tag{11.15}$$

which is a more general way to express (11.6), necessary for the detailed description of distributed amplifiers. It must be noted that the factor R_g is a measure of the signal power variation (normally called the *signal excursion*) along the transmission line.

From this we find that

$$\langle E\rangle = E_s\cdot R_g \tag{11.16}$$

where E_s is the signal pulse energy at the beginning of the amplification period. Considering the distributed amplifier as a discrete amplifier in the overall transmission line, the noise energy in the signal mode after each amplification period is

$$\tilde{E} = \tilde{n} \cdot h \cdot \nu_s \tag{11.17}$$

where $\tilde{n}$ is the number of photons in the signal mode generated by spontaneous emission, and $h \cdot \nu_s$ is the photon energy at the signal wavelength ν_s. After the passage of $N = L_{tot}/L$ amplification periods, the noise energy in the signal mode is

$$\tilde{E} = \frac{1}{2} \cdot L_{tot} \cdot \alpha_{ASE} \cdot h \cdot \nu_s \tag{11.18}$$

The key parameter α_{ASE} is related to the noise figure by $\alpha_{ASE} = 2\tilde{n}/L = (F - 1)/L$. The factor of 2 arises as the generated ASE photons are equally divided in the two orthogonal polarizations of the fiber. Assuming that the receiver consists of an optical filter of bandwidth B_0 followed by a photodetector and an integrator, the energy within one time slot ΔT is measured [28]. The number of communication channels M to which the receiver is sensitive is then $M = K \cdot B_0 \cdot \Delta T$, where $K = 1$ if the detector is sensitive to one polarization, and $K = 2$ if the detector is sensitive to two polarizations.

As will be shown in Section 11.1.4, the electrical signal-to-noise ratio for a mark in a digital signal may be improved by the minimization of the product $\tilde{E} \cdot R_g$. In the following section, we shall see that besides the noise power error, another major error source is timing jitter. This error occurs when spectral noise is added to the signal, resulting in fluctuations of the group velocity of a transmitted pulse train, which result in timing jitter, often called the *Gordon Haus effect* [26]. Also, this error source may be reduced by minimizing $\alpha_{ASE} \cdot R_g$.

It is therefore evident that the product $\alpha_{ASE} \cdot R_g$ is an essential figure in the description of distributed EDFAs, and the dEDFA is optimally designed with respect to the influence of the noise by minimizing the product $\alpha_{ASE} \cdot R_g$. The factor α_{ASE} is, as mentioned, the number of generated ASE photons per hertz, per second, and per unit length of the amplifier.

In the following section, we will describe the noise properties in more detail, and the focus will therefore be on the generation of ASE photons. In the present section, however, it is important to remember that the factor R_g, which describes the signal excursions, is also an essential figure of merit for distinguishing between discrete and distributed amplification. The difference is illustrated qualitatively with an example in Figure 11.3, where it is shown that the signal excursions (and thereby the factor R_g) are significantly larger in the case of lumped gain.

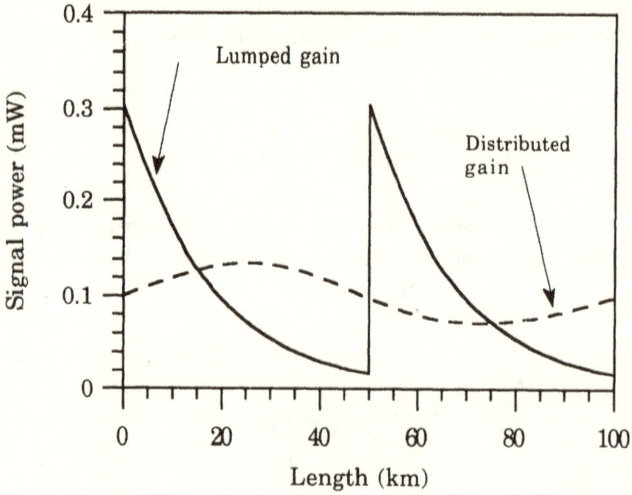

Figure 11.3 Signal power excursions over a distance of 100 km as a function of the length. The solid curves correspond to a transmission link applying lumped gain, and the dashed is for a distributed amplifier. The pump power consumption for both systems is 100 mW supplied by 1480-nm sources.

11.1.4 Noise Performance of Distributed Fiber Amplifiers

In this section, we will discuss the expressions for the two most significant noise contributions in soliton communication systems. These are amplitude noise and variance in the pulse arrival time at the detector, and they originate from the ASE added by the fiber amplifiers.

It should at this point be noted that one of the first treatments of the subject of noise in distributed fiber amplifiers was presented in [30]. However, for the sake of continuity in the use of parameters, the present description is based on the results of [29].

In discussing the amplitude noise properties of dEDFAs, note that, in a real receiver, there is always more than just the signal mode received. The number of signal modes to which the receiver is sensitive is denoted by $M = KB_0\Delta T$. Here B_0 is the detector bandwidth and ΔT is the time constant of the integrator in the detector [28]. If $K = 1$, the detector is sensitive to one polarization, and, if $K = 2$, the detector is sensitive to two polarizations. In [29] it is shown that the electrical signal-to-noise ratio for a mark is

$$\frac{S}{N} \cong \frac{E_s + 2M\tilde{E}}{2\tilde{E}} = \frac{\langle E(1)\rangle + 2R_gM\tilde{E}}{2R_g\tilde{E}} \tag{11.19}$$

where (11.16) has been used. Since $\tilde{E}$, according to (11.18), is proportional to α_{ASE}, we find, as mentioned in Section 11.1.3, that the product $\alpha_{\text{ASE}} \cdot R_g$ is central in the evaluation of the signal-to-noise ratio.

We will now turn our attention to the variance in pulse arrival time (or timing jitter), which is the other major noise contribution in soliton communication systems. This error occurs when spectral noise is added to the signal, resulting in fluctuations of the group velocity of a transmitted pulse train. The variance for the arrival time for a pulse may, according to [29, 31], may be expressed as

$$\langle \delta t^2 \rangle = \frac{1.763 \cdot h}{9} \cdot n_{ke}^{l} \cdot \frac{D \cdot L_{\text{tot}}^3}{\tau \cdot A_{\text{eff}}} \cdot \frac{\alpha_{\text{ASE}}}{2} \cdot R_g \tag{11.20}$$

where the parameters are as described in Section 11.1.3. In [29] it is found that the timing jitter reduces with increasing signal power and decreasing noise power. Equation (11.12) shows that the higher the A_{eff}, the higher the signal power level that is required to support soliton operation. Thus, from this point of view, fibers with high A_{eff} are desirable. An increasing A_{eff} may, as we shall see in Section 11.2, be achieved by reducing the cutoff wavelength for a fixed NA. However, this results in a reduced dispersion that for NA higher than 0.15 is negative. For soliton operation, a positive mean dispersion is necessary for the transmission of optical solitary pulses [23]. In lumped-gain systems, the mean dispersion is the average of the dispersion in the passive fiber section and the following active fiber section. Therefore, in lumped-gain systems, the restrictions to NA, cutoff wavelength, and A_{eff} are not as important as in distributed active fibers. Consequently, it must be emphasized that timing jitter is reduced by reducing the noise power relative to the signal power level, but in a transmission line based on distributed amplification, at the same time, it is necessary to ensure a positive mean dispersion. This tradeoff will be the subject of the following section, where we will discuss how the distributed amplifier is optimally designed. It must also be pointed out that the product $\alpha_{\text{ASE}} \cdot R_g$ should be minimized in order to reduce the timing jitter. This means that with respect to both major noise contributions, the object is to reduce $\alpha_{\text{ASE}} \cdot R_g$ as much as possible to achieve optimal noise performance.

11.2 DESIGN OF DISTRIBUTED EDFA LINKS FOR LONG-DISTANCE TRANSMISSION

The thorough description of central parameters for the analysis of distributed EDFAs presented in the previous section now allows us to move a step further, to the design of dEDFAs. Therefore, in this section, we will focus on the special properties that have to be considered in the planning of soliton transmission systems applying dEDFAs. Before we go into details concerning the design, however, a few limitations to the description should be made. These are primarily with respect to the choice of the pump wavelength. It has been demonstrated that the 1480-nm pump wavelength for the distributed EDFA is more efficient compared to the 980-nm pumping band due to the low intrinsic losses at 1480 nm, even though this results in increased noise [32]. The fact that the pump light has to propagate over long distances therefore becomes decisive, and in the following analysis we have chosen to use this pumping band. Furthermore, a fixed pump power of 100 mW will be applied unless specified differently. Finally, we have for the results presented in Section 11.2 and 11.2.1 used a signal wavelength equal to 1535 nm, where the EDFA has a peak in its emission cross-section spectrum.

As a first step towards a useful design of a distributed erbium-doped amplifier, we will first establish relationships between waveguide parameters, dispersion, and effective core area. Calculation of the dispersion for a fiber with a step index profile as a function of the cutoff wavelength for the LP_{11} mode and with the NA as parameter leads to the results in Figure 11.4(a). Figure 11.4(b) shows the effective core area A_{eff} for the same refractive index profiles calculated by integration of the electrical field in the fiber as those described in (11.11). From Figure 11.4(a), NA clearly has to be kept below 0.3 to ensure positive dispersion when the cutoff wavelength has to be low enough to ensure that the fiber is operated with a single

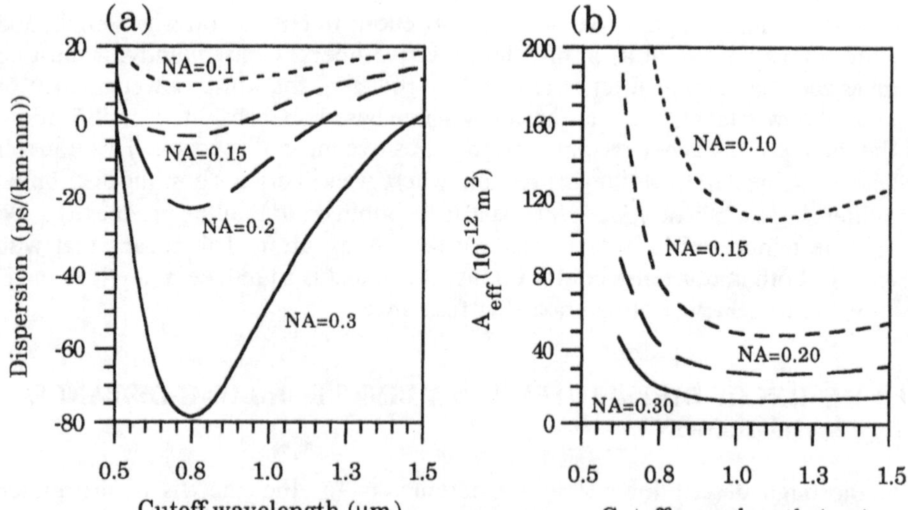

Figure 11.4 (a) The total fiber dispersion as a function of the LP_{11} cutoff wavelength. (b) The effective core area, calculated for a step-index profile as a function of the LP_{11} cutoff wavelength with the NA of the fiber as parameter. (Data from [33].)

pump mode. For NA equal to 0.2, the dispersion changes from positive to negative values for cutoff wavelengths lower than 1160 nm. Having NA equal to 0.1, the dispersion is positive for any cutoff wavelength. The curves show a minimum dispersion for λ_c between 700 and 800 nm where the waveguide dispersion is most pronounced. For any fixed value of the NA, the total fiber dispersion is dominated by the material dispersion for small values of the cutoff wavelength, since the electrical field in this situation is poorly guided. For large values of the cutoff wavelength, the electrical field is broadened, so again the total fiber dispersion is dominated by the material dispersion. In the region between these two boundary cases, the waveguide dispersion has an influence, and therefore an extremum is formed.

As illustrated in Figure 11.4(b), the effective core area decreases with increasing NA. This is due to the guiding properties of the high refractive index obtained in the fiber core, the higher index difference (and hence the higher NA), the better guiding property, and therefore the lower effective core area. For the various values of cutoff wavelength, there exist a minimum effective core area near cutoff wavelengths equal to 1000 nm. For small cutoff wavelengths, the electrical field is poorly guided, and there is hence a large A_{eff}. For larger values of the cutoff wavelengths, the electrical field becomes well guided, and therefore A_{eff} is decreased. For even larger cutoff wavelengths, the electrical field is broadened, and therefore A_{eff} is increased.

As described previously, the product $\alpha_{ASE} \cdot R_g$ is an important figure when the two system errors, noise power error, and timing jitter are considered (see (11.19) and (11.20)). Calculation of this product with the NA and the cutoff wavelength as parameter gives the results shown in Figure 11.5. This is calculated for a 100-km-long distributed EDFA with equal backward- and forward-propagating pumping. The confinement factor, defined as the erbium dopant radius relative to the fiber core radius, is equal to 0.8. Through these fixed values, further calculations have shown that the figure is qualitatively valid for other values of the above parameters. The solid curve represents the zero dispersion limit. Above this limit, the total fiber dispersion is positive, whereas it is negative below. The figure clearly demonstrates that the optimum fiber is achieved using as high NA as possible, but at the same time the NA should be chosen in the region where the total fiber dispersion is positive.

In a long transmission line consisting of cascade coupled dEDFAs, it is, as mentioned earlier, essential to maintain a small and positive dispersion for propagation of solitary pulses. Therefore, we have for the following discussion chosen to use a fiber with the NA equal to 0.2 and the cutoff wavelength equal to 1220 nm. According to Figure 11.4, this gives a total fiber dispersion equal to 3.2 ps/(nm · km) and an effective core area equal to 27.7 $(\mu m)^2$. The total ASE power co-propagating with the signal was in all the simulations on the order of 10^{-6}W, which is far from the saturation level. This fiber gives, as shown in Figure 11.5, an $\alpha_{ASE} \cdot R_g$ product

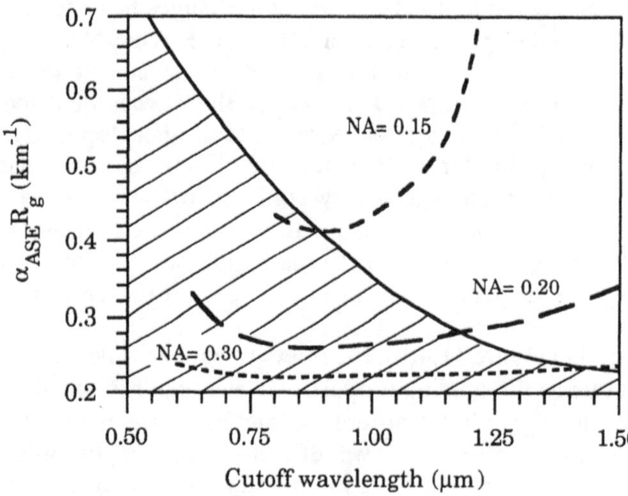

Figure 11.5 The $\alpha_{ASE} \cdot R_g$ product as a function of the LP_{11} cutoff wavelength for different values of the NA. In the hatched area, the total fiber dispersion is negative. (Data from [33].)

equal to 0.29 km^{-1} for 50-mW pump power co-propagating and counter-propagating with the signal.

It should finally be mentioned that even though the investigations presented here are made for a step-index fiber, the results may, with minor modifications, be applied for other profiles. The purpose of these modifications, not surprisingly, is to ensure positive dispersion together with reliable fabrication of the fibers. However, due to the large amount of similarity with the results for step-index fibers, further results are omitted here.

11.2.1 Optimum Pump Scheme for Distributed Fiber Amplifiers

In the previous analysis, the pump power has been distributed with equal power from both ends of the fiber section. To evaluate this choice, and, furthermore, get a better understanding of how to pump the distributed EDFAs optimally, we will look at this problem in more detail. Figure 11.6(a) shows the $\alpha_{ASE} \cdot R_g$ product as a function of the ratio between forward- and backward-propagating pumping for different lengths of the distributed EDFA. It can be found that the product $\alpha_{ASE} \cdot R_g$ is different, depending on whether backward or forward pumping is used, and lowest for the shortest fiber.

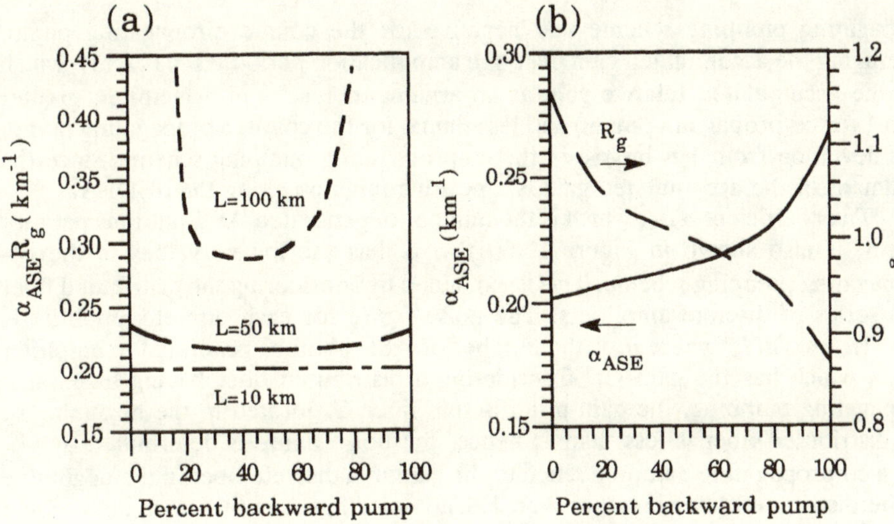

Figure 11.6 (a) The $\alpha_{ASE} \cdot R_g$ product as a function of the backward-propagating pump power relative to the forward-propagating pump power for three different fiber lengths. (b) α_{ASE} and R_g as a function of the backward-propagating pump power relative to the forward-propagating pump power in a 50km-long fiber with NA = 0.2 and cutoff wavelength equal to 1220 nm. (Data for both parts of the figure from [29].)

It is generally obvious that the lowest $\alpha_{ASE} \cdot R_g$ product is achieved for the shortest length of fiber. It is also worth noting that the curves show a minimum near an equal ratio between backward- and forward-propagating pumping. The precise minimum for backward-propagating pump powers is located a little higher than the forward-propagating pump power. The displacement from an equal ratio, though, is negligible; therefore, we conclude that the optimum pumping scheme is equal pumping of the fiber from both ends. The curves are not symmetric; they show an advantage in 100% counter-propagating pump and signal compared to 100% co-propagating pump and signal. This is explained by the two factors represented in the $\alpha_{ASE} \cdot R_g$ product. The coefficient R_g, which expresses the mean power level deviation from the initial power level, is larger than 1 for the co-propagating pump and less than 1 for the counter-propagating pump. This is illustrated in Figure 11.6(b), which shows the factors R_g and α_{ASE} as a function of the ratio between forward- and backward-propagating pump. The used fiber length for this figure is 50 km, with an NA for the fiber equal to 0.2 and a cutoff wavelength equal to 1220 nm. The curve is not symmetric around $R_g = 1$, which is due to the accumulated relative gain described by an exponential relation. The accumulated relative gain is a function that begins and ends at zero, but is positive for the co-propagating pumping scheme and negative for the counter-propagating pump. Averaging the accumulated gain over one amplification period, as in (11.15), which has the accumulated relative gain as an argument, results in a factor R_g greater than 1 for co-propagating pump and less than 1 for the counter-propagating pump. The deviation from 1 is largest in the co-propagating pumping scheme, since the argument of the accumulated gain is a positive function along the dEDFA.

The coefficient α_{ASE}, which is the number of generated ASE photons per unit length, is also shown in Figure 11.6(b). α_{ASE} has the lowest values in the co-propagating pumping scheme. This is explained by considering the distributed fiber as a series of discrete amplifiers. The noise figure for each discrete amplifier is $F_n = (n_n + 1)/G_n$, where n_n is the number of ASE photons generated in amplifier No. n which has the gain G_n. Considering a distributed fiber having backward-propagating pumping, the gain in a discrete fiber G_n located in the beginning of the distributed fiber is less than 1; hence, the noise figure is determined by G_n. For a co-propagating pumping scheme, the gain of a discrete fiber in the beginning of the distributed fiber is larger than 1. The noise figure is therefore determined by both n_n and G_n, and so is less than in the counter-propagating situation. Looking at the total distributed fiber, the overall noise figure is higher than a noise figure in any infinitesimal part of the fiber. The noise in a forward-pumped distributed fiber is therefore lower than in a backward-pumped distributed fiber. Fundamentally, α_{ASE} is above the limit given by $\alpha_i \cdot 1.74 = 92.31 \cdot 10^{-3}\ \text{km}^{-1}$, as previously mentioned.

11.2.2 Optimum Signal Wavelength for a Distributed EDFA

In the major part of this book, the focus has been the design of discrete or lumped amplifiers, and as we have seen in Chapters 5 to 8, the specific spectral properties of erbium play a central role in the way the amplifier operates. It is therefore an obvious task to examine the influence of the emission and absorption cross sections on the properties of dEDFAs. This has been done in [24,34], and we will here go through the primary results.

We consider two fibers that are both based on a GeO_2/SiO_2 host glass composition. One fiber is purely co-doped with germanium, which is the Ge/Er fiber. The other fiber has an additional low aluminum co-dopant concentration, which is the Al/Ge/Er fiber and ensures that the cross-section spectra are dominated by the aluminum spectra, as shown in [35]. The cross-section spectra are similar to those presented in [21]. The analysis is based on an accurate numerical model for the dEDFA [21]. Throughout the description in this section, the length of the dEDFA is 100 km and the fiber design is as described above, with an NA of 0.2 and a cutoff wavelength of 1220 nm. The dEDFA is bi-directionally pumped at 1480 nm and the pump power values are the total launched pump power.

Operating at a fixed signal wavelength, the necessary pump power for transparency decreases with increasing erbium dopant concentration until the signal absorption introduced by the erbium results in an overall signal attenuation. This defines an optimum erbium concentration where the pump power is minimized. Figure 11.7(a) displays the necessary pump power for transparency and the corresponding noise figure at this erbium concentration. From the figure we conclude that the minimum pump power for transparency is achieved for the signal wavelengths 1534 and 1554 nm for the Ge/Er fiber and 1562 nm for the Al/Ge/Er fiber. In Figure 11.7(a), the noise figure is also shown where the pump power exhibits minimum. However, the noise figure increases with increasing erbium concentration. Therefore, the minimum noise figure is achieved when only the erbium concentration is sufficient in the case where an infinite pump power is used to obtain transparency. This defines a concentration described in [21] as the critical erbium concentration. Thus, for a fixed signal wavelength, the erbium concentration has to be in the range between the critical erbium concentration and the optimum erbium concentration. A reasonable analysis of the noise figures is therefore based on a fixed pump power level. Calculations of the noise figure based on an equal pump power level are shown in Figure 11.7(b) (for a total of 100 mW pump power). We here find that the most efficient signal wavelengths coincide with the wave-

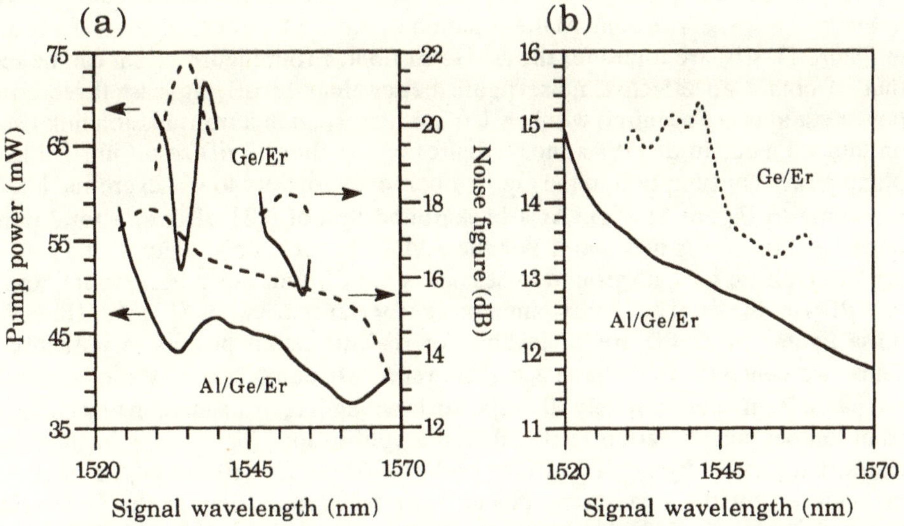

Figure 11.7 (a) Minimum pump power for transparency (solid) and the corresponding noise figure (dashed) versus signal wavelength in the Ge/Er and the Al/Ge/Er fiber. (b) Noise figure versus signal wavelength for a total pump power of 100 mW, calculated for the Ge/Er and the Al/Ge/Er fiber. (Data from [24].)

lengths, where the minimum in the necessary pump power for transparency occurs [35]. Furthermore, the noise figure in the Ge/Er fiber shows an absolute minimum around the 1554-nm signal wavelength, where the noise figure is 13.1 dB. In comparison, the noise figure at the signal wavelength 1534 nm is 14.4 dB. Therefore, the following analysis is concentrated on the optimum signal wavelengths 1554 nm for the Ge/Er fiber and 1562 nm for the Al/Ge/Er fiber. Further calculations have shown that the optimum signal wavelengths concluded above are identical to the signal wavelengths resulting in the lowest signal excursions. Additionally, they are independent of the fiber length and the erbium dopant radius relative to the fiber core radius (normally known as the confinement factor).

In dEDFAs, the background loss is essential due to the relatively long length of the fiber. For the above conclusions, it is assumed that the background loss in both fibers is 0.23 dB/km. For the dEDFA purely co-doped with germanium, this is a realistic background loss [36]. However, introducing aluminum may lead to an additional background loss, even for a very low aluminum co-dopant concentration. Furthermore, the background loss of a dEDFA may be increased due to introduced impurities in the fabrication process [2]. In the case of transparency along the whole fiber, an increased background loss causes an increase in both the necessary pump power for transparency and in the generation of ASE [22]. Calculating the necessary pump power for transparency as a function of the background loss with fixed noise figures, the results displayed in Figure 11.8 are achieved. Here, the curves shown in Figure 11.8(a) correspond to the situation of the Ge/Er fiber, whereas the results in Figure 11.8(b) are found for the Al/Ge/Er fiber. From Figure 11.2 it can be seen that to obtain an effective noise figure better than 13 dB, at least three pump power stations are required within a 100-km fiber span in a transmission link based on lumped gain. In dEDFAs, noise figures of less than 13 dB result in significant pump power consumption in the Ge/Er fiber, even for very low background losses, according to Figure 11.8(a). At a background loss of 0.23 dB/km, a total pump power of 110 mW is necessary. For the Al/Ge/Er fiber, noise figures of 13 dB or less are realistic for background losses below 0.3 dB/km. For a background loss of 0.29 dB/km, the total necessary pump power for transparency is 150 mW. However, noise figures of 13 dB are achievable for realistic pump powers in both fibers. Thus, we conclude that the spacing between two pump power stations may be increased from approximately 30 to 100 km in a long-haul transmission link changing from lumped amplification to distributed amplification. Figure 11.8 further illustrates that a noise figure of 13 dB in both the Al/Ge/Er and the Ge/Er fiber may be achieved for the same pump power if the background loss of the Ge/Er fiber is approximately 0.05 dB/km less than the background loss of the Al/Ge/Er fiber.

The other property that may be evaluated from Figure 11.8, is how the pump power necessary to obtain a given noise figure is affected by the erbium confinement factor. This parameter expresses, as discussed in Chapter 5, the ratio between the erbium dopant radius and the fiber core radius. The curves in Figure 11.8 show

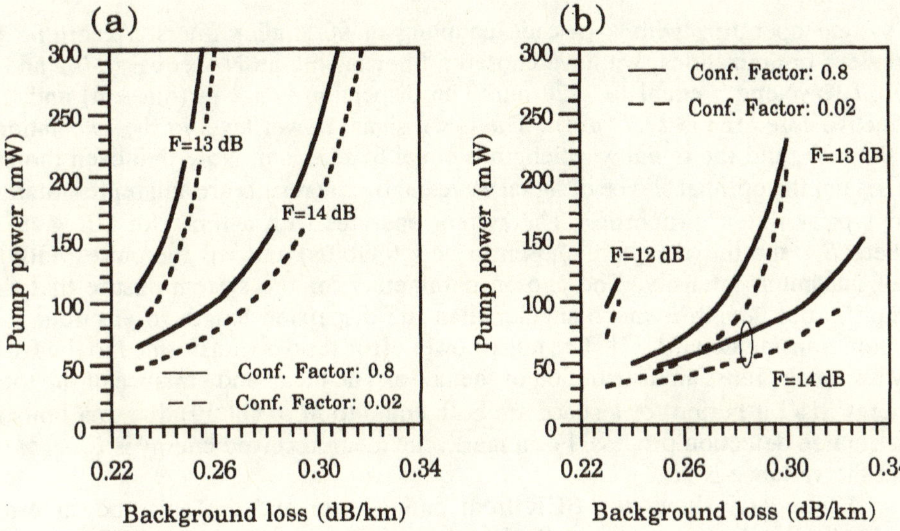

Figure 11.8 Necessary pump power for transparency versus background loss at the optimum signal wavelength. The noise figure F in decibels is parameter. The two sets of curves correspond to a confinement factor of 0.8 (solid curves) and 0.02 (dashed curves), respectively. (a) Results for the Ge/Er fiber. (Data from [24].) (b) Curves are calculated for the Al/Ge/Er fiber.

that in the case of the Ge/Er fiber, the pump power required to obtain a noise figure of 13 dB may be lowered by more than 30 mW by changing the confinement factor from 0.8 to 0.02. For an Al/Ge/Er fiber, a relative pump power decrease of more than 20% may be expected by changing to distributed amplifiers fabricated by the seed fiber technique (see Chapter 2). All these improvements naturally provide that the background loss is not increased simultaneously with the stronger erbium confinement.

11.2.3 Noise Evaluation in a 9000-km System Employing Distributed Gain

Having discussed the waveguide design, the co-dopant choice, and the erbium confinement properties of a distributed fiber amplifier, it is also interesting to quantify the sizes of the different types of noise described in Section 11.1.4. For this purpose, we follow the description of [29] and focus on the performance of an ultralong transmission line, which may be evaluated according to (11.19) and (11.20) for the noise power error and the timing jitter, both represented through BER.

As an example of a 9000-km-long transmission system based on distributed amplification without any ASE- or timing-jitter reducing components, we consider a system operating with a cascade coupling of 90 similar fibers. According to previous optimizations, we have chosen a fiber having an NA equal to 0.2 and a cutoff wavelength equal to 1220 nm. The dispersion is 3.2 ps/(nm·km) and the effective core area is 27.7 (μm)^2. The mean signal power level in the calculations is 0.1 mW, and the signal wavelength is equal to 1535 nm. Note that even though this is not the optimal choice of signal wavelength, the results are still representative for typical system properties. The system operates with a time slot $1/B = 5 \cdot \tau$, where B is the bit rate (here chosen to be 2.6 Gbit/s) and τ is the pulse width at half maximum intensity. The chosen parameters for the system ensure that the amplification length is much smaller than the dispersion length to ensure stable soliton transmission [31,37]. The noise power error ratio is transformed to the BER by assuming Gaussian distribution of the noise. The mean and variance of the total energy in a bit period for a space are both equal to M in (11.19) due to a Poisson distributed detection process. For a mark, the mean received energy is $E_s + M \cdot \tilde{E}$ and the variance is $2E_s\tilde{E}$.

Figure 11.9 shows the BER from timing jitter and the noise power error versus the product of α_{ASE} and R_g. For the noise power error, the BER is shown for three different values of M. We conclude from this figure that for the 2.6-Gbit/s system considered here, the BER is limited by timing jitter for any value of $\alpha_{\text{ASE}} \cdot R_g$. Calculations performed for other bit rates, furthermore, leads to the same conclusion. This implies that the timing jitter is the essential property, and for BER less than 10^{-12} we have to ensure that $\alpha_{\text{ASE}} \cdot R_g$ is less than 0.3 for the 2.6-Gbit/s system.

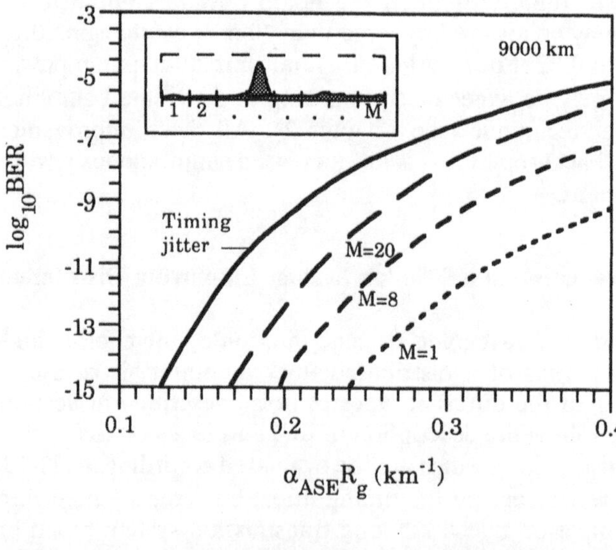

Figure 11.9 The BER for a 2.6-Gbit/s system with the $\alpha_{\text{ASE}} \cdot R_g$ product as parameter. (Data from [33].)

Returning to Figure 11.5, we find that $\alpha_{ASE} \cdot R_g$ may be reduced either by using a fiber with a high NA, and hence a high cutoff wavelength, or by reducing the spacing between two pump stations. The system we have investigated here had an $\alpha_{ASE} \cdot R_g$ product equal to 0.3 for 100-km spacing between each pump station and equal to 0.2 for 10-km spacing. Figure 11.5 shows that $\alpha_{ASE} \cdot R_g$ may be reduced to lower than 0.22 for NA = 0.3 and a spacing between the pump stations between 50 km and 100 km. It is, though, very important to notice that the above examples have a ratio τ/D equal to 24 (nm·km). For lumped gain systems, it is shown in [31] that the timing jitter is the limiting phenomenon for this value of the pulse width/dispersion ratio. In future work, it is therefore important to investigate the fraction τ/D for distributed fibers in ultralong transmission lines.

11.2.4 Lengthwise Distribution of Erbium Concentration for an Attenuation-Free Fiber

In this chapter we have discussed how both lumped and distributed gain may be used in optical communication systems employing the nonlinear Kerr effect for optimum transmission. This description has shown that systems with distributed EDFAs are superior to the lumped-gain system in the respect that the required distance between pump sources may be improved considerably.

For the purpose of demonstrating the optimum performance of the distributed fiber amplifiers, we have considered transmission link properties such as waveguide design, erbium concentration, choice of index-raising co-dopant, and optimum signal wavelength. However, in all of the previous description, it has been assumed that the distributed fiber amplifier had a uniform erbium dopant concentration along the entire fiber length. Aside from the reason that this clarifies the description, it is, furthermore, the simplest to achieve in practice. In [7] it was, however, suggested that a more complex distribution of the doping concentration be used, such as linear grading of dopant along the full length of the fiber. Even though a thorough analysis of the possible advantages of this approach has not yet been performed, the subject of this section will be a short presentation of a simplified approach in order to demonstrate the advantages of nonuniform erbium distribution.

As described in [7], an approximation of a specific grading function of the erbium concentration may be made by the interconnection of fiber sections with a different erbium concentration. A simple example of approximate grading of the fiber doping concentration is realized by connecting three sections of fiber to form an 80-km transmission link. This link is made by splicing two weakly doped fibers (each 20 km long) at the ends of a fiber 40 km long with a higher erbium concentration. The whole link is then pumped bi-directionally by launching 50 mW of 1480-nm light at each of the free ends of the 20-km sections.

To see the advantage of the lengthwise variation of the erbium concentration, the signal energy relative to the one coupled into the signal input end of the link is shown in Figure 11.10. Also illustrated is the relative signal energy for a uniformly doped dEDFA of equal length (80 km) and for similar pump power consumption. From Figure 11.10 it can be clearly seen that the signal level fluctuations along the length of the three section fiber link are lower than the fluctuations in the uniformly doped fiber. This lower signal excursion is, according to the discussion of Section 11.1.3, advantageous for the pulse transmission.

This improved performance of the multisection fiber may be understood better by imagining the situation in which the full fiber length consisted of the most dilutely doped fiber. In this case, the number of erbium ions would be too small to absorb enough pump energy and therefore too small to supply enough energy for the signal to overcome the background attenuation. However, a partial amplification takes place as represented by the curves in Figure 11.10, showing the relative signal energy at the ends of the transmission link. Besides this, there will be a significant residual pump power emitted from such a dilutely doped fiber. In order to absorb this residual pump, the more highly doped center section is introduced, whereby

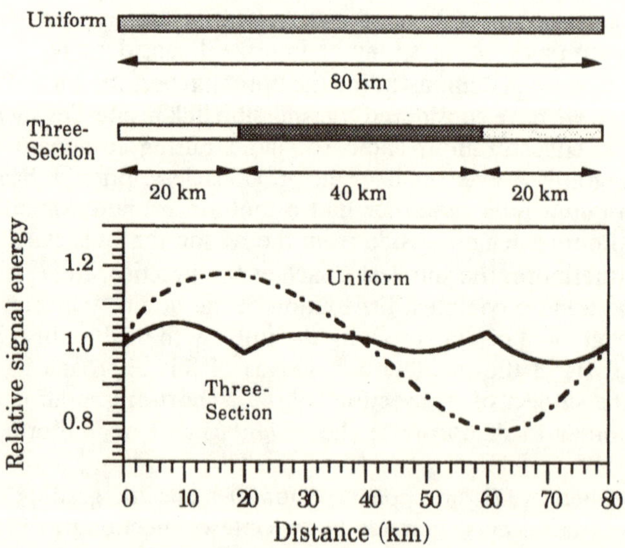

Figure 11.10 Relative signal energy fluctuation along 80-km-long transmission links consisting of either a uniformly doped dEDFA (with erbium concentration of $1.07 \cdot 10^{20}$ m^{-3}) or a three-section dEDFA (of which the end fibers have an erbium concentration of $0.8 \cdot 10^{20}$ m^{-3} and the center section an erbium concentration of $1.4 \cdot 10^{20}$ m^{-3}). (Data for the lower sketch from [7].)

the signal again undergoes a positive gain. Note, however, that if the higher erbium concentration was chosen in the entire length, the pump power would be insufficient to overcome the signal attenuation introduced by the erbium itself.

To conclude this multistage distributed amplifier suggestion, note that further improvement may be obtained with additional subsections of a different erbium concentration. The obvious advantage of this approach is that the pump power requirement for a given transmission link may be weakened or, alternatively, that the pumping distance may be extended for equal pump power consumption.

11.3 BUS NETWORKS WITH UNIFORMLY DISTRIBUTED GAIN

For the final subject of this chapter, we turn to a different application of the distributed erbium-doped fiber. The concern in this case is not with the problems of signal transmission over ultralong distances but instead with the wish of extending optical bus network architectures, both as a means of switching and as a means of broadcast distribution. The fundamental problem has until now been the accumulation of loss through the many tapping points such passive architectures has, which has limited these networks to a modest size. However, the advances of optical fiber amplifiers have also provided new possibilities for the network design for this application by extending the available power budget. As explained in [38], a single distributed EDFA is particularly suited to this application because the gain is distributed along the entire length of the network and is not specific to a particular location.

The intent here to give a short presentation of this alternative application of the dEDFA as a distribution and amplifying medium in bus networks. The idea behind such networks is to have a system in which the active fiber spine of the bus is kept at a high intensity and the individual tapping coefficients are low. The overall operation of the bus is thereby relatively insensitive to any single tap, thus providing great flexibility in terms of online reconfiguration of the network [38]. Of course, obtaining this ideal system is not without problems, since it might require the extraction of the signal at strategic points along the bus, and, more critically, the pump power remains in the backbone fiber in order to pump the erbium ions further along the fiber.

So far, the experiments and investigations on systems of this type are in a preliminary stage, but theoretical work indicates that bus systems of several kilometers are possible with practical pump powers [8]. Furthermore, transparent optical bus networks have been demonstrated using 825 m of lightly doped erbium fiber as the amplifying backbone of a bus network with five equally spaced tapping points [38]. The need for pump power was modest, since only 25 mW of 980-nm pump power was used to successfully balance the tapped signal power at each port. In addition, it must be noted that studies of the accumulation of spontaneous noise along the distributed erbium-doped fiber shows no degradation in the performance of a 140-Mb/s system.

This type of distributed amplifier has recently been combined with D-fiber technology [39] in order to provide increased tapping points [40] within a racking-type system. In a D-fiber, a part of the cladding has been removed on one side of the core, resulting in a D-shaped fiber cross section. Also, 980 nm pump sources were used to obtain amplification of a 1530-nm signal. It is also important to note that careful design of the D-fiber obtained a rejection of the pump-to-signal power at the evanescent tapping points larger than 30 dB [41].

11.4 SUMMARY

This chapter has dealt primarily with the central problems of ultralong transmission systems in which the nonlinear effects are used constructively. Such systems, which can also be called soliton transmission systems, may be realized either by the use of discrete amplifiers or by the application of distributed EDFAs. To evaluate these possibilities, basic properties of transmission through systems with discrete or lumped gain were described. It was shown in this connection that pulse transmission over more than 500 km requires the use of many amplifiers, each with limited gain. Over shorter distances, such as 100 km, the use of higher order ($N = 2.0$) solitons may also be advantageous, since the number of optical amplifiers may therefore be reduced.

In order to further quantify the problems combined with the use of lumped amplification, the central properties, such as accumulation of ASE and signal power fluctuation along the transmission line, were studied, which showed that to obtain an overall noise figure at an acceptable level (of around 13 dB for a 100-km transmission link) the maximum spacing of the amplifiers may be no more than 30 to 40 km. This further relates to the need of limiting the signal power excursions or fluctuations, since large spacing leads to more significant energy variations along the line followed by the possible problems of exciting higher order solitons or adding power to dispersive waves, which acts like noise on the overall system.

To overcome these problems, the distribution of erbium along the entire transmission line has been shown to be advantageous, since the ASE generation and the signal excursions may be reduced by this approach. The attractive properties of these distributed amplifiers have been demonstrated through the investigation of different design parameters analyzed in 1480-nm pumping schemes. The evaluation was performed through the inclusion of noise properties. From the analysis, it may be concluded that the optimum distributed EDFA is achieved using a step-index profile with a large index difference. However, in contrast to the situation of erbium-doped fibers for discrete amplifiers, it is not only a question of making the NA as high as possible. It is also important to maintain a positive total fiber dispersion, and a combination of optimum values of the NA and the cutoff wavelength is found. For cutoff wavelengths where the pump power is single-moded, the optimum NA is close to 0.3, with the cutoff wavelength in the range of 1450 to 1480 nm. However, the design tolerances of such a fiber might be too restrictive, and a more realistic suggestion is, therefore, a fiber with NA equal to 0.2 and a cutoff wavelength equal to 1220 nm. This results in a total fiber dispersion of 3.2 ps/(nm · km) and an effective core area equal to 27.7 $(m)^2$ at the signal wavelength equal to 1535 nm.

The optimum pumping scheme is bi-directional pumping with equal co- and counter-propagating pumping. Using a distributed EDFA with only one pump, the counter-propagating pump configuration is most advantageous. Furthermore, it should be noted that the ASE level in our analysis is found to be below the saturation level of the EDFA, even after 9000-km transmission. For realistic systems, among which an example is the investigated 9000-km-long transmission line, the timing jitter is found to be the most critical noise contribution compared to signal power fluctuations.

The comparison between lumped and distributed amplification shows that it is possible to reduce the required number of pump power stations by approximately 60%. This may be achieved by changing from lumped to distributed amplification in long-haul transmission with signal powers large enough to generate solitary pulses. Furthermore, two erbium-doped distributed fibers were compared: one co-doped with germanium and another that was additionally co-doped with a weak aluminum concentration high enough to ensure aluminum dominated emission and absorption cross-section spectra. It was found that the latter fiber provides the best performance, considering both pump power requirements and noise figures. This conclusion is, however, conditioned by the requirement that the background loss in the aluminum-doped fiber not exceed the background loss in the purely germanium-co-doped fiber by more than 0.05 dB/km for a noise figure of 13 dB.

The description of fiber amplifiers for ultralong transmission applications is concluded with the presentation of an approach [7] to limit the signal power excursions in the fiber. This idea is based on the use of a lengthwise variable erbium concentration along the transmission line. However, due to the extra difficulties of splicing different fibers together, the value of the approach is yet to be seen.

One very important factor restricting the application of distributed amplifiers should be pointed out here, that is, that these systems have to be installed completely from scratch rather than upgrade existing links with lumped amplifiers. Furthermore, it needs to be added that in systems with distributed amplification, the fibers are opaque unless pumped. This is unlike present links, but is fundamentally the case for all erbium-doped fibers.

Finally, the chapter gives a short survey of optical distribution networks based on lightly doped erbium fiber amplifiers. These are likely to enable high-capacity, multi-tap optical buses to be realized in the near future. It has been shown in this connection [38] that the pump power requirements and noise accumulation in such networks will not preclude the demonstration of useful optical distribution systems. It is furthermore expected that future use of small tapping coefficients will enable many hundreds of taps, thereby opening up many potential application areas, from physically small inter-rack distribution systems to large scale loop applications with a length of many kilometers.

REFERENCES

[1] Simpson, J. R., L. F. Mollenauer, K. S. Kranz, P. J. Lemaire, N. A. Olsson, H. T. Shang, and P. C. Becker, "A Distributed Erbium Doped Fiber Amplifier," *Proceedings of Optical Fiber Communication Conference OFC '90*, San Francisco, 22–26 Jan. 1990, postdeadline paper PD19.

[2] Simpson, J. R., H. T. Shang, L. F. Mollenauer, N. A. Olsson, P. C. Becker, K. S. Kranz, P. J. Lemaire, and M. J. Neubelt, "Performance of a Distributed Erbium-Doped Dispersion-Shifted Fiber Amplifier," *IEEE Journal of Lightwave Technology*, Vol. 9, No. 2, 1991, pp. 228–233.

[3] Olsson, N. A., "Lightwave Systems With Optical Amplifiers," *IEEE Journal of Lightwave Technology*, Vol. 7, No. 7, 1989, pp. 1071–1082.

[4] Hasegawa, A., and Y. Kodama, "Signal Transmission by Optical Solitons in Monomode Fiber," *Proceedings of the IEEE*, Vol. 69, No. 9, 1981, pp. 1145–1150.

[5] Hasegawa, A., *Plasma Instabilities and Nonlinear Effects*, New York: Springer-Verlag, 1975, p. 195.

[6] Nakazawa, M., Y. Kimura, K. Suzuki, and H. Kubota, "Wavelength Multiple Soliton Amplification and Transmission With an Er^{3+}-Doped Optical Amplifier," *Journal of Applied Physics*, Vol. 66, No. 7, 1989, pp. 2803–2812.

[7] Mollenauer, L. F., "Distributed Amplification for Lightwave Transmission System," European patent application, No. 90310604.5, 27/9-1990.

[8] Urquart, P., and T. J. Whitley, "Long Span Fiber Amplifiers," *Applied Optics*, Vol. 29, No. 24, 1990, pp. 3503–3509.

[9] Nakazawa, M., K. Suzuki, E. Yamada, and Y. Kimura, "5 Gbit/s, 250km Error-Free Soliton Transmission With Er^{3+}-Doped Fibre Amplifiers and Repeaters," *IEE Electronics Letters*, Vol. 26, No. 8, 1990, pp. 551–553.

[10] Rottwitt, K., A. Bjarklev, and J. H. Povlsen, "Optimum Use of Cascade Coupled In-Line Amplifiers for Soliton Communication Systems," *Optics Communications*, Vol. 84, No. 56, 1991, pp. 339–342.

[11] Cotter, D., "Observation of Stimulated Brillouin Scattering in Low-Loss Silica Fibre at 1.3 μm," *Electronics Letters*, Vol. 18, No. 12, 1982, pp. 495–496.

[12] Cotter, D., "Transient Stimulated Brillouin Scattering in Long Single-Mode Fibres," *Electronics Letters*, Vol. 18, No. 12, 1982, pp. 504–506.

[13] Doran, N. J., and K. J. Blow, "Solitons in Optical Communications," *IEEE Journal of Quantum Electronics*, Vol. 19, No. 12, 1983, pp. 1883–1888.

[14] Marcuse, D., Light Transmission Optics, New York: Van Nostrand Reinhold, 1982.

[15] Chiao, R. Y., E. Garmire, and C. H. Townes, "Self-Trapping of Optical Beams," *Physical Review Letters*, Vol. 13, 1964, p. 479.

[16] Hasagawa, A., and F. Tappert, "Transmission of Stationary Nonlinear Optical Pulses in Dispersive Dielectric Fibres, I Anomalous Dispersion," *Applied Physics Letters*, Vol. 23, 1981, pp. 142–144.

[17] Agrawal, G. P., *Nonlinear Fiber Optics*, London: Academic Press, 1989.

[18] Satsuna, J., and N. Yajima, "Initial Value Problems of One-Dimensional Self Modulation of Nonlinear Waves in Dispersive Media," *Supplement of the Progress of Theoretical Physics*, No. 55, 1974, pp. 284–306.

[19] Mollenauer, L. F., M. J. Neubelt, S. G. Evangelides, J. P. Gordon, J. R. Simpson, and L. G. Cohen, "Experimental Study of Soliton Transmission Over More Than 10, 000 km in Dispersion-Shifted Fiber," *Optics Letters*, Vol. 15, No. 21, 1990, pp. 1203–1205.

[20] Yevick, D., and B. Hermansson, "Soliton Analysis With the Propagating Beam Method," *Optical Communications*, Vol. 47, No. 2, 1983, pp. 101–106.

[21] Pedersen, B., A. Bjarklev, J. H. Povlsen, K. Dybdal, and C. C. Larsen, "The Design of Erbium-Doped Fiber Amplifiers," *IEEE Journal of Lightwave Technology*, Vol. 9, No. 9, 1991, pp. 1105–1112.

[22] Chen, D. N., and E. Desurvire, "Noise Performance Evaluation of Distributed Erbium-Doped Fiber Amplifiers With Bidirectional Pumping at 1.48 μm," *IEEE Photonics Technology Letters*, Vol. 4, No. 1, 1992, pp. 52–55.

[23] Mollenauer, L. F., S. G. Evangelides, and H. A. Haus, "Long-Distance Soliton Propagation Using Lumped Amplifiers and Dispersion Shifted Fiber," *IEEE Journal of Lightwave Technology*, Vol. 9, No. 2, 1991, pp. 194–197.

[24] Rottwitt, K., J. H. Povlsen, A. Bjarklev, O. Lumholt, B. Pedersen, and T. Rasmussen, "Noise in Distributed Erbium Doped Fibers," *IEEE Photonics Technology Letters*, Vol. 5, No. 2, 1993, pp. 218–221.

[25] Hasagawa, A., "Amplification and Reshaping of Optical Solitons in a Glass Fiber-IV: Use of the Stimulated Raman Process," *Optics Letters*, Vol. 8, No. 12, 1983, pp. 650–652.

[26] Gordon, J. P., and H. A. Haus, "Random Walk of Coherently Amplified Solitons in Optical Fiber Transmission," *Optics Letters*, Vol. 11, No. 10, 1986, pp. 665–667.

[27] Stolen, R. H., and C. Lin, "Self-Phase-Modulation in Silica Optical Fibers," *Physical Review A*, Vol. 17, No. 4, 1978, pp. 1448–1453.

[28] Shimoda, K., H. Takahasi, and C. H. Townes, "Fluctuations in Amplification of Quanta With Application to Maser Amplifiers," *Journal of the Physical Society of Japan*, Vol. 12, No. 6, 1957, pp. 686–700.

[29] Rottwitt, K., A. Bjarklev, J. H. Povlsen, O. Lumholt, and T. Rasmussen, "Fundamental Design of a Distributed Erbium Doped Fiber Amplifier for Long Distance Transmission," *IEEE Journal of Lightwave Technology*, Vol. 10, No. 11, 1992, pp. 1544–1552.

[30] Walker, G. R., D. M. Spirit, D. L. Williams, and S. T. Davey, "Noise Performance of Distributed Fibre Amplifiers," *IEE Electronics Letters*, Vol. 27, 1991, pp. 1390–1391.

[31] Gordon, J. P., and L. F. Mollenauer, "Effects of Fiber Nonlinearities and Amplifier Spacing on Ultra-Long Transmission," *IEEE Journal of Lightwave Technology*, Vol. 9, No. 2, 1991, pp. 170–173.

[32] Rottwitt, K., J. H. Povlsen, and A. Bjarklev, "Demand for Accuracy of the Attenuation Constant in Distributed Active Fibers," *SPIE OE/Fibers '91,* part: Fiber Laser Sources and Amplifiers III, Proc. Vol. 1581, paper. 20, 1991.

[33] Rottwitt, K., A. Bjarklev, O. Lumholt, J. H. Povlsen, and T. Rasmussen, "Design of Long Distance Distributed Erbium Doped Fibre Amplifier," *IEE Electronics Letters*, Vol. 28, No. 3, 1992, pp. 287–288.

[34] Rottwitt, K., J. H. Povlsen, A. Bjarklev, O. Lumholt, B. Pedersen, F. B. Pedersen, and T. Rasmussen, "Detailed Analysis of Distributed Erbium Doped Fiber Amplifiers," *Proc. Optical Amplifiers and Their Application OAA '92*, Santa Fe, 1992, paper FB6, pp. 178–181.

[35] Craig-Ryan, S. P., J. F. Massicott, M. Wilson, B. J. Ainslie, and R. Wyatt, "Optical Study of Low Concentration Er^{3+} Fibres for Efficient Power Amplifiers," *Proc. European Conference on Optical Communication ECOC '90*, Amsterdam, 1990, Vol. 1, pp. 571–574.

[36] Tanaka, D., A. Wada, T. Sakai, and R. Yamauchi, "Attenuation Free, Dispersion Shifted Fiber Doped With Distributed Erbium," *Proc. Optical Amplifiers and Their Application OAA '90*, Monterey, 1990, paper TuB4, pp. 138–141.

[37] Mollenauer, L. F., J. P. Gordon, and M. N. Islam, "Soliton Propagation in Long Fibers With Periodically Compensated Loss," *IEEE Journal of Quantum Electronics*, Vol. 22, No. 1, 1986, pp. 157–173.

[38] Whitley, T. J., C. A. Millar, S. P Craig-Ryan, and P. Urquhart, "Demonstration of a Distributed Optical Fibre Amplifier BUS network," *Proc. Optical Amplifiers and Their Application OAA '90*, 1990, paper WB2, pp. 236–239.

[39] Dakin, J., and B. Culshaw, *Optical Fiber Sensors: Principles and Components, Volume 1*, Artech House, 1989.

[40] MacKenzie, F., S. A. Cassidy, S. T. Davey, T. K. White, and D. L. Williams, "Dichroic Tap for Optical Multiple Access Systems With Distributed Amplification," *IEE Electronics Letters*, Vol. 27, No. 2, 1991, pp. 181–183.

[41] Ainslie, B. J., "Recent Advances in Discrete and Distributed Fibre Amplifiers," *Proc. European Conference on Optical Communication ECOC '91*, Paris, 1991, Vol. 2, invited paper, pp. 47–50.

Chapter 12
Amplifiers for Analog-Modulated Systems

It is at this point evident that the erbium-doped amplifiers are promising for future long-haul digital lightwave transmission system applications, and the EDFA is going to be an important factor in the development of digital system architecture. However, the mass deployment of single-mode cable and the availability of DFB lasers have opened up a new application: broadband analog networks [1]. The most widely deployed analog networks today are operated for *cable television* (CATV) systems, which are realized as coaxial cable broadband networks. However, the demand for new video services challenges suppliers of transmission technology to expand the bandwidth and performance of these coaxial networks. It is noteworthy that, according to [1], a practical bandwidth limit is reached around 600 MHz. The single-mode fiber-optic technology, which is characterized by extremely low signal loss and high bandwidth, therefore becomes a very interesting alternative to the coaxial technology. However, the digital applications have dominated the fiber technology, and at present no simple, economical interface exists for delivering video over digital networks. Therefore, analog FDM optical transmission systems, which eliminate the interface problem, are of interest, and there has been intensive research and development over the last decade [2–8].

The analog signals transported on optical broadband links are fundamentally different from those encountered in classical digital telephony, and it has therefore been necessary to develop an understanding of the critical parameters that affect the performance of these systems. From these investigations it appears that several parameters, which are of only secondary importance in traditionally baseband digital applications, are of prime importance in the design of analog links. We will, therefore, in Section 12.1 discuss the modulation formats used in analog FDM lightwave transmission and point out the important parameters relevant for their description. Thereafter, the aim of Section 12.2 is to provide a short review of some of the most central properties of the EDFA with respect to applications in analog communication systems. Finally, Section 12.3 describes recent reports comparing different modulation approaches in CATV distribution systems employing EDFAs. It is at this point important to stress that the present description in no way attempts to give a complete description of all aspects of the design and application of EDFAs for analog applications. The idea is merely to provide a short review of some of the latest highlights of a rapidly developing application area. In addition, the most widely used terms and key parameters for AM systems are defined.

12.1 PARAMETERS FOR CHARACTERIZATION OF ANALOG SYSTEMS

CATV networks, using analog FDM optical transmission systems, are becoming very popular because of their reliability and high quality [9]. In this section, we will briefly review the considered modulation formats and define the advantageous features normally used in connection with analog modulation.

12.1.1 AM or FM Formats

Two types of formats, AM and FM, are being considered for CATV networks. The photocurrent resulting from a received optical AM signal may be expressed as follows:

$$x_c(t) = A_c(1 + x(t)) \cos \omega_c t = A(t) \cos \omega_c t \tag{12.1}$$

where $|A(t)|$ is the envelope of the modulated carrier and $\omega_c = 2\pi f_c$ is the angular frequency of the carrier. Two unique features of the AM signal are that a carrier frequency component is present and that the envelope $|A(t)|$ of the modulated carrier has the same shape of $x(t)$ as long as the signal frequency $f_s \ll f_c$ and as long as $A(t) = A_c(1 + x(t))$ does not go negative. An important parameter of an AM signal is its modulation index m, which is defined as

$$m = \frac{[A(t)]_{\max} - [A(t)]_{\min}}{[A(t)]_{\max} + [A(t)]_{\min}} \tag{12.2}$$

If m exceeds 1, the carrier is said to be overmodulated, resulting in envelope distortion.

It should be mentioned that many large-bandwidth message signals, such as television video and high-speed data signals, are often transmitted using the *vestigial sideband* (VSB) modulation scheme. This modulation is derived by filtering double sideband or AM signals in such a fashion that one sideband is passed almost completely, while only a trace of the other sideband is included. The purpose is thereby to obtain the best compromise between bandwidth conservation, improved low-frequency response, and improved power efficiency. It is not the purpose of this description to go further into the advantages of this approach, but we will only note that the AM-VSB signal may be obtained by filtering the AM signal of (12.1). For further details, the reader is referred to [10].

Another often used term when CATV signal transmission is discussed is NTSC, which is an abbreviation for *National Television System Committee*. Video signals that obey these requirement standards have a bandwidth of 6 MHz, and the lower edge of the NTSC signal band is 1.25 MHz below the picture carrier. Other discrete-frequency components in a color television channel are the sound carrier and the color sub-carrier, which are located 4.5 and 3.58 MHz above the picture carrier, respectively. These properties are described in [11], but it should here also be pointed out that the amplitudes of the sound carrier and the color sub-carrier are much smaller than that of the picture carrier (at least 15 dB below). Hence, potential intermodulation products are negligible.

AM-VSB is attractive for distributing multi-channel television signals to subscribers because it is compatible with both existing CATV networks and current *very-high-frequency* (VHF) and *ultrahigh-frequency* (UHF) NTSC television receivers. However, the total loss budget of multi-channel AM-VSB FDM video transmission systems is small. An example of this has been mentioned in [9], where a loss budget as low as 5 to 7 dB is stated as the allowed limit for 40 channels. Therefore, video transmission systems without optical amplifiers have severely limited properties, and only trunk line applications are realistic.

The FM modulated signal can be written as

$$x_c(t) = A_c \cdot \cos[\omega_c t + k \int_{-\infty}^{t} x(\tau)\,d\tau] \tag{12.3}$$

which means that the phase of the signal consists of one contribution that increases linearly with time corresponding to a carrier angular frequency ω_c, and another contribution that varies according to the modulated signal $x(t)$. Where $x(t)$ is assumed to be sinusoidal (or a tone) ($x(t) = A_m \cos \omega_m t$), the instantaneous phase deviation of the modulated signal is $\phi(t) = (kA_m/\omega_m) \cdot \sin \omega_m t = \beta \cdot \sin \omega_m t$. The parameter β, which is called the modulation index, represents the maximum phase deviation produced by the modulating tone.

The FM-FDM video transmission system is incompatible with both existing CATV networks and current VHF and UHF NTSC television receivers. To connect an FM-FDM video transmission system to existing CATV networks, therefore, FM/AM converters are needed, which naturally results in a cost increase. However, FM optical networks are still attractive, since they are compatible with *broadcast satellite intermediate frequency* (BS-IF) receivers, and the total loss budget is increased compared to the AM systems. Note, however, that although the increased loss budget is 20 dB for 16-channel transmission and 12 dB for 60-channel transmission [12], this is still insufficient for video distribution [9].

12.1.2 Carrier-to-Noise Ratios in Fiber-Optic AM Links

One of the important features of AM links for CATV applications is the *carrier-to-noise ratio* (CNR), defined as the ratio of the peak carrier power for a given channel to the noise floor near the carrier. Note that, according to NTSC standards, a noise bandwidth of 4 MHz is assumed, although a standard AM-VSB video channel occupies a total bandwidth of 6 MHz. The noise is specified in terms of a 4MHz bandwidth centered within the 6-MHz channel (see [11]). Note also, however, that these standards are not specified internationally, and they therefore vary among different references.

According to [11], the received signal CNR can be calculated from the following equation:

$$\mathrm{CNR}^{-1} = \frac{\mathrm{RIN}\cdot B}{1/2\cdot m^2} + \frac{2q\cdot(I_{\mathrm{ph}} + I_{\mathrm{dark}})\cdot M^2\cdot F(M)\cdot B}{1/2\cdot m^2M^2I_{\mathrm{ph}}^2} + \frac{\langle i_{\mathrm{ckt}}^2\rangle\cdot B}{1/2\cdot m^2M^2I_{\mathrm{ph}}^2} + \frac{\sigma_{IM}^2\cdot B}{1/2\cdot m^2M^2I_{\mathrm{ph}}^2} \tag{12.4}$$

where B is the signal bandwidth, m is the optical modulation index, I_{ph} is the received photocurrent, I_{dark} is the dark current of the photodiode, M is the gain of an avalanche photodiode ($M = 1$ for a pin diode), $F(M)$ is the excess noise factor of an avalanche photodiode, σ_{IM}^2 is the intermodulation product spectral noise density, and $\langle i_{\mathrm{ckt}}^2\rangle$ is the equivalent input noise current spectral density of a receiver. The term RIN denotes the *relative intensity noise* [13], defined as $\langle i^2\rangle/\langle i\rangle^2$, where $\langle i^2\rangle$ is the noise current spectral density and $\langle i\rangle$ is the detected dc photocurrent.

The first term in (12.4) accounts for the CNR limitation caused by system intensity noise, and laser intensity noise is particularly important here. The second term accounts for shot noise, dark current, and the excess noise limitation. The third term describes the influence of the thermal noise of the circuit, and the last term accounts for the degradation due to intermodulation noise. In [11] it is pointed out that the last term usually can be ignored for multi-channel FM video transmission due to the tolerance of FM video signals to noise. However, it can be a relevant factor in limiting CNR performance for multi-channel AM video transmission systems. The avalanche gain of an APD can improve the CNR by overcoming the receiver thermal noise as shown by the third term. However, it can also degrade the CNR performance by increasing the excess noise as shown by the second term. This defines an optimum APD gain for which the CNR is maximized.

12.1.3 Intermodulation Noise

For multi-channel signal transmission through a nonlinear device such as a laser diode, intermodulation distortions (IMDs) occur. These are normally separated in second-order IMDs ($\omega_1 - \omega_2$ or $\omega_1 + \omega_2$ type, where ω_1 and ω_2 are the angular frequency of two arbitrary *radio frequency* (RF) signals) and triple-beat products (especially $\omega_1 + \omega_2 - \omega_3$ type, which occurs within signal bands most frequently). Also the number of two-tone IMDs ($2\omega_1 - \omega_2$) increases with $N_c(N_c - 1)$, and the number of triple-beat products increases with the factor $N_c(N_c - 1)(N_c - 2)/2$, where N_c is the number of channels. Therefore, triple beat nonlinear products are always the dominant third-order nonlinear distortion products for large numbers of channels ($N_c > 7$ according to [14]). For RF and microwave signals occupying a bandwidth of more than one octave, the second order nonlinear distortions must be taken into account. Otherwise, only third-order nonlinear distortions have to be considered. The high CNR requirements for AM-FDM signals will, according to [4], limit the bandwidth of each AM-FDM channel within one octave.

The mentioned distortions cause cross modulation, which is the transfer of modulation from one or more carriers to another in a multi-channel system. The intermodulation noise is normally described by the two most dominant terms for second-order and third-order IMD, respectively.

The first feature is the *composite second-order* (CSO) component, which is defined as the ratio of the peak carrier to the peak power in the composite second-order intermodulation tone [15]. For standard and *incrementally related carrier* (IRC) frequency plans, the CSO appears at the carrier frequency ±1.25 MHz. For the *harmonically related carrier* (HRC) frequency plan, the CSO beats appear under the carriers.

The second key parameter in the representation of intermodulation noise is the *composite triple beat* (CTB), which is defined as the ratio of the peak carrier to the peak power in the composite third-order intermodulation tone. The CTB may for many frequency plans occur at the carrier frequency [15]. Note that if the spectrum of a large number of FDM channels is plotted, then we will find the CTB to be the highest in the central channels, whereas the CSO is the highest at low- and high-frequency edges [11].

To give an idea of the typical orders of magnitude of the involved parameters in a CATV network, we have in Table 12.1 reproduced a set of data from an AM-VSB system [15]. The data are for a 42-channel link with a transmission loss of 2 dB, and all noise contributions are reduced to an equivalent noise current of the electrical input of the optical receiver. Performance superior to the results of Table

Table 12.1
Data From AM-VSB System

Parameter	*Value*
Number of channels	42
Transmitted power	+2 dBm optical
Modulation index	0.04 per channel
CNR (4 MHz)	51 dB
CTB	−65 dBc
CSO	−60 dBc
Noise sources	
Shot noise	13 $pA/(Hz)^{1/2}$
RIN	12 $pA/(Hz)^{1/2}$
Front-end noise	14 $pA/(Hz)^{1/2}$

Note: values represent typical results and are not necessarily representative of system specifications or best achievable results [15].

12.1 can be obtained by the careful choice of the involved components, but it is not included here. For further information, the reader may consult [15].

From Table 12.1 we find that the three noise components are of comparable magnitude and that both CTB and CSO are larger than 60 dBc (values larger than 70 dBc may very well be achieved [15]). Note that dBc denotes the power in decibels measured relative to the carrier.

12.2 PROPERTIES OF FIBER AMPLIFIERS FOR ANALOG-MODULATED SIGNALS

As was discussed in the previous section, analog transmission of optical signals suffers from a very limited loss budget. This means that the inclusion of fiber amplifiers, which may help us to overcome these problems, becomes very attractive, and optical distribution networks can become realistic. Such systems have the advantage of supplying improved video quality while the maintenance cost will be reduced. Furthermore, the reliability improvements are expected because all electrical amplifiers can be replaced. However, even though the first reports on AM-FDM transmission experiments with optical amplifiers did not show sufficient results for video quality improvements [16], later reports promise greatly improved properties in AM-FDM systems [9].

There are, however, several points to consider when the EDFA is employed in CATV network applications. One of the primary concerns is that the application of EDFAs only allows the use of the 1550-nm wavelength band and not the 1300-nm band, which is popularly used in AM-FDM optical transmission systems for

CATV [9]. Therefore, existing transmitter laser diodes must be replaced with 1550-nm devices. Furthermore, if standard optical fibers with zero dispersion in the 1300-nm band are used, a single-mode laser diode must be used to prevent transmission quality deterioration due to chromatic dispersion. Bearing this in mind, we will now turn to the properties of the EDFA in analog FDM systems.

12.2.1 Gain Linearity of EDFAs

One of the first concerns that may come to mind is the question of gain linearity. For output powers above 0 dBm, the dc gain becomes dependent on input power (see, for instance, Figure 7.6). This might at first lead us to believe that the fiber amplifier is unsuitable for analog transmission. However, as shown in Figure 12.1, ac amplification properties are quite different, and the measured linearity was good in a specific case [9]. The pump wavelength was 1480 nm, and an erbium-doped fiber with a core radius of 4 μm and an erbium concentration of 50 ppm was used in this experiment.

The reason for the demonstrated linear behavior is that the saturated gain characteristic does not respond to input signal variations at speeds faster than a

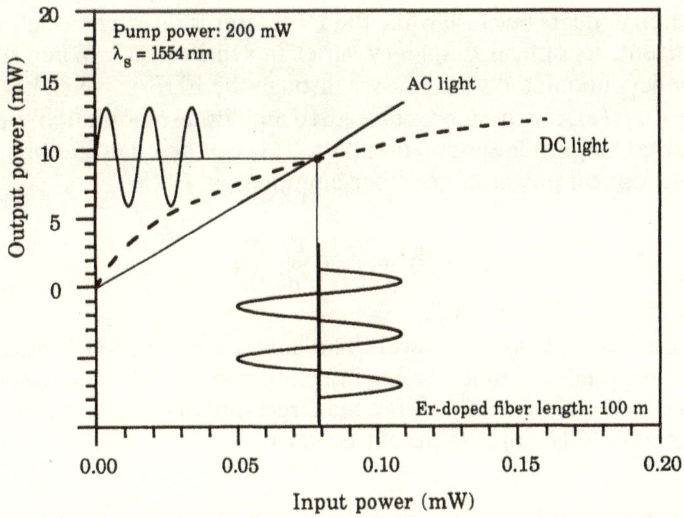

Figure 12.1 Instant signal output power from EDFA shown versus instant signal input power. The two curves represent dc operation and constant biased ac light at a frequency of 1 MHz. (Data from [9].)

few kilohertz. This is because of the small absorption and stimulated emission cross sections and the long metastable lifetime (10 to 14 ms) of erbium ions [17, 18]. The ac period of the 1-MHz signal is therefore expected to show a very fine linearity such as that found in Figure 12.1. Note that this contrasts with semiconductor amplifiers, whose saturated gain dynamics are on the order of hundreds of picoseconds [19], and the semiconductor amplifier would therefore create distortion in AM CATV systems.

12.2.2 EDFA Second-Order Distortion in Analog Links

Even though the gain linearity with signal input power of optical-fiber amplifiers according to the results of Section 12.2.1 seems very promising, distortions may still appear in an optically amplified lightwave analog AM CATV transport system. These distortions depend critically on the operating conditions of the EDFA, and they can be due to the interaction between the frequency chirping of the light source and the nonuniform gain coefficient within the amplification bandwidth. To describe these, we will review the results of [20], in which the second-order distortion is investigated for the case in which an EDFA is used to amplify the analog optical AM CATV multiple carrier signal from a directly modulated DFB semiconductor laser.

The distortion results from FM-to-AM conversion in the fiber amplifier. The FM modulation appears because while the DFB laser is directly modulated through injection current, its optical frequency varies in synchronism. When this inadvertently frequency-modulated signal passes through the EDFA, which has frequency-dependent gain $G(\nu)$, it experiences unwanted amplitude modulation superimposed upon its desired amplitude modulation [20]. This can be quantified by writing the instantaneous optical power of the fiber amplifier as

$$p_i(t) = \overline{P} + \frac{dp}{di}\, i(t) \tag{12.5}$$

where $\overline{P}$ is the average optical power. The analog information is encoded in the optical power through the time varying injection current $i(t)$ to the laser, assuming negligible second-order distortion from the direct modulation. If the fiber amplifier gain is denoted by $G(\overline{P}, \nu)$, the output power is

$$p_o(t) = G(\overline{P}, \nu(t) + \nu_0)p_i(t) = [G(\overline{P}, \nu_0) + \frac{\partial G}{\partial \nu} \cdot \nu(t)]p_i(t) \tag{12.6}$$

where $\nu(t) \approx (d\nu/di)\,i(t)$ is the instantaneous frequency deviation from the steady-state frequency ν_0, and the gain has been expanded to the first order. Because the speed of the dynamics of $i(t)$ far exceeds the time scale of the erbium-doped fiber dynamics, the gain $G(\overline{P}, \nu_0)$ may be considered time invariant. The derivative $d\nu/di$ is the frequency chirping, which, more generally, is a function of the modulation frequency [21]. However, as we follow the description of [20] here, $d\nu/di$ is considered constant for clarity of the discussion. Combining (12.5) and (12.6), we obtain the following expression for the output power:

$$p_o(t) = \overline{P}\cdot G(\overline{P}, \nu_0) + \left[G(\overline{P}, \nu_0)\frac{dp}{di} + \frac{\partial G}{\partial \nu}\frac{d\nu}{di}\overline{P}\right]\cdot i(t) + \left[\frac{dp}{di}\frac{\partial G}{\partial \nu}\frac{d\nu}{di}\right]\cdot i^2(t) \tag{12.7}$$

The second term in the first bracket represents the change of carrier intensity by the FM-to-AM conversion. In practice, it is about 1000 times smaller than the first term and may therefore be omitted for most applications [20]. For a single-carrier modulation, $i(t)$ is expressed as $I_m e^{-j\omega t}$, with I_m being the modulation amplitude. The *second-order distortion* (2HD) is the ratio of the third term to the second term in (12.7), where $i(t)$ is replaced by I_m:

$$2\text{HD} \approx \frac{\frac{\partial G}{\partial \nu}\frac{d\nu}{di}I_m}{G(\overline{P}, \nu_0)} \tag{12.8}$$

The 2HD is, from (12.8), seen to be proportional to the slope of the gain, the frequency chirping, and the modulation current. It is therefore critical to use lasers with little frequency chirping (e.g., multi-QW lasers) to reduce the 2HD. The relationship between 2HD and CSO is described below.

As we have already seen, the important performance parameter of an AM-VSB modulated system is the CNR, which is best maintained using the highest possible input levels to the EDFA [20]. Therefore, for the calculation of 2HD, the saturated gain as a function of wavelength should be used. For the representative example shown in Figure 12.2, the input power was −5.0 dBm, which corresponded to a saturated gain of 16.6 dB at 1548 nm. Further parameters are the laser chirping of 300 MHz/mA, and a modulation current of 2.6 mA for $m = 0.04$. Figure 12.2 illustrates that the 2HD varies in proportion to the slope of the gain curve, and it can be positive or negative depending on the slope of G with respect to the wavelength.

The 2HD is related to the CSO distortion. According to [20], it is a common criterion in CATV standards that the CSO may be calculated as the squared 2HD multiplied by the proper count of the distortion components for the particular

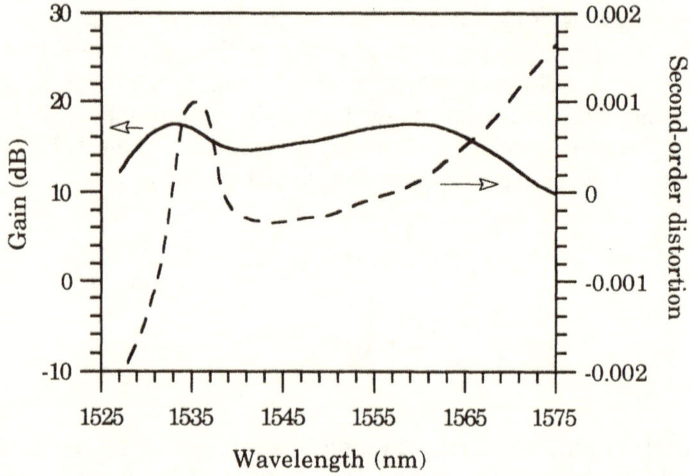

Figure 12.2 The calculated gain and the 2HD as a function of wavelength at an input power of −5 dBm and a gain of 16.6 dB at 1548 nm. (Curves from [20].)

frequency plan. Evaluations in [20] show that even by limiting the laser wavelength to narrow ranges near 1554 to 1565 nm, CSOs below 60 dB are very difficult to obtain. The second-order nonlinearities in EDFA CATV transport systems are therefore a significant problem, which may suggest the need for gain equalization in EDFAs. Another possibility exists, however, as suggested in [20], where a low-cost electronic predistorter is demonstrated to compensate for the distortion incurred by the EDFA. This scheme is reported to efficiently suppress the distortion for most operating conditions.

12.2.3 Increased System Power Budgets with the Use of Fiber Amplifiers

Most current analog optical systems do not have as high a power margin as baseband digital systems [11]. This is due to the poor sensitivity of 50 Ω front-end receivers used in most of the reported experiments, where the cost of the receiver is an essential parameter. Exceptions may, however, be found (see Table 12.1) where a better receiver is used, leading to comparable RIN, shot-noise, and receiver noise contributions.

The differences in input power requirements for analog and digital communication systems may be evaluated from the minimum required amplifier input signal power for a single in-line amplifier and a preset CNR (or BER for digital systems). This quantity can, according to [22], be estimated from the EDFA input power that causes 1-dB power penalty in the receiver sensitivity, as compared to the case where no amplifiers are used.

Analyses of [22] show that the minimum required input power to the amplifiers is about −10 dBm for an AM system with a modulation index per channel of 5%. This is for a system where the CNR requirement is 45 dB. It is also concluded that optical filtering has very little influence on the attainable input power results for AM modulation.

In the case of FM modulation, the required CNR is 17 dB in the typical example discussed in [22]. The important results are that an optical filter can significantly reduce the required amplifier input power for FM systems and that the FM system for the same filter bandwidth can be made comparable to that of a digital system by adjusting the optical modulation index per channel. Amplifier input power values even below −40 dBm are possible in these systems, and it is therefore obvious that the EDFA input power of an AM system is incompatible with that of an FM or digital system.

To give a better impression of the involved improvements, which are obtainable with optical amplifiers, Figure 12.3 has been included. It is important to note that the present experimental results of CNR versus the total transmission loss are obtained by *traveling wave* (TW) semiconductor laser amplifiers. The reason for this choice is to underline that the first demonstrations of optical amplifiers in analog communication systems were performed with TW laser amplifiers. Figure 12.3 shows that the power margin of a 90-channel FM video *subcarrier multiplexed*

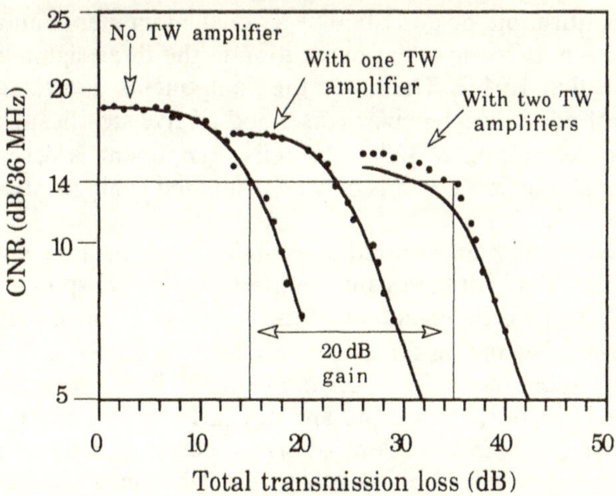

Figure 12.3 CNR per channel versus total transmission loss for a lightwave system that employed two in-line laser amplifiers to distribute 90-channel FM video signals. (Figure from [11].)

(SCM) system can be increased by 20 dB using two stages of amplifiers. Each FMchannel reaches a CNR of 14 to 15 dB, which is close to a studio-quality requirement of about 16.45 dB [23].

The TW laser is also mentioned here to include some comparative comments concerning this and the EDFA. According to [24], the TW laser amplifier can be used only for a channel frequency range within one octave, primarily because of the second-order nonlinear IMD caused by signal-induced carrier density modulation [19]. On the other hand, it has been shown that the interchannel crosstalk introduced by the EDFA is below 40 dB for modulation frequencies greater than 100 kHz [17]. This is mainly due to the long fluorescence lifetime in the upper laser level (see Chapter 3). Consequently, the EDFA is the best candidate for potential applications in multi-channel analog systems [24].

The EDFA has shown successful results as a power booster, especially in multi-channel AM-VSB applications, which are attractive due to compatibility with current NTSC TV receivers. A good example is reported in [24], where an EDFA (operated 10 dB into saturation) showed no additional IMD or intensity noise penalty. The signal-to-noise ratio of 43 ± 3 dB of each received TV channel was limited by nonlinearities of the DFB laser, and the achieved 8-dB power budget improvement may be expected to be even bigger if the amplifier output power is increased.

To get a clearer picture of what the limiting factors are of an AM-VSB modulated system employing EDFAs, we will review here some of the results of [25]. More specifically, we will focus on the CNR and study the relative contribution of each noise component to the total system noise. The CNR may, in a system with optical amplification, be described by a signal-spontaneous and a spontaneous-spontaneous beat noise component in addition to the three significant components described in Section 12.1.2. These are the components due to relative intensity noise, RIN, shot noise, and receiver noise, and all five significant components are included in the description of [25]. The RIN component is described by (12.4), and the additional four components may be evaluated analogously with the expressions listed in (10.7).

The following comments should be included in connection with the individual CNR components. First, for high-gain amplifiers, the signal-spontaneous beat noise component, $CNR_{sig\text{-}sp}$, is independent of the signal gain and it scales linearly with input signal power. Second, in the case where a single EDFA is followed by a loss, the spontaneous-spontaneous beat component, $CNR_{sp\text{-}sp}$, is negligible (i.e., very large) compared to $CNR_{sig\text{-}sp}$ (see the analogy in Figure 10.8(b)). Third, the RIN component, CNR_{RIN}, is independent of signal power and amplifier gain, and is therefore also independent of pump power. Fourth, the circuit noise component, CNR_{th}, depends quadratically on the received power, which in turn depends linearly on pump power. Fifth, the shot noise component, CNR_{sh}, depends linearly on the received power, and it will therefore vary linearly with pump power. Finally, note that all CNR components mentioned here depend quadratically on the modulation index.

An example of the dependence of each CNR component on pump power is shown in Figure 12.4. Also shown is the total CNR. For the calculations that formed the basis of this illustration, [25] used the signal input power of 2 mW, the modulation index of 5%, the circuit noise of 5 pA/(Hz)$^{1/2}$, and the RIN of −155 dB/Hz. A postamplifier loss of 22 dB was assumed, which for a loss margin of 10 dB leaves a loss of 12 dB. This accounts for four stages of 3-dB couplers, thus implying a capability to divide the signal 16 ways. Note that $CNR_{sp\text{-}sp}$ is not shown in Figure 12.4, since it is much larger than the other CNR components. For each pump power, the fiber amplifier length is chosen such that the CNR, after 22 dB of post-amplifier loss, is maximized.

Even though Figure 12.4 only represents an example of CNR components, some general conclusions can be drawn from it [25]. At high pump powers, the dominant noise sources are RIN and signal-spontaneous beat noise. These two noise mechanisms will either together or individually set an ultimate upper limit on the achievable CNR. As the pump power decreases, shot noise and eventually thermal noise become the dominant noise mechanisms. The relative onset of dominance will naturally depend on the specific system parameters.

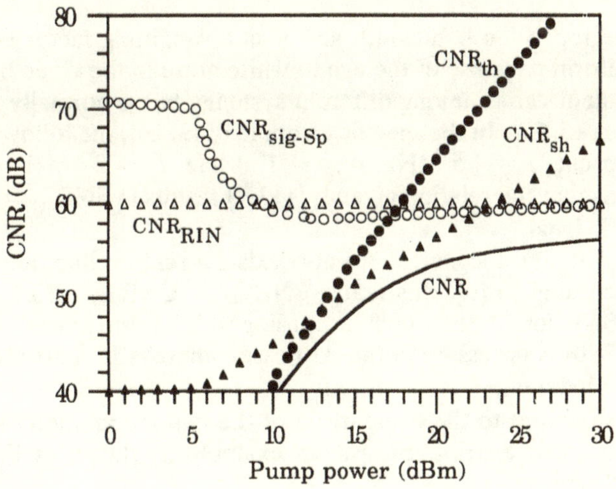

Figure 12.4 The dependence of each CNR component on pump power. Also shown is the total CNR. It is assumed that the signal input power is 2 mW, the modulation index m = 5%, the effective rms receiver noise current is 5 pA/$\sqrt{\text{Hz}}$, and RIN = −155 dB/Hz. A post-amplifier loss of L = 22 dB is assumed, and at each pump power the fiber amplifier length is chosen such that the CNR, after 22 dB of post-amplifier loss, is maximized. (Figure from [25].)

12.3 DESIGN CONSIDERATIONS FOR ANALOG OPTICAL COMMUNICATION SYSTEMS

To give an idea of the tradeoffs involved in the design of analog-modulated CATV distribution systems, we will here review the evaluation of AM-FDM or FM-FDM signals presented in [9, 26]. As previously discussed, the CNR is often used instead of the weighted signal-to-noise ratio in analog FDM system design, but their relationship should, however, be briefly discussed, starting with the FM-FDM case.

The FM signals exchange a wide bandwidth for a considerable *signal-to-noise ratio* (SNR) improvement [11]. The SNR at the output of the FM detector is much larger than the CNR at the detector input, provided that the CNR is above a threshold value characteristic for that detector. According to [11], an appropriate value for the required bandwidth B is given by Carson's rule, $B = \Delta f_{pp} + 2f_m$, where Δf_{pp} is the peak-to-peak frequency deviation of the modulator, and f_m is the top audio subcarrier frequency. The corresponding output SNR improvement over the input CNR is given by [11]

$$\mathrm{SNR}_{\mathrm{out}} = \mathrm{CNR}_{\mathrm{in}} + 10\cdot\log\left[\frac{3}{2}\cdot\frac{B}{f_v}\cdot\left(\frac{\Delta f_{pp}}{f_v}\right)^2\right] + w \tag{12.9}$$

where f_v is the top video bandwidth and w is a weighting factor used to account for the nonuniform response of the eye to white noise in the video bandwidth. The total improvement varies among different systems, but is normally in the range of 35 to 44 dB [11, 15, 27]. In the specific example of [9, 26], the following parameters have been chosen: f_v = 4.5 MHz, Δf_{pp} = 17 MHz, B = 27 MHz, and w = 14.3 dB. This means that the relation given by (12.9) results in $\mathrm{SNR}_{\mathrm{out}} = \mathrm{CNR}_{\mathrm{in}} + 35.5$ dB for the FM signal.

For the AM-FDM signal, there also exists a relationship between SNR and CNR and, according to [9], this reads: $\mathrm{SNR}_{\mathrm{AM}} = \mathrm{CNR}_{\mathrm{AM}} + 0.3$ dB. To obtain the weighted SNR for 52 dB, which, according to [9], ensures studio video quality, a CNR of 49.7 dB is necessary in the AM case, whereas 16.5 dB is needed for the CNR in the FM system.

We will now turn to the comparison of the two modulation formats, and for this purpose a video distribution system example as shown in Figure 12.5 was proposed in [9, 26].

The evaluation parameter is chosen as the obtainable distribution number as a function of the length of the trunk transmission. The results of the analysis are shown in Figure 12.6, and it should be noted that the calculation conditions are based on experimental values, which may be found in [9, 26]. We will not, however, include the specific amplifier data at this point, since Figure 12.6 primarily serves as a typical example of the potential performance improvement, which is obtainable

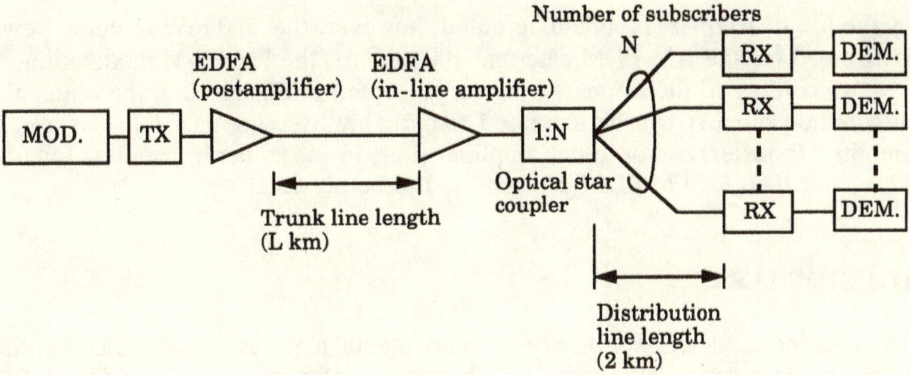

Figure 12.5 Schematic illustration of CATV distribution system using EDFAs in analog FDM. (Figure reproduced from [26].)

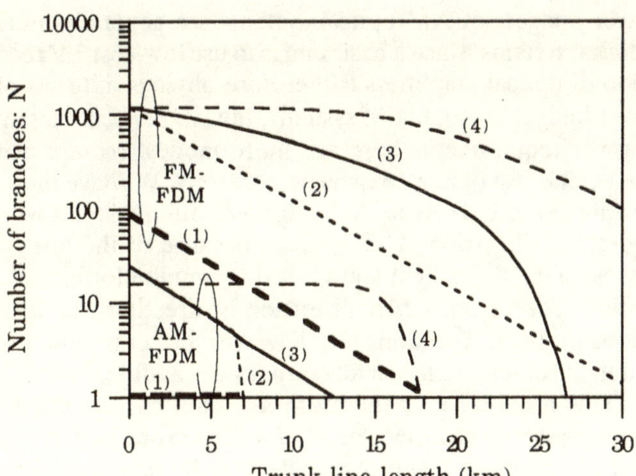

Figure 12.6 Obtainable distribution number for the communication system shown in Figure 12.5. The curve legends have the following significance: (1) without amplifier; (2) with post-amplifier; (3) with in-line amplifier; (4) with post- and in-line amplifier. (Figure from [26].)

by the use of EDFAs. It should be noted, however, that 200 mW of pump power is assumed for the AM-FDM case and 100 mW for the FM-FDM modulation.

According to the estimated results presented in Figure 12.6, the obtainable distribution number can be increased remarkably by using an EDFA as a post-amplifier (booster) and/or in-line amplifier (i.e., 16-subscriber distribution for AM-FDM and 1000 for FM-FDM are expected to be possible).

12.4 SUMMARY

This chapter has described some of the fundamental features of analog FDM optical transmission systems. This includes a short description of the AM and FM signals. It was pointed out that the AM signals are particularly interesting for the distribution of television signals due to their compatibility with current TV receivers. FM signals, on the other hand, are compatible with broadcast satellite intermediate-frequency receivers, and their total loss budget is increased compared to the AM systems. However, FM/AM converters will be needed if the FM signals are to be connected to existing TV networks. A key parameter for analog systems is the CNR, and typical requirements for a video channel are CNR ≈ 50 dB for AM signals and CNR ≈ 16 dB for FM signals. Other features are the CSO and the CTB components, which dominate the intermodulation noise properties of analog systems. Typical required magnitudes are CSO ≈ 60 dBc and CTB ≈ 65 dBc.

The power budgets of AM optical systems are generally much tighter than in baseband digital systems, since a basic aim is to use low-cost TV receivers directly. The application of optical amplifiers is therefore obvious in terms of achieving an increased power budget. Also, in FM systems, the EDFA becomes interesting even though the power requirements here are more modest (comparable to those of digital systems) at the cost of more expensive receivers. We have therefore discussed the special properties of EDFAs in AM communication systems with special focus on the second-order distortion. This appears because of the interaction between the frequency chirping of the light source and the nonuniformity of the gain (i.e., spectral gain tilt). The second-order distortion is directly related to the CSO and may be a serious problem. Solutions are, however, either the use of gain-flattened amplifiers and light sources with a small chirp, or the addition of low-cost electronic predistorters. This property may therefore not limit the potential application of EDFAs in analog systems, and theoretical as well as experimental evaluations show that significant power budget improvements may be expected. Improvements of 10 to 20 dB are readily demonstrated today. It may also be added that the CNR properties of AM systems employing an EDFA at high pump powers are dominated by RIN and signal-spontaneous beat noise. At low pump powers, the CNR is dominated by thermal and shot noise.

Finally, results on the allowable distribution number of TV receivers were reviewed. This showed that the obtainable distribution number can be increased remarkably using the EDFA as post- and/or in-line amplifier. A 16-subscriber distribution for AM-FDM and 1000 for FM-FDM is possible. Although these numbers indicate that the AM-FDM system, especially, operates within a few decibels of several fundamental physical limits, both represent a cost-competitive and technically advanced alternative to coaxial-based CATV networks, and they deliver improved reliability and performance compared to the latter. Optical-fiber amplifiers are therefore expected to play a significant role in the next generation of CATV systems.

REFERENCES

[1] McGrath, C. J., "AM Lightwave Technology," *AT&T Technical Journal*, January/February 1992, pp. 23–30.

[2] Nagano, K., Y. Takahashi, Y. Takasaki, M. Maeda, and M. Tanaka, "Optimizing Optical Transmitters and Receivers for Transmitting Multichannel Video Signals Using Laser Diodes," *IEEE Transactions on Communication*, Vol. 29, 1981, pp. 41–45.

[3] Sato, K., "Intensity Noise of Semiconductor Laser Diodes in Fiber Optic Analog Video Transmission," *IEEE Journal of Quantum Electronics*, Vol. 19, 1983, pp. 1380–1391.

[4] Olshansky, R., V. A. Lanzisera, and P. M. Hill, "Subcarrier Multiplexed Lightwave Systems for Broad-Band Distribution," *IEEE Journal of Lightwave Technology*, Vol. 7, No. 9, 1989, pp. 1329–1342.

[5] Way, W. I., and M. M. Choy, "Optical Feedback on Linearity Performance of 1.3-μm DFB and Multimode Lasers Under Microwave Intensity Modulation," *IEEE Journal of Lightwave Technology*, Vol. 6, No. 1, 1988, pp. 100–108.

[6] Petermann, K., and G. Arnold, "Noise and Distortion Characteristics of Semiconductor Lasers in Optical Fiber Communication Systems," *IEEE Journal of Quantum Electronics*, Vol. 18, No. 4, 1982, pp. 543–555.

[7] Fujito, K., T. Uno, T. Ichida, and H. Serizawa, "Low Noise Wideband Analog Optical Link Using a DFB Laser Diode," *Optical Fiber Communications OFC '88*, New Orleans, 1990, Proc. Vol. 1.

[8] Chiddix, J., "Application of Optical Fiber Transmission Technology to Existing CATV Networks," FOC/LAN '88, Atlanta, 1988, Proc. Vol. 1, pp. 98–105

[9] Yoneda, E., K. Kikushima, T. Tsuchita, and K.-I. Suto, "Erbium-Doped Fiber Amplifier for Video Distribution Networks," *IEEE Journal on Selected Areas in Communications*, Vol. 8, No. 7, 1990, pp. 1249–1256.

[10] Shanmugam, K. S., *Digital and Analog Communication Systems*, New York, Chichester, Brisbane, Toronto, Singapore: John Wiley & Sons, 1979, 1985.

[11] Way, W. I., "Subcarrier Multiplexed Lightwave System Design Considerations for Subscriber Loop Applications," *IEEE Journal of Lightwave Technology*, Vol. 7, No. 11, 1989, pp. 1806–1818.

[12] Olshansky, R., and E. Eichen, "Microwave-Multiplexed Wideband Lightwave Systems Using Optical Amplifiers for Subscriber Distribution," *IEE Electronics Letters*, Vol. 24, No. 15, 1988, pp. 922–923.

[13] Sato, K., "Intensity Noise of Semiconductor Laser Diodes in Fiber-Optic Analog Video Transmission," *IEEE Journal of Quantum Electronics*, Vol. 19, No. 9, 1983, pp. 1380–1391.

[14] Gündner, H. M., R. Stannard, and U. Scholz, "Optical Fiber CATV Distribution System for 12 TV and 12 FM Stereo Radio Channels," *European Conference on Optical Communication ECOC '83*, Geneva, 1983, pp. 559–563.

[15] Lipson, J., L. Chainulu Upadhyayula, S.-Y. Huang, C. B. Roxlo, E. J. Flynn, P. M. Nitzsche, C. J. McGrath, G. L. Fenderson, and M. S. Schaefer, "High-Fidelity Lightwave Transmission of Multiple AM-VSB NTSC Signals," *IEEE Transactions on Microwave Theory and Techniques*, Vol. 38, No. 5, 1990, pp. 483–493.

[16] Way, W. I., M. M. Choy, A. Yi-Yan, M. Andrejco, M. Saifi, and C. Lin, "Multi-Channel AM-VSB Television Signal Transmission Using an Erbium-Doped Optical Fiber Power Amplifier," *IOOC '89*, Kobe, Japan, 1989, paper 20PDA-10.

[17] Laming, R. I., L. Reekie, and D. N. Payne, "Multichannel Crosstalk and Pump Noise in an Er-Doped Fibre Amplifier Pumped at 980 nm," *IOOC '89*, Kobe, Japan, 1989, paper 20A4-2.

[18] Giles, C. R., E. Desurvire, and J. R. Simpson, "Transient Gain and Crosstalk in Erbium-Doped Fiber Amplifiers," *Optics Letters*, Vol. 14, 1989, p. 880.

[19] Saleh, A. A., T. E. Darcie, and R. Jopson, "Nonlinear Distortion Due to Optical Amplifiers in Subcarrier-Multiplexed Lightwave Communications Systems," *IEE Electronics Letters*, Vol. 25, No. 1, 1989, pp. 79–80.

[20] Kuo, C. Y., and E. E. Bergmann, "Erbium-Doped Fiber Amplifier Second-Order Distortion in Analog Links and Electronic Compensation," *IEEE Photonics Technology Letters*, Vol. 3, No. 9, 1991, pp. 829–831.

[21] Koch, T. L., and Bowers, J. E. "Nature of wavelength chirping in directly modulated semiconductor lasers," *IEE Electronics Letters*, Vol. 20, No. 25/26, 1984, pp. 1038–1040.

[22] Way, W. I., "Optical Fiber Amplifiers For Multichannel Video Transmission and Distribution," *Techn. Dig. of Optical Amplifiers and Their Applications OAA '91*, Snowmass, CO, 24–26 July 1991, paper FB1, pp. 206–209.

[23] Way, W. I., R. S. Wolff, and M. Krain, "A 1.3-μm 35-km fiber-optic Microwave Multicarrier Transmission System for Satellite Earth Station," *IEEE Journal of Lightwave Technology*, Vol. 5, No. 9, 1987, pp. 1325–1332.

[24] Way, W. I., M. M. Choy, A. Yi-Yan, M. Andrejco, M. Saifi, and C. Lin, "Multichannel AM-VSB Television Signal Transmission Using an Erbium-Doped Optical Fiber Power Amplifier," *IEEE Photonics Technology Letters*, Vol. 1, No. 10, 1989, pp. 343–345.

[25] Habbab, I. M. I., and L. J. Cimini, Jr., "Optimized Performance of Erbium-Doped Fiber Amplifiers in Subcarrier Multiplexed Lightwave AM-VSB CATV Systems," *IEEE Journal of Lightwave Technology*, Vol. 9, No. 10, 1991, pp. 1321–1329.

[26] Nakagawa, K., S. Nishi, K. Aida, and E. Yoneda, "Trunk and Distribution Network Application of Erbium-Doped Fiber Amplifier," *IEEE Journal of Lightwave Technology*, Vol. 9, No. 20, 1991, pp. 198–208.

[27] Mendis, F. V. C., and P. A. Rosher, "C/N Requirements for Subcarrier Multiplexed Multichannel Video FM Transmission in Optical Fibre," *IEE Electronics Letters*, Vol. 25, No. 1, 1989, pp. 72–74.

Acronyms

AGC	Automatic gain control
AM	Amplitude modulation
APD	Avalanche photodiode
ASE	Amplified spontaneous emission
BER	Bit error rate
BS-IF	Broadcast satellite intermediate frequency
CATV	Cable television
CNR	Carrier-to-noise ratio
CSO	Composite second order
CTB	Composite triple beat
dB	Decibel
dEDFA	Distributed erbium-doped fiber amplifier
DS	Dispersion shifted
DFB	Distributed feedback
EDFA	Erbium-doped fiber amplifier
Er	Erbium
ESA	Exited-state absorption
FDM	Frequency division multiplexing
FET	Field-effect transistor
FM	Frequency modulation
FWHM	Full width half maximum
GSA	Ground-state absorption
IMD	Intermodulation distortion
IRC	Incrementally related carriers
LD	Laser diode
LED	Light-emitting diode
MCVD	Modified chemical vapor deposition
NA	Numerical aperture

Nd	Neodymium
NDFA	Neodymium-doped fiber amplifier
NLS	Nonlinear Schrödinger equation
NTSC	National Television System Committee
OTDR	Optical time domain reflectometry
OVD	Outside vapor deposition
PDFA	Praseodymium-doped fiber amplifiers
Pr	Praseodymium
QCE	Quantum conversion efficiency
QW	Quantum well
RIN	Relative intensity noise
RBS	Rutherford backscattering spectrometry
RNF	Refracted near field
SEM	Scanning electron microscope
SIC	Saturation-induced crosstallk
SLJ	Russel-Saunders term
TW	Traveling wave
VAD	Vapor axial deposition
VSB	Vestigial sideband
WDM	Wavelength division multiplexing
YAG	Yttrium-aluminum-garnet ($Y_3Al_5O_{12}$)
Yb	Ytterbium
YLF	$YLiF_4$
ZBLAN	ZrF_4-BaF_2-LaF_3-AlF_3-NaF
ZBLANP	ZrF_4-BaF_2-LaF_3-AlF_3-NaF-PbF_2
2HD	Second-order distortion

Symbols

a	Constant used in expression for resulting transition probability
a	Fiber core radius
$a(z)$	Emission factor of the LP_{01} signal mode
$a_d = a_{dot}/a$	Confinement factor
a_{dot}	Erbium dopant radius
$att(\nu)$	Attenuation in dB/m as a function of the optical frequency
$a_{1050}(z, \nu)$	Emission factor of the 1050-nm ASE in PDFA
$a_{1310}(z, \nu)$	Emission factor of the 1310-nm ASE in PDFA
A	Fiber core area
A	Soliton amplitude
$A(\lambda)$	Attenuation as a function of the wavelength
A_c	Amplitude factor of analog-modulated signal
A_e	Spontaneous emission rate
A_{eff}	Effective core area
A_d	Photoresponsivity of detector
b	Outer fiber radius
B	Bit rate
B	Equivalent spontaneous emission bandwidth
B	Required bandwidth of an FM system
B_e	Electrical bandwidth
B_o	Optical bandwidth
C_0	Scaling constant used in ESA determination
CNR_{AM}	Input carrier-to-noise ratio in an AM system
CNR_{in}	Input carrier-to-noise ratio in an FM system
CNR_{RIN}	Relative intensity noise component to the carrier-to-noise ratio
CNR_{sh}	Shot noise component to the carrier-to-noise ratio
$CNR_{sig\text{-}sp}$	Signal-spontaneous beat noise component to the carrier-to-noise ratio

$CNR_{sp\text{-}sp}$	Spontaneous-spontaneous beat noise component to the carrier-to-noise ratio
CNR_{th}	Circuit noise component to the carrier-to-noise ratio
D	Fiber dispersion
D_s	Fiber dispersion
c	Vacuum velocity of light
E	Electric field in fiber
E	Energy of scattered particles
$E(r, \phi, \nu)$	Normalized transverse electrical field distribution
$\langle E \rangle$	Mean energy
E_{ij}	Difference in energy between the jth and the lowest component of level i
E_i	Energy separation between the lowest component of manifold i
E_O	Impact energy of α-particle
E_s	Signal energy
$f(t)$	Time-dependent electrical field
f_c	Carrier frequency of analog-modulated signal
f_m	Top audio subcarrier frequency
f_s	Signal frequency of analog modulated signal
f_ν	Video top bandwidth in an FM system
F	Noise figure
F	Three-way overlap factor between pump mode, signal mode, and fiber core
$F(M)$	Excess noise factor of an avalanche photodiode
$\mathcal{F}$	Coefficient used to distinguish between co- and counter-propagating pump and signal
F_{I}	Total noise figure of communication link realized by power amplifier-configurated blocks
F_{II}	Total noise figure of communication link realized by preamplifier-configurated blocks
$F_{\lim}$	Asymptotic value of effective noise figure
F_p	Overlap factor between pump mode and fiber core
F_s	Overlap factor between signal mode and fiber core
F_1	Noise figure of remote pumped amplifier
F_2	Noise figure of preamplifier
$g(x)$	Relative gain at a position x along the transmission length
$g(\nu)$	Gain in dB/m at the optical frequency ν
$g_s(z)$	Local gain coefficient
g_i	Degeneracy of level i
g_i	Gain in section i of a Nd-doped fiber with ASE filters
$g_p^{lp}(z)$	Pump absorption factor of the LP_{lp} mode

g_p	Attenuation factor in Nd- or Pr-doped fiber amplifiers
$g_{1050}(z, \nu)$	Gain factor of the 1050-nm ASE in PDFA
$g_{1310}(z, \nu)$	Gain factor of the 1310-nm ASE in PDFA
G	Gain
G_{Hp}	High-pump small-signal gain for idealized 4-level system
G_i	Saturated gain of amplifier i in a cascade
G_{Lp}	Low-pump small-signal gain for idealized 4-level system
G_{opt}	Maximum gain of Nd-doped fiber amplifier
$G_{\max}$	Maximum gain
G_0	Unsaturated gain of amplifier in a cascade
G_{10}	Small-signal gain
h	Planck's constant
$h_e(t)$	Current response of a single electron in the photodetector
$H_e(f)$	Detector transfer function
$H_T(f)$	Normalized spectral dependence of the transfer impedance from $I_p(f)$ to $V_s(f)$
$\langle i \rangle$	Detected dc photocurrent
$\langle i^2 \rangle$	Noise current spectral density
$\langle i^2_{ckt} \rangle$	Equivalent input noise current spectral density of receiver
$i_p(t)$	Photocurrent as a function of time
$I(r, \nu, z, \nu)$	Spatial intensity distribution
$I^{01}(\nu, r)$	Normalized LP_{01} mode intensity
$I_p^{lp}(r, \phi, z)$	Mode intensity of the LP_{lp} pump mode
I_s^{sat}	Signal saturation intensity
$I_s^{01}(r, z)$	Normalized signal mode intensity
$I_a(\nu)$	Spectrally dependent absorption
I_{dark}	Dark current of photodiode
$I_e(\nu)$	Spectrally dependent fluorescence
$I_p(f)$	Fourier transform of the photocurrent $i_p(t)$
I_{ph}	Received photocurrent
$I_s(0)$	Photocurrent equivalent of the input power for a space
$I_s(1)$	Photocurrent equivalent of the input power for a mark
J	Total angular momentum
k	Boltzmann constant
K	The number of polarization directions to which the receiver is sensitive
K_M^2	Kinematic factor
l_{tot}	Total transmission distance in direct detection system
l_1	Pumping distance in a remotely pumped DD system
L	Attenuation in a direct detection system
L	Total amplifier length

L	Orbital angular momentum
L_A	Reference fiber length in spectral attenuation measurement
L_B	Fiber length in spectral attenuation measurement
L_{opt}	Maximum gain length of Er-doped fiber
L_{tot}	Total transmission length
L_s	Original transmission distance in a remotely pumped DD system
m	Modulation index for AM signal
M	Gain of an avalanche photodiode
M	Number of communication channels
M_1	Mass of α-particle
M_2	Mass of target nuclei in RBS measurements
n	Refractive index
$n(t)$	Function representing the electrical field launched at the detector
$\langle n \rangle$	Static mean value of $n(t)$
$\langle \overline{n} \rangle$	Time averaged mean value of $\langle n \rangle$
$\tilde{n}$	Number of photons in the signal mode
$n_{\text{ASE},b}$	Number of ASE photons after one block of a communication link
$n_{\text{ASE},t}$	Total number of ASE photons after a communication link
n_{ke}	Nonlinear Kerr coefficient
n_{ke}^{I}	Nonlinear Kerr coefficient, expressed in terms of intensity
n_{sp}	Spontaneous-emission factor
n_0	Linear part of refractive index
n_2	Nonlinear part of refractive index
n_7	Population in ground level of Pr-doped fiber amplifiers
$n_1(r, \phi, z)$	Steady-state population in the ground state of erbium
$n_2(r, \phi, z)$	Steady-state population in the upper state of erbium
$n_2^{\max}$	Maximum population in an Er-doped fiber
n_1	Population in upper laser level of Pr-doped fiber amplifiers
N	Number of blocks of a communication link
N_c	Number of channels in an analog multichannel transmission system
N_{th}	Electrical spectral density of the circuit noise
NF_{I}	Effective noise figure of power amplifier configurated transmission link
NF_{II}	Effective noise figure of preamplifier configurated transmission link
p	Radial mode number
$p(z)$	Quantity used in derivation of noise figure expression
$\langle P \rangle$	Path average signal power
$P(z)$	Power distribution along the fiber

p_i	Probability of finding an electron at the sublevel number i
$p_l(l)$	Conditional Poisson probability distribution involved in photodetection process
$\underline{p}_m(m)$	Probability density involved in photodetection process
$\overline{P}$	Average optical power in an analog system
P_{ai}	Absorbed pump power in fiber section no. i for NDFA with filters
P_{in}	Total available pump power to NDFA with filters
P_{ASE}	ASE power
$P_A(\lambda)$, $P_B(\lambda)$	Transmitted optical power as functions of the wavelength
P_p^{abs}	Absorbed pump power
P_p^{in}	Pump input power
P_s^{in}	Signal input power
P_s^{out}	Signal output power
P_s^{sat}	Signal saturation power
$P^-_{\text{ASE},i}$	Output power of backward-traveling ASE at amplifier no. i in a cascade
$P^+_{\text{ASE},i}$	Output power of forward-traveling ASE at amplifier no. i in a cascade
$P_{\text{in},i}$	Total input power of amplifier no. i in a cascade
P_{I}, P_{II}	Launched signal power
P_{out}	Output power from transmitter
P_p	Pump power
$P_{p1,th}$	Pumping threshold
P_s	Signal power
$P_{s,i}$	Output signal power of amplifier no. i in a cascade
P_{sat}	Saturation power of amplifier
P_{th}	Threshold pump power
P_{tr}	Transparency power
P_0	Launched signal power in soliton communication link
$P^{\pm}_{900}$	Spectrally dependent power of forward (+) and backward (−) propagating ASE in the 900-nm band of Nd-doped fibers
$P^{\pm}_{1050}$	Spectrally dependent power of forward (+) and backward (−) propagating ASE in the 1050-nm band of Nd-doped fibers
$P^{\pm}_{1300}$	Spectrally dependent power of forward (+) and backward (−) propagating ASE in the 1300-nm band of Nd-doped fibers
$P^{\pm}_{1050}(z, \nu)$	Spectrally dependent power of forward (+) and backward (−) propagating ASE in the 1050-nm band of Pr-doped fibers
$P^{\pm}_{1310}(z, \nu)$	Spectrally dependent power of forward (+) and backward (−) propagating ASE in the 1310-nm band of Pr-doped fibers
q	Electron charge
q_1	Electrostatic charges of an α-particle

Q	"Quality" parameter used in BER calculation
r	Extinction ratio
r	Radial coordinate in a cylindrical coordinate system
R_c	Coupler reflection coefficient
R_g	Measure of the pulse energy fluctuation along a communication link
R_I	Isolator reflection coefficient
R_n	Excitation/deexcitation rate ($= n_2/n_1$)
$R_{pa}(r, \phi, z)$	Pump absorption rate of erbium
$R_{pe}(r, \phi, z)$	Pump emission rate of erbium
R_s	Splice reflection coefficient
R_T	Normalization factor of the transfer impedance from $I_p(f)$ to $V_s(f)$
S	Total spin momentum
S_{ASE}	Total amplified spontaneous emission power spectral density
S^-_{ASE}	Backward-propagating amplified spontaneous emission power spectral density
S^+_{ASE}	Forward-propagating amplified spontaneous emission power spectral density
$S_n(f)$	Intensity spectrum of $n(t)$
$\mathrm{SNR}_{\mathrm{AM}}$	Signal-to-noise ratio of an AM system
$\mathrm{SNR}_{\mathrm{in}}$	Signal-to-noise ratios at amplifier input end
$\mathrm{SNR}_{\mathrm{out}}$	Signal-to-noise ratios at amplifier output end
S_0	Photocurrent for a space
S_1	Photocurrent for a mark
t	Time
τ_{FWHM}	Full-width half-maximum optical pulse intensity
t_p	Physical time coordinate
$t_1, t_2, \ldots, t_l$	Time variables
T	Temperature
T	Time interval length
T_{on}	Transmission factor of pump probe for pumped fiber
T_{off}	Transmission factor of pump probe when background pump is turned off
U_2, U_4, U_6	Tabulated quantities used in the Judd-Ofelt analysis
v_g	Group velocity
V	Normalized frequency
V_s	Variance of signal power
$V_s(f)$	Fourier transform of the receiver output voltage
w	Weighting factor used in the description of an FM system
$W_p^{\mathrm{abs}}(r, z)$	Absorption rate of the 1015-nm pump of praseodymium
$W_p^{\mathrm{ems}}(r, z)$	Emission rate of the 1015-nm pump of praseodymium

$W_{sa}(r, \phi, z)$	Signal absorption rate of erbium
$W_{sa}(r, \phi, z)$	Signal absorption rate of erbium
$W_{1050}^{abs}(r, z)$	Absorption rate of the 1050-nm wavelength band of praseodymium
$W_{1050}^{ems}(r, z)$	Emission rate of the 1050-nm wavelength band of praseodymium
$W_{1310}^{ems}(r, z)$	Emission rate of the 1310-nm wavelength band of praseodymium
W_{ESA}	Excited-state absorption efficiency for the ASE in neodymium fibers
W_{SESA}	Excited-state absorption efficiency for the signal in neodymium fibers
W_{SSE}	Stimulated emission efficiency for the signal in neodymium or praseodymium fibers
W_{900}	Stimulated emission efficiency for the ASE around 900 nm in neodymium fibers
W_{1050}	Stimulated emission efficiency for the ASE around 1050 nm in neodymium fibers
W_{1300}	Stimulated emission efficiency for the ASE around 1300 nm in neodymium fibers
x_{rep}	Spacing of lumped amplifiers in a nonlinear communication system
z_p	Physical place coordinate
z	Longitudinal position along a fiber
Z	Normalization factor (in Boltzmann distribution)
Z_T	Transfer impedance from $I_p(f)$ to $V_s(f)$
$^1D_2$	Signal upper ESA level of praseodymium ions
$^4F_{3/2}$	Upper state of the gain transition of neodymium ions
$^3F_4$	Upper GSA level of praseodymium ions
$^1G_4$	Upper laser level of praseodymium ions
$^3H_4$	Ground level of praseodymium ions
$^3H_5$	Lower laser level of praseodymium ions
$^4I_{9/2}$	Ground state of neodymium ions
$^4I_{13/2}$	Upper laser level of erbium ions
$^4I_{15/2}$	Ground laser level of erbium ions
α	Attenuation of Er-doped fiber
α	Constant used in description of solutions to numerical difficulties
α_{ASE}	The number of generated ASE photons per hertz, per second, and per unit length of the amplifier
α_k	Small signal absorption coefficient
α_p	Loss constant of passive fiber at the pump wavelength

α_s	Loss constant of passive fiber at the signal wavelength
α_{13}	Attenuation due to the pump GSA transition in erbium
α_{24}	Attenuation due to the pump ESA transition in erbium
β	Modulation index for FM signal
β_o''	Second derivative of the propagation constant with regard to the angular frequency
δ	Saturation power ratio
$\langle \delta t^2 \rangle$	Pulse arrival time variance
$\delta\lambda$	Laser linewidth
ΔE_i	Energy difference between sublevel *i* and the lowest lying sublevel in the manifold
Δf_{pp}	Peak-to-peak frequency deviation of the modulator in an FM system
Δn	Refractive index difference
ΔT	Difference in transmission factors
ΔT	Time slot
$\Delta\beta_\mu$	Spatial frequency
$\Delta\beta_{1p}$	Spatial frequency
$\Delta\nu$	Optical bandwidth interval
$\Delta\tau$	Pulse broadening
ϵ	Temperature-dependent excitation energy
ϕ	Azimuthal coordinate in a cylindrical coordinate system
γ	Ratio of absorption and emission cross section at pump and signal wavelengths
Φ	Pulse envelope
γ	Attenuation constant of fiber
γ_a	Absorption factor
γ_e	Emission factor
Γ	Loss term used in NLS
η	Quantum efficiency of photodiode
η_c	Coupler efficiency
η_{in}	Input coupling efficiency
η_I^-	Isolator backward efficiency
η_I^+	Isolator forward efficiency
η_{out}	Output coupling efficiency
η_s	Splice coupling efficiency
κ	Normalized pump power
λ	Wavelength
λ_ψ	Cutoff wavelength
λ_π	Pump wavelength
λ_σ	Signal wavelength
λ_μ	Grating period

λ_μ Grating period
ν Optical frequency of signal light
ν_ψ Crossing frequency used in McCumber theory
ν_π Optical pump frequency
ν_σ Optical signal frequency
θ Scattering angle
Ω Steradian
Ω_2, Ω_4, Ω_6 Judd-Ofelt parameters (determined from absorption spectra)
$\rho(\mathrm{r})$ Radial density distribution of neodymium ions
$\rho_{\mathrm{Er}}(r)$ Radial density distribution of erbium ions
ρ_0, ρ_0' Maximum density of erbium ions
σ_a Absorption cross section
σ_e Emission cross section
σ_{ESA} ESA cross section in erbium
σ_{ij} Cross-section value corresponding to a transition from level *i* to level *j*
σ^2_{IM} Intermodulation products spectral noise density
σ_{pa} Absorption cross section of pump
σ_{pe} Emission cross section of pump
$\sigma_{\mathrm{s\text{-}sp}}$ Signal-spontaneous beat noise component
$\sigma_{\mathrm{sh,s}}$ Signal shot noise component
$\sigma_{\mathrm{sh,sp}}$ Amplified spontaneous emission shot noise component
$\sigma_{\mathrm{sp\text{-}sp}}$ Spontaneous-spontaneous beat noise component
σ' Cross-section ratio used in simplified noise expression of Nd-doped fibers
σ_0 Root-mean-squared noise term for a space
σ_1 Root-mean-squared noise term for a mark
σ_{13} GSA cross section in erbium
σ_{24} ESA cross section in erbium
σ_{43} Signal emission cross section in neodymium
σ_{46} ESA cross section in neodymium
τ Pulse width (time) at half maximum intensity
τ Radiative lifetime
τ_e Radiative lifetime of erbium
τ_{ij} Lifetime related to the transition between level *i* and level *j*
ω Spot size
ω_c Angular frequency of carrier wave
ω_s Fiber spot size at the signal wavelength
б(r, θ) Modal distribution of the fundamental fiber mode field

Index